NICHE CONSTRUCTION

NICHE CONSTRUCTION

How Life Contributes to Its Own Evolution

JOHN ODLING-SMEE

The MIT Press
Cambridge, Massachusetts
London, England

The MIT Press would like to thank the anonymous peer reviewers who provided comments on drafts of this book. The generous work of academic experts is essential for establishing the authority and quality of our publications. We acknowledge with gratitude the contributions of these otherwise uncredited readers.

This book was set in Adobe Garamond and Berthold Akzidenz Grotesk by Jen Jackowitz. Printed and bound in the United States of America.

Library of Congress Cataloging-in-Publication Data

Names: Odling-Smee, J. C. (John C.), author.
Title: Niche construction: how life contributes to its own evolution / John Odling-Smee.
Description: Cambridge, Massachusetts : The MIT Press, [2024] | Includes bibliographical
     references and index.
Identifiers: LCCN 2023035586 (print) | LCCN 2023035587 (ebook) |
     ISBN 9780262548168 (paperback) | ISBN 9780262378895 (epub) |
     ISBN 9780262378888 (pdf)
Subjects: LCSH: Evolution (Biology) | Adaptation (Biology) | Bioenergetics.
Classification: LCC QH366.2 .O33 2024 (print) | LCC QH366.2 (ebook) |
     DDC 576.8—dc23/eng/20231211
LC record available at https://lccn.loc.gov/2023035586
LC ebook record available at https://lccn.loc.gov/2023035587

10 9 8 7 6 5 4 3 2 1

This book is dedicated to my wife, Ros. I dedicate it to her in gratitude for giving me so much affection and support during my years of blindness, as well as for her indispensable help in the writing of this book.

Contents

Preface

This book was written in strange times. I shared many of these strange times with everyone else. They comprised the 2019 COVID-19 pandemic and its associated disruptions and lockdowns in the following two years and then, in 2022, the war in Ukraine began after the invasion by Russia.

It was also a strange time for me personally. After enjoying a lifetime of excellent vision, shortly before my seventy-eighth birthday, I suddenly became totally blind in less than ten days. It was a considerable shock for both myself and my wife and family, as well as for our friends and colleagues. What was I to do? Give up, or start a new kind of life? With a great deal of encouragement and concerned affection from so many friends and colleagues and above all from my family, I eventually decided to start life again. But how? A UK charity called Blind Veterans came to my rescue. They equipped me with machines and gadgets that allowed me to read and write again, and they trained me how to use them. I then resumed the work that I had been in the middle of before I went blind, most of it in collaboration with my colleagues. Subsequently, I continued to collaborate with my colleagues on new papers for the next six years.

I then received an unexpected email from the life sciences editor of the MIT Press, Robert Prior, asking me if I would consider writing a new book about niche construction. I emailed him back saying I would not be able to manage it because I was now blind. To my amazement and lasting gratitude, Prior pushed back. I remember him saying something like, "If you can write emails, then surely you can write a book." That was how this book came to be written. I still had a lot of unfinished business in my head, in the form of ideas about evolution. Prior's challenge was all I needed to start putting some of these ideas on paper as best as I could.

Nonetheless, I fear that blind authorship has its limitations and consequences. One of its consequences is that I have been unable to cite the work of many authors that I should have done but couldn't because I can no longer track the scientific literature in the ways that I used to do. I owe all these authors an apology and ask for their understanding. Another consequence of my blind authorship is that, with the exception of a single diagram, I have been unable to use figures, graphs, or even photographs to support my text. I have also cited a minimal number of equations and have avoided discussing mathematical models, partly because my reading machine goes berserk if I ask it to read mathematics. That may prove to be an asset rather than a deficit, given that allegedly so many readers are put off by equations.

I should also like to acknowledge the many people who have in different ways helped me to write this book. I would particularly like to thank Kevin Lala (formerly Laland) at the University of St Andrew's, Scotland, who more than anyone else helped me to prepare my manuscript by reading and editing drafts and writing endnotes. In addition, Kevin, with the help of his assistant Linda Hall, did one job for me that I now find so difficult: tracking down and compiling the list of references that I have managed to cite in this book. I would also like to thank Tobias Uller, at Lund University, Sweden, who likewise gave me critical but very helpful feedback on all my chapters. I would like to thank several other scientists as well, who have helped me with particular chapters. They include Addy Pross, Ben Gurion University, Israel; David Deamer and Bruce Damer, Santa Cruz University; Sonia Sultan, Wesleyan University; and Philip Kreager, Oxford University. I would like to thank three anonymous referees too, who read part of or all my manuscript. They gave me encouragement as well as specific criticisms, which further improved my text. Finally, I should like to thank Anne-Marie Bono and her team of editors at the MIT Press for coping with my difficult circumstances so well. As always, any remaining faults are mine and no one else's.

I should also like to acknowledge the help that I received from Helen Krajicek, who read aloud to me numerous articles from *Nature* and other journals over the course of many years. By doing so, she helped me to keep abreast of at least some of the things that were going on in contemporary science. I also owe a debt to both my daughters, Anne and Lucy Odling-Smee, who helped me in different ways. Anne runs a design company in London, Design Science, concerned with communicating science to lay-people, as well as to other scientists. She has greatly helped me with the presentation of my book. Lucy is one of *Nature*'s senior editors. That allowed her to criticize some of my chapters, which she did with filial severity. I also need to thank the two blind charities in the UK, which have helped me so much. One is the Blind Veterans Charity, which I've already mentioned. The second is the Royal National Institute of the Blind, who were particularly helpful during my first months of blindness. With my wife's help, I intend to set up a trust to enable at least some of the proceeds that may arise from the sales of this book to be given to these two charities.

I want to dedicate this book to my wife, Ros. Many authors dedicate their books to their wives. It's almost a convention. But I need to add something extra about her being the hidden coauthor of this book since almost every word in it started by my dictating sentences to my wife, who then turned it into text. She often did so by turning my sentences into better English. She also insisted that even though she is neither a biologist nor a scientist, she had to understand each sentence before she wrote it down. I think and hope that she improved my text considerably.

John Odling-Smee
September 2023

I LIFE ON ANY PLANET

1 WHAT IS LIFE?

We are incredibly improbable systems. The second law of thermodynamics dictates that everything in the universe inevitably travels toward a state of greater disorder. Yet somehow human beings, along with the rest of life, buck this trend. We are made of trillions of cells, packed densely together and organized into functional tissues and organs in our bodies. For instance, there are roughly a hundred billion interconnected neurons in a human brain, about as many as there are stars in the known universe. In virtually every cell in our bodies, there are innumerable molecules, including proteins and nucleic acids, interacting with a complexity greater than any human-made artifact. If the fate of all things is an inevitable journey toward greater disorder, how can such an immense concentration of order and complexity be packed together in a single human body? Come to think of it, how can any living organism exist? Also, once life appeared on Earth, how could more complex organisms evolve? And how has the biosphere continued to grow in complexity?

The laws of physics and chemistry alone cannot answer these questions, nor can contemporary biology. For example, orthodox evolutionary theory cannot explain how life got started. Nor can it explain how purposeful, goal-directed organisms could evolve out of an apparently purposeless universe. To understand these issues, I suggest we are eventually going to have to develop a more comprehensive theory of evolution than the current orthodox theory. Among other things, we are going to need a theory that recognizes that living organisms are active, purposeful agents that contribute in fundamental ways to their own and each other's evolution.

The questions that I want to focus on in this book are closely related to another question, which was asked by the famous physicist Erwin Schrödinger in 1944, in the title of his now-celebrated book *What Is Life?*[1] Schrödinger was primarily concerned with a subordinate question, which I will paraphrase here: *What else is needed beyond the known laws of physics and chemistry to understand life?* He indicated that these questions can be answered only in the context of the laws of thermodynamics. Life has to obey these laws.

There are three laws of thermodynamics, which were progressively discovered by physicists during the second half of the nineteenth and the beginning of the twentieth century. Initially, I will focus only on the first two. The first law states that energy is conserved throughout the universe, while the second law specifies that entropy—that is, the amount of disorder or randomness in a system—is always increasing. The second law implies that everything in the universe is traveling from a state of relatively higher organization and order to a state of greater disorder. This implies that any system that happens to be more organized or more ordered than its immediate surroundings will dissipate over time and continue to do so until it has returned to thermodynamic equilibrium with its surroundings.

Everything in the universe obeys the first and second laws of thermodynamics. However, the ways in which living and nonliving systems obey the second law differ in important ways. Leaving aside human artifacts for the moment, inanimate objects are relatively simple. They are also passive and purposeless. They obey the second law of thermodynamics reactively. Forces act on them, but they do not act of their own accord. This is true on multiple scales. For example, on a large scale, the Sun responds reactively to the laws of physics, including the laws of thermodynamics. If you know these laws, it is possible to predict the future of the Sun with some accuracy. On a smaller scale, the iceberg that sank the *Titanic* was responding passively to the laws of physics. If the crew of the *Titanic* had known enough about where and when the iceberg had originated and the subsequent prevailing weather and sea conditions that it encountered, they could have predicted its position accurately. The tragedy was that they did not. At a smaller scale still, the molecules of metal in the head of a pin change their state only through random motion unless acted on by physical forces.

In contrast, living organisms are far more complicated. Crucially, they are active, purposeful systems, which makes them less predictable. Because of their complexity and high degree of organization, organisms are very far from thermodynamic equilibrium systems relative to their abiotic surroundings. If living organisms did nothing, if they were passive in the same way that inanimate systems are passive, they would rapidly dissipate until they were back in thermodynamic equilibrium with their surrounding environment. That is, they would die. Organisms have to be active, purposeful systems to oppose the second law and stay alive. However, at all times and places, organisms are still subject to the second law. Somehow, organisms have to "cheat" the second law of thermodynamics, while still obeying it, to stay alive. How is it possible for organisms to pull off this trick? Unfortunately, the laws of thermodynamics are notoriously difficult to understand. Even contemporary physicists sometimes struggle with them.[2] Yet we encounter them every day of our lives, usually without realizing it.

Before going any further, I will give some examples of the ways in which we meet the first and second laws in everyday life by focusing on three increasingly complicated yet familiar examples. Let's start with a simple armchair experiment, one that you can easily do yourself. Take a glass of water, and then add to the water a drop of dye, such as red ink. As soon as you've added the drop, the situation in the glass changes. It becomes more complicated because the drop of ink is out of thermodynamic equilibrium with the surrounding water. The ink will immediately start to dissipate by diffusing into the water around it. A few hours later, you will have a glass of slightly pink water. The ink drop has obeyed the second law of thermodynamics by dissipating. It was initially concentrated in one dense area rather than spread evenly around. Such dissipation is an everyday experience. It's equivalent to making yourself a hot cup of coffee and pouring sugar in it, which by dissipating will sweeten your entire drink rather than staying as one very sweet mouthful. Likewise, after making your coffee, if it is still too hot for you to drink, you can rely on the second law again to cool it until it's closer to the temperature of its surrounding environment.

There is an important twist here that is not so obvious. When you first add the drop of red ink into the glass of water, you have not only introduced a more complicated system in the glass, but you have also turned the entire glass of liquid into a less probable system. The system will then gradually travel toward a more probable state as the ink diffuses. Another way of characterizing the second law of thermodynamics is to say that everything in the universe is traveling from a less probable state to a more probable state. This is illustrated by the fate of the ink in the water. It travels from the less thermodynamically stable state of the concentrated ink drop to the more thermodynamically stable state of the well-mixed solution.

At the molecular level, this journey toward greater probability is driven by background thermal agitation, called "Brownian motion." It happens by chance. But because it only happens by chance, it is still possible for the unexpected to happen instead of the expected. Purely by chance, it is possible that the drop of red ink that you added to the water might start to dissipate but then return to a discrete drop of red ink surrounded by water again. That is incredibly unlikely but it is a statistical possibility. To give an even crazier example, it's also possible that, purely by chance, your drop of red ink might, for a fleeting moment, cause a picture of your face to appear in the water. But I wouldn't wait around for this to happen. That is so improbable that you might have to hang around for several times the age of the universe before it occurred. Nonetheless, it is statistically still a possibility, however remote. In contrast, you may have to wait for only a couple of hours for the red ink to be wholly dissipated in your glass of uniform pink water. That is a vastly more probable outcome. It was another famous physicist, Ludwig Boltzman,[3] who first established the underlying statistical nature of the second law of thermodynamics. It should not be forgotten that, even in everyday life, you just might be taken by surprise by a highly improbable event one day.

My second example is more complicated. Suppose that you are a carpenter and want to make an object such as a wooden table. To make the table, you have to oppose the second law of thermodynamics by investing in active, purposeful, physical work. Once you have made it, the table will be even more out of thermodynamic equilibrium with its surroundings than the wood was before you started to work on it. To make the table, you also have to know in advance what you have to do to the wood and how to do it. You start by preparing the wood. You do so by sawing the wood to the size and shape that you're going to need. You then smooth down the wood by hand with sandpaper to eliminate rough edges. You have to supply the energy needed to do this work from your own body, drawing on energy extracted from the food that you have eaten. Your work of sawing and smoothing the wood necessarily generates friction and heat. The heat dissipates in the surrounding atmosphere; you are unlikely to notice it, but nevertheless, it happens. The energy that you used to do your work is conserved as per the first law of thermodynamics, but it has now dissipated in the surrounding atmosphere by being transformed into other forms of energy (heat and sound), in accordance with the second law. This second example is more complex than the first because active, purposeful work is being done on a system. That is what allows a highly improbable object, the table, to be created in your workshop against the wider background of constantly increasing entropy.

My last example is still more complicated. Imagine that you are a potter and want to make some coffee mugs. But this time, you also want to teach other people how to do the same. Once again, you will need the raw materials: some china clay, a supply of water and electricity, and perhaps paint and glaze to decorate the mugs after you have made them. You also need some tools, such as a potter's wheel and a kiln to fire the mugs after they are made. Potters have to buy both their raw materials and their tools from others, who must in turn invest work and energy to manufacture or gather them and transport them to where they can be purchased. The clay would not have been out of thermodynamic equilibrium with its surroundings before it was dug out of the earth. But it would have taken energy from other people to dig it out.

As you work on the clay, showing your pupils how to make a mug, you turn it into an increasingly improbable object. The clay is also becoming increasingly far from thermodynamic equilibrium with its surroundings. In this case, in the same way as when you were building your table, you are opposing the second law, to create the order that you are putting into each of your mugs by your work. After you have made the mugs, they will be put to use and start to dissipate. Over time, cracks and chips may gradually appear in them. Then one day you drop a mug and it smashes into fragments. Your carelessness has assisted the mug on its journey to greater thermodynamic stability and a more probable state, in accordance with the second law. What is interesting this time is that the information or know-how that you were transmitting to your pupils about how to make the mug will not have dissipated when you break the mug. Rather, the mug-making knowledge will have spread to all the other people you were teaching. That knowledge about how to make another mug will remain in your head too. It will not have dissipated either, just because you've broken your mug, nor has the net amount of mug-making knowledge remained constant. It has increased.

The relationship between information and the laws of thermodynamics is still not fully understood,[4] but nonetheless there is an intriguing difference between what happens to the energy that you have been using to make your mugs, as opposed to the information or knowledge that you also needed to make your mugs and that you passed on to the others you were teaching. In everyday life, you will probably take this difference for granted. But physicists cannot do that.[5] Neither will it be possible to take it for granted if we want to understand the biology of life, including the biology that lies behind the different ways in which living organisms contribute to their own and each other's evolution. The distinction between physical resources and informational resources is highly relevant to the way that evolution works. It is not an overstatement to suggest that it may hold the key to the very existence of life.

That's enough examples. Because they are all prosaic, everyday human experiences, it may still be hard to recognize the generality of the natural phenomena that they are illustrating. If the physicists are correct, the laws of thermodynamics are as universal as the laws of gravity. The falling apple that inspired Isaac Newton was equally a prosaic event. However, because Newton realized that it was an example of a universal phenomenon, he was able to discover so much about the universal laws of gravity. Similarly, if you want to understand the answers to Schrödinger's "What is life?" question, you have to start by trying to understand how life copes with the universal laws of thermodynamics, no matter how humdrum the examples are with which you begin.

Let's return to considering the basic physics, chemistry, and biology of these laws. Schrödinger emphasized there were two outstanding problems that biologists had to solve, each of which relates to a basic requirement of life. The first is that organisms must actively oppose the second law of thermodynamics to stay alive. Organisms have to maintain their own very-far-from-thermodynamic-equilibrium status, relative to their external environment, by active, purposeful, energy-consuming, and physical work. If organisms did nothing, they would dissipate and die. For every moment they are alive, organisms must actively oppose the flow of energy and matter between themselves and their environments, which is favored by the second law.

Organisms must continuously import resources that are relatively high in free energy[6] from their environments. They must also export the detritus that they generate by living, and that are lower in free energy, back to their environment. This exchange of physical resources between organisms and their environments is primarily understood by biologists in terms of the metabolic activities of organisms. However, the metabolic activities of organisms depend in turn on the capacity of organisms to interact with their external environments by two-way-street interactions. Via these interactions, organisms must take energy and matter resources from their environments and also dump their detritus back into their environments. It follows that organisms must cause some changes in their local external environments by interacting with them, as well as causing changes in their own internal environments by their metabolic activities. However, the physical changes that organisms cause in their external environments will not just be by-products of their metabolic activities. They will also be the consequences of other kinds of activities and the choices that organisms make, which are now referred to as "niche construction."[7]

In organisms, niche construction promotes "dynamic kinetic stability (DKS)"[8] (discussed further in chapter 6), which for organisms refers to the interactive niche relationships between the phenotypic traits of organisms and the multiple variables that they encounter in their local external environment. The niche-constructing activities of organisms are likely to increase their DKS, not just their thermodynamic stability. Ultimately, the niche-constructing activities of organisms cause them to coevolve with their external environments instead of just evolving reactively in response to their allegedly autonomous environments.[9]

The second basic requirement of life can then be derived from the first. Life cannot oppose the second law of thermodynamics by chance. The odds of it doing so effectively by chance are so improbable that it's not worth considering in practice. Neither can organisms resist the second law through random activity. To survive, all living organisms have to be informed by "meaningful" information. They must have the requisite kind of knowledge, or "adaptive know-how," about their local external environment and the nature of the work that they need to do to retain their far-from-thermodynamic equilibrium status. Living organisms cannot harvest the energy and matter resources that they need from their external environment to stay alive if they don't know what those resources are, where they can be found, or how they can be extracted and processed. Neither can they dump the detritus that they have to get rid of back into their external environment without appropriate adaptive know-how, concerned with how to excrete their waste products. These considerations dictate that life must have certain key properties. Living organisms not only have to be active, fuel-consuming systems, but they must also be informed, purposeful, goal-seeking agents to oppose the second law.[10]

The knowledge of how to oppose the second law is the second basic requirement of life. As we saw in my third example, there is a fundamental asymmetry between the availability of the two kinds of resources that organisms need to stay alive. They are energy and matter resources, which I'll call the "bioenergetics resources" of life, as opposed to informational resources, which I'll call the "bioinformatics resources" of life. Although biologists usually think of the bioenergetic resources of life in terms of the metabolic activities of organisms, here I'm going to use the term "bioenergetics" primarily in an ecological sense, to refer to the energy and matter that flow through ecosystems and that organisms have to tap into to stay alive.

With respect to the bioenergetics of life, raw energy and matter resources exist in the external environments of organisms. Energy and matter resources even exist on lifeless planets, in the total absence of organisms. But with respect to the bioinformatics of life, raw informational resources, in the form of adaptive know-how or knowledge, do not exist in the external environments of organisms. Raw informational resources exist only in the form of the information carried by other organisms, but they don't exist anywhere else in their environment. The knowledge that organisms require to oppose the second law has to be made by life itself. In practice, adaptive know-how is made by evolutionary processes as a consequence of evolving populations of organisms interacting with their external environments and the natural selection that ensues from their interactions.

Genetic variations, which were previously adaptive for ancestral organisms relative to their environment in the past, can accumulate in evolving populations. Individual organisms may then inherit a sample of adaptive information as a function of their membership of their particular evolving populations. Individual organisms may also gain additional adaptive know-how for themselves as a function of a variety of developmental processes, such as adaptive immunity, epigenetic processes, or learning in animals. By using whatever adaptive know-how they acquire, individual organisms may be able to control the energy and matter flow between themselves and their external environment in ways that allow them to oppose the second law of thermodynamics and survive. Throughout this book, we are going to be preoccupied with the asymmetries between these two kinds of resources.

Schrödinger was well aware of both the bioenergetics requirements and the bioinformatics requirements of life, when he asked his question "What is life?"[11] Largely thanks to him, we can now see that both of these basic requirements of life are closely related to the laws of thermodynamics. The first requirement, the bioenergetics of life, stems directly from the second law of thermodynamics. The second requirement, the bioinformatics of life, stems directly from the first requirement. Life cannot solve the first bioenergetics of life problem unless it can also solve the second bioinformatics of life problem, and vice versa.

Schrödinger explicitly addressed the first bioenergetics of life problem as follows: "A living organism. . . . feeds upon negative entropy [by] attracting. . . . a stream of negative entropy upon itself, to compensate [for] the entropy increase it produces by living and thus to maintain itself on a stationary and fairly low entropy level."[12] His answer to this first bioenergetics of life problem introduces two key concepts that we need: "negative entropy" and "entropy." Entropy equates to greater disorder and is reasonably familiar. However, "negative entropy," or "negentropy," is less intuitive: it means a greater order.

As yet, there is probably no such thing as a common sense version of either of these concepts. For the moment, I will continue to use the term "entropy" but I will often substitute the terms "free energy" or "order" for the awkward term "negative entropy." Free energy is any kind of energy that can be utilized by a physical system to do work.[13] In crude terms, the greater the amount of order in a resource, the more free energy it possesses.

Schrödinger was less explicit about the second bioinformatics requirement of life. He simply assumed that organisms are informed by the processes of evolution, and primarily by the process of natural selection discovered by Charles Darwin.[14] However, the acquisition of adaptive know-how by organisms, as a function of their evolution, is not cost-free. The processes whereby organisms acquire, encode, store, transmit, and decode adaptive know-how, or even erase or forget it,[15] all cost energy and matter. They all require some kind of physical platform, which must be built and maintained.

By the time Schrödinger was writing, Darwin's theory of evolution by natural selection had been combined with Gregor Mendel's laws of genetic inheritance to provide the first truly coherent theory of evolution. In the title of a now well-known book, Julian Huxley called it the "modern synthesis."[16] In light of the modern synthesis theory, evolutionary biologists came to understand that for natural selection to be able to inform individual organisms in evolving populations, physically encoded "memories" of some kind had to exist. These memories also had to be transmitted between parent organisms and their offspring across successive generations of their populations. Specifically, memories of the historical outcomes of prior natural selection, acting on ancestral organisms, had to be transmitted to contemporary organisms.

In 1944, the concepts of genetic inheritance and gene-based "memories" already existed. It was assumed that successive generations of organisms could inherit adaptive know-how from ancestral generations via whatever "particles" that underlay Mendel's laws of inheritance. But no one knew how memories of the past were physically registered and stored in the current generation, nor how that generation could transmit those memories to subsequent generations, because at that time, no one knew what the physical bases of gene-based memories were.

It was in these circumstances that Schrödinger put forward a hypothesis. He suggested that the physical basis of genetic memories should be some kind of quasi-aperiodic crystal. At the time, he was probably thinking that a protein molecule might do this job rather than a nucleic acid. But the proposal was prescient. The breakthrough came only nine years later, when James Watson and Francis Crick (1953), aided by others, most notably Rosalind Franklin, discovered the real memory molecule, deoxyribonucleic acid (DNA), which indeed has the properties of a quasi-aperiodic crystal. Their discovery was inspired, at least in part, by Schrödinger.

Since Schrödinger's time, orthodox evolutionary theory, today usually referred to as "neo-Darwinism," has made huge advances in understanding the bioinformatics of life problem. In contrast, orthodox evolutionary theory has paid far less attention to the bioenergetics of life, which has been largely left to ecologists. Orthodox evolutionary theory has also failed to pay much attention to the interactions between the bioenergetics and the bioinformatics requirements of life, which are considerable. As a result, contemporary evolutionary theory doesn't address how life contributes to its own evolution. We will be giving these issues far more attention in this book.

Given that Schrödinger was writing seventy-five years ago, why are we still going back to him and the physics of his time rather than building on contemporary physics? This question has recently become more pertinent to biologists because a contemporary biophysicist, Stuart Kauffman, has been asking similar questions to the ones that we have been asking here about the nature of life, while also acknowledging his debt to Schrödinger. For instance, in his intriguingly named book *A World beyond Physics*,[17] Kauffman asks how life evades but does not avoid the second law. However, Schrödinger's physics allows me to use conventional or even common sense versions of the flow of time, from past to present to future. It also allows me to use conventional ideas about cause-and-effect relationships in nature, which I couldn't use if I based my approach on more modern understandings of those concepts by contemporary physicists, such as Ravelli's *The Order of Time*.[18] That would demand placing my account of the thermodynamics of life on quantum field theory, which is beyond me and I imagine most others (but see Carroll, 2019). For these reasons, I will stick with Schrödinger's physics, combined primarily with the classical physics of the nineteenth and twentieth centuries.

MAXWELL'S DEMON

Where have the physicists left us? What can they tell us about the capacity of organisms to resist the second law of thermodynamics? Instructive here is the work of a father figure in the field of thermodynamics from an earlier generation: James Clerk Maxwell. In the nineteenth century, Maxwell was interested in the laws of thermodynamics for two reasons. First, he wanted to know whether it was really impossible for any system to violate the second law of thermodynamics in a closed system (i.e., a system closed off from the rest of the universe). Can anything cheat the second law simply by violating it? At the time, that was still a sensible question to ask since the laws of thermodynamics were not as well established then as they are now. Implicitly, Maxwell also asked a second question: What minimal properties are needed for any system to resist the second law? He was interested in the efficiency of Victorian steam engines and, more generally, in human-designed and manufactured fuel-consuming machines. Such machines are also out of thermodynamic equilibrium with their surroundings, although not nearly so far out of equilibrium as any living organism. Maxwell wondered whether it might be possible to build a machine that was so efficient that it could resist the second law by violating it in a closed system. One consequence of his second question was that Maxwell inadvertently provided biologists with an extremely helpful starting point for considering Schrödinger's "What is life?" question.[19]

In his *Theory of Heat* (1871), Maxwell described a hypothetical "intelligent," conceptually subversive, second-law-opposing imaginary creature, which subsequently became known as "Maxwell's demon." Maxwell equipped his demon with the ability to monitor individual molecules of a gas in a cartoon environment, usually reduced to a boxlike container filled with a gas. Typically, the container is divided into two compartments, A and B, by a partition with a molecular-sized hole in it. A and B are initially in thermodynamic equilibrium with each other because the temperature and pressure of the gas are the same in both compartments. The container is shut off from the rest of the universe and is therefore a closed system.

Maxwell wanted to know whether in theory, a suitably energized and structurally and functionally informed demon could drive compartments A and B out of thermodynamic equilibrium by doing nonrandom, second-law-opposing work. He was therefore exploring the opposite problem to the one that I illustrated in my first example. Instead of introducing something that was deliberately out of thermodynamic equilibrium with its surroundings (as in my example of the red ink in the glass of water), to see whether it would achieve thermodynamic equilibrium by passively reacting to the second law, Maxwell started with the opposite problem. He started with compartments A and B in his cartoon environment that were already in thermodynamic equilibrium with each other (equivalent to the pink water), and he investigated whether his demon could drive compartments A and B out of thermodynamic equilibrium relative to each other by doing active, informed work (equivalent to producing a concentrated drop of dye surrounded by clear water in the glass). Would this be possible? That was Maxwell's question.

Let's consider the two kinds of resources (labeled R) that the demon would require to do its job. First, it needs sufficient physical energy and matter to fuel its second-law-opposing work, and to supply it with the material ingredients that it needs to accumulate negative entropy or order to build its body. From here on, this physical resource will be notated as R_p, where the subscript p representing physical energy and matter. The demon also needs sufficient information in the form of meaningful, adaptive know-how in order to do its work "correctly" relative to its local external environment. From here on, this informational resource will be notated as R_i, where the subscript i represents information in the form of adaptive know-how relative to the demon's task in its environment. These two kinds of resources, R_p and R_i, are identical to the basic requirements that living organisms need to oppose the second law, and it's these equivalences that make the questions that Maxwell was asking so illuminating. First, we will consider how Maxwell's demon acquires both its bioenergetic (or R_p requirements) and its adaptive know-how (or R_i requirements), and then we will return to consider how living organisms acquire these same resources.

Let's start with the informational requirements of Maxwell's demon. There was no problem for Maxwell's demon in acquiring its adaptive know-how in advance: Maxwell simply gave the demon the R_i that it needed to do its work, drawing on his scientific knowledge to design, inform, and instruct it. Maxwell also gave the demon a goal by instructing it to open and shut the door between compartments A and B selectively so as to cause it to oppose the second law of thermodynamics. For instance, Maxwell told his demon to sort all the fast-moving molecules into compartment A and all the slow-moving molecules into compartment B. In doing so, he effectively turned the demon into a purposeful agent, or goal-seeking system. Maxwell also gave his demon sufficient intelligence, or know-how, to perceive, remember, and demarcate between fast and slow molecules and to assign different values to different molecules. Their values were derived from his demon's goal or purpose. Maxwell then gave his demon the physical capacity to open and shut the door between the two compartments in its environment selectively. In this way, his demon acquired both its necessary structural design and functional adaptations to do its second-law-opposing work. Maxwell also designed his demon's environment. To ask whether it was theoretically possible for a demon to oppose the second law in a closed system, he placed it in a closed environment (i.e., the box with two compartments and a small partition).

Now let's consider the physical resources (R_p) that the demon also needed to oppose the second law. Here, it had a problem—Maxwell did not give it any energy to fuel its work. This was a deliberate decision on Maxwell's part because the acquisition of sufficient energy (R_p) by the demon from its closed environment was the principal problem that Maxwell was investigating. The demon had to take the energy that it needed from within the closed environment (i.e., within the box). A subsidiary problem is that the demon had to dump the entropy that it was generating by its work, presumably in the form of heat, back into its closed environment without destroying the capacity of its environment to go on supplying it with energy.

What did Maxwell conclude? Was it possible for the demon to cheat the second law in its closed environment, at least in theory? Could it do so by creating free energy, by sorting between the faster and the slower molecules? If it could, the demon would succeed in violating the second law of thermodynamics in a closed system. But the answer was, famously, "No." It was only the presence of variant molecules in its environment that allowed the demon to accumulate negative entropy by driving compartments A and B out of thermodynamic equilibrium by its intelligent work. In the absence of any environmental variance, the demon's task would not have been possible, not even in theory.[20] Moreover, the demon could only sort molecules effectively because it was fully informed. It had been given all the relevant information that it needed by Maxwell.

Maxwell, and later other physicists, demonstrated that no conceivable demon could oppose the second law by doing nonrandom work in a closed environment. Subsequently, many physicists challenged Maxwell's findings by imagining ever more ingenious and minimalist versions of his demon, typically demons that needed either more R_i or less R_p than did Maxwell's.[21] Their efforts have been extremely enlightening, but none changed the basic result. No one has ever come up with a Maxwell's demon that could violate the second law of thermodynamics in a closed environment by generating more free energy than it consumed.

Could the demon work in an open environment? Physicists have demonstrated another much more mundane result that is of considerable interest to biologists. It turns out that it is relatively easy to design a "demon," such as a fuel-consuming engine, that can oppose the second law of thermodynamics, temporarily and locally, without violating the law, provided that the demon is placed in an open environment, which can supply it with the extra free energy that it needs to fuel its work. The demon's open environment must also be able to absorb the increased entropy that its work generates.

The second law determines that the net entropy generated by the demon's work must always be equal to or greater than the net free energy that the demon produces by its work. Maxwell demonstrated that although it is not possible for his demon to oppose the second law in a closed system, it is possible for a suitably informed demon to assemble highly improbable, very-far-from-thermodynamic-equilibrium systems in open environments, which contain sufficient R_p to sustain them, at the cost of increasing entropy somewhere else in the open environment. That is a less dramatic result than the violation of the second law, but it is nonetheless of major significance to the questions that interest us here. Maxwell's ingenious thought experiment takes us at least part of the way toward satisfying Schrödinger's criteria for life.

THE NECESSARY REQUIREMENTS FOR LIVING ORGANISMS

Now we can compare the (R_p and R_i) requirements of living organisms to those of Maxwell's demon. There is a strong convergence between the bio-energetic (R_p) requirements of Maxwell's demon and those of organisms. By demonstrating that it is impossible for any system to oppose the second law in a closed system, Maxwell demonstrated that it is impossible for a living organism to oppose the second law in a closed system. Both Maxwell's demon and living organisms can acquire the energy and matter (R_p) resources that they need to do their second-law-opposing work only in open environments.

However, the requirement that organisms must live in open environments is easily satisfied. Life on Earth (or any conceivable extraterrestrial life) can exist only in open environments. More specifically, organisms can live only in those open environments that can supply them with enough free energy to build their improbable bodies and fuel their second-law-opposing work. Organisms must also live in environments that can absorb the entropy or detritus that they generate through their work, without destroying the environment's capacity to continue supplying them with their energy and matter needs. All organisms live in open heterogeneous ecosystems that are themselves open to wider ecosystems, up to and including Earth's biosphere, which is open to the solar system. The last point is relevant because it means that even though the physical energy and matter resources available on Earth are finite, the radiant energy from the Sun is effectively unlimited relative to life on Earth. All Earth's life forms exist in the so-called Goldilocks zone of the Sun, which refers to regions of the solar system that are neither too cold or too hot, but, as in the children's story, are just right. The Earth satisfies this criterion.

The Sun too obeys the second law of thermodynamics, but it is such a vast source of free energy and it lasts so long that, in practice, it is legitimate for biologists to treat it as an external source of unlimited free energy. Consequently, all organisms on Earth do live in open environments that can supply them with sufficient R_p to fuel their second-law-opposing work. The Earth can also supply different living organisms with the diverse physical ingredients that they may need to allow them to assemble and maintain their highly improbable bodies. In addition, it allows some organisms to assemble other highly improbable systems in their environments, such as animal and human artifacts.

Another convergence between the requirements of Maxwell's demon and living organisms concerns the properties that Maxwell gave his demon to do its second-law-opposing work. Organisms must possess the same minimal set of properties that Maxwell gave his demon.[22] It follows that organisms like Maxwell's demon have to be purposeful, goal-seeking systems. They must possess at least the minimal goal of actively opposing the second law to stay alive. In conventional biology, this basic goal of organisms corresponds to the survival component of fitness.

There is no comparable convergence between the demon and living organisms with respect to the bioinformatics requirements of organisms for two reasons. A minor reason for this lack of convergence is that organisms need far more adaptive know-how (R_i) than Maxwell's demon needed to oppose the second law because of the sheer complexity of living organisms' internal and external environments compared with those of Maxwell's demon. The demon's environment was sufficiently simple that it did not need to "know" much to pursue its goals.

The opposite is true of organisms, whose internal and external environments vary in a very large number of dimensions. For example, the external environments of organisms always include other second-law-opposing organisms, including multiple predators, prey, parasites, competitors, and other species, with whom they share mutualistic relationships. Maxwell's demon's task was to drive a very simple system out of equilibrium to a minimal degree. In contrast, organisms have to build and maintain massively improbable, far-from-equilibrium living bodies relative to extremely complex external environments. For these reasons, organisms need far more adaptive know-how (R_i) than Maxwell's demon needed. But this difference between Maxwell's demon and living organisms is only a matter of degree.

The major reason for the divergence between Maxwell's demon and living organisms, with respect to the bioinformatics requirements of life, is much more fundamental. If organisms were really designed and informed by an omniscient, transcendental deity or god, in the same way as Maxwell was able to design and inform his demon on the basis of his own scientific knowledge, there would have been a much stronger convergence between living organisms and Maxwell's demon. But nature tells us that that is not what happened. Organisms were not "designed," nor spontaneously created, to fill their respective niches by a divine, transcendent, intelligent Creator. Instead, a huge amount of evidence tells us that all organisms owe their adaptive know-how to the natural processes of evolution. In particular, they owe it to the processes of natural selection, acting on populations of variant organisms over eons. Unlike Maxwell's demon, organisms must be supplied with all the knowledge they need to do their second-law-opposing work by evolutionary processes. But how?

NATURAL SELECTION AND INFORMATION

Assuming that natural selection is the primary source of adaptive know-how (R_i) for living organisms, we immediately run into two major limitations of natural selection. The first concerns the inability of natural selection to prepare organisms in advance for their second-law-opposing work, in the same way that Maxwell was able to prepare his demon in advance to do its work. If organisms did not carry adaptive R_i from the outset of their lives, they would not be viable. Their initial interactions with their environments would be chance-based and probably fatal. However, natural selection does not work prospectively, but only retrospectively. It cannot inform organisms a priori about their futures, in advance of their lives. It can only inform them a posteriori, after natural selection has already occurred relative to their ancestors and parents.

For this reason, natural selection has often been described as a "blind" process. It is blind to the future. Natural selection cannot predict new changes either in organisms or in their environments ahead of time. Instead, natural selection works by selecting those memories of past historical interactions between ancestral organisms and their ancestral environments that happened to be adaptive in the past. These memories, which we now assume are primarily encoded in genes, are then transmitted via inheritance systems to contemporary organisms in each successive generation of evolving populations. But this process cannot guarantee that the R_i that was adaptive for ancestral organisms in the past will still be adaptive for contemporary organisms in the present.

The best that natural selection can do is to inform contemporary organisms with know-how (R_i) that was adaptive in the past. Unlike the demon, which was given all the knowledge it needed by Maxwell a priori, natural selection cannot guarantee that organisms will possess the adaptive know-how (R_i), that they need to survive. Natural selection informs organisms by an inductive gamble, which amounts to a bet that the present and future environments of contemporary individual organisms will be similar to the past environments of their ancestors.

If there is little or no change either in organisms or their environments across generations, this inductive gamble is likely to work by producing well-adapted contemporary organisms. However, if there are any changes either in the organisms or in their environments, the gamble will be much more likely to fail. Later generations of organisms, informed by what is now out-of-date adaptive know-how, may not possess the knowledge that they need to survive. The persistence of life depends on endless probabilistic inductive gambles. The fact that species survive for substantial periods of time shows that these inductive gambles can be effective. At the same time, the observation that all species eventually go extinct highlights the fact that these gambles are not foolproof.

What kinds of changes are likely to undermine the capacity of natural selection to supply organisms with adaptive know-how? Some components of the external environments of organisms do not change. For example, the laws of gravity and thermodynamics don't change, at least on the timescales that interest us here, and neither do the structures of atoms unless they become negative or positive ions by gaining or losing electrons. Subatomic elementary particles do not change either, nor do the laws of chemistry. It is a safe bet that all these laws were the same at the origin of life on Earth approximately four billion years ago as they are today. Insofar as natural selection is sensitive to these unchanging components of organisms' environments, it should be able to supply contemporary organisms with the adaptive know-how that they need to survive.

Other components of the external environments of organisms do change, but only slowly. For example, the radiant energy from the Sun takes millions of years to change. The prior natural selection of ancestral organisms is likely to have a considerable capacity for preparing contemporary organisms with appropriate adaptations in response to the radiant energy of the Sun.

Difficulties arise when components of the external environments of organisms change more rapidly, such as climate changes and evolutionary changes in other organisms. Moreover, because almost all organisms modify their environments by their niche-constructing activities, these activities are also likely to cause some changes in their own and in each other's environments. Some of these changes may be rapid, and they may also be significant relative to the adaptations of contemporary organisms. They may demand further evolution from evolving populations.[23] In these circumstances, we might expect natural selection to favor any biological trait that can offset this first limitation of natural selection. For example, natural selection might favor organisms that live only very short lives, thereby limiting the amount of change that they are likely to have to cope with in each generation.

Natural selection might also favor organisms that invest in large numbers of diverse offspring in order to increase the chances that a fraction of their descendants will happen to be well adapted to whatever environmental changes do occur. Alternatively, in longer-lived organisms, natural selection may favor the evolution of supplementary developmental information gaining processes that update adaptive know-how (R_j), in individual organisms. These supplementary developmental processes include immune systems, epigenetic processes, and learning in individual animals. Natural selection may also favor the evolution of sociocultural information-gaining processes in social animals, including humans (see chapter 7).

There is also another way in which natural selection can favor traits in organisms that compensate for natural selection's limitation of only being able to confer knowledge that was adaptive in the past. Natural selection may favor organisms that partly determine their own environment, either by physically modifying it by their niche-constructing activities or by actively relocating to environments that have properties that resemble their ancestors' environments.[24]

Natural selection should also favor organisms that choose, modify, or regulate the environments in which their offspring develop in order to maximize the chances that their offspring's environments are well suited to their inherited adaptive know-how. For instance, birds' nests, termite mounds, and mammal burrows reduce the range of temperatures that these organisms and their offspring experience in their environments.[25] The transmission of environments, modified by the niche-constructing activities of ancestral and parent organisms to their descendants, is referred to as an "ecological inheritance"[26] in niche construction theory (NCT). Living organisms frequently inherit not just their genes, but an environment that has been shaped by their parents' niche construction, in an effort to ensure that the adaptive know-how (R_i) that they inherit via genetic inheritance will be adaptive.

The second major limitation of natural selection is that it cannot inform individual organisms with adaptive know-how (R_i) directly. It can inform evolving populations directly, but it can inform individual organisms in populations only indirectly, as a function of the membership of their particular populations. Let's first consider how natural selection informs evolving populations. Then I'll consider why natural selection cannot do the same for individual organisms. Evolving populations comprise multiple organisms, manifesting a variety of diverse phenotypes in each generation. This diversity is partly due to the various adaptive know-how (R_i) and resources (R_p) that each individual organism in a population inherits from its ancestors via its parents. The population will then undergo natural selection arising from the interactions of each individual organism in the population, with its own local external environment. Each individual organism in each generation is exposed to its own idiosyncratic selection pressures.

The different selective fates of individual organisms in each generation then allow natural selection to sort between better- or worse-adapted individuals. The least well adapted, or the least fit, are likely to be eliminated from their population, while the best adapted, or most fit, are likely to make the greatest contribution to the next generation. The net effect should be to update the adaptive know-how (R_i), carried collectively by the individual organisms in each successive generation of their populations. In this manner, natural selection is able to inform evolving populations directly.

Natural selection cannot do the same for individual organisms. Individual organisms are similar to populations in that they can express a large number of diverse phenotypic traits, loosely analogous to the large number of diverse phenotypes expressed by individual organisms in populations. What may be confusing is that biologists use the same word, "phenotype," to describe both each different individual organism in a population and each separate phenotypic trait in each individual organism. For example, a spaniel is a phenotype in a population of dogs, while a spaniel's characteristic floppy ears are a particular phenotypic trait in individual spaniels. The difference is that the diverse phenotypic traits in individual organisms cannot be separately evaluated by natural selection, in the same way that individual organisms in populations can be separately evaluated.

Natural selection cannot act within individual organisms by selecting between their diverse phenotypic traits. For example, if the spaniel has a left ear slightly bigger than its right ear, even if that makes it hear better in the left, natural selection cannot favor the left ear over the right. The fates of all an individual spaniel's traits are tied together. Natural selection can only act directly on the whole organism that carries them, in this case selecting both ears together. Neither natural selection nor artificial selection can affect the floppy ears of an individual spaniel during its lifetime. It can change the ears of a population of spaniels only by imposing a selective bias, either in favor of or against bigger or smaller floppy ears. The phenotypic traits expressed by any individual organism are bound to share the same selective fate as all the other phenotypic traits carried in that particular organism. Sometimes this may have unfortunate consequences for an organism. For instance, if one of its phenotypic traits is severely maladapted, it may cause the whole organism to be eliminated by natural selection, even though all its other phenotypic traits may be well adapted. The net result is that individual organisms cannot be directly informed by natural selection during their lifetimes, in the same way that evolving populations can be directly informed by natural selection over successive generations.

This second limitation of natural selection was noted by Darwin when he originally discovered natural selection. Darwin could not explain how natural selection could inform individual organisms with adaptive structural and functional know-how (R_i), until he had invented population biology. Only then could he explain how natural selection could inform individual organisms as a function of their membership of their particular evolving populations.[27]

Now we can take a closer look at how natural selection is able to inform individual organisms in populations indirectly. It starts with the legacies that newly born organisms inherit from their ancestors via their parents. We know that offspring inherit genes from their ancestors, and we have also seen that they frequently inherit an ecological legacy in the form of an ancestrally modified local environment. For the moment, I will assume that this initial inheritance comprises no more than a new organism's genetic inheritance and ecological inheritance, relative to each other. This package I call "niche inheritance."[28] In reality, it may involve some additional nongenetic (R_i) inheritance systems, as well as the inheritance of some initial energy and matter (R_p) resources. But I'll ignore these complications for the moment.

The ecological inheritance of any neonate organism will always include an initial address in space and time. The new organism then has to immediately start translating its niche inheritance into the assembly and growth of its very-far-from-thermodynamic-equilibrium internal environment and body. It has to do so in opposition to the second law of thermodynamics by immediately importing energy and matter from, and exporting detritus to, its local environment. Natural selection will be evaluating the new individual organism's capacity for survival throughout this time. Only organisms equipped with the appropriate properties, including Maxwell demon–type properties relative to their local selective environments, will survive and grow into adulthood. This accounts for the first classical component of fitness: survival.

But survival alone is not enough. The only organisms that can exist in successive generations of a population are those that happen to be instructed by the adaptive know-how (R_i) that they inherit, in order to pay back the gift of their initial adaptive know-how that they receive from their ancestors. They pay for the adaptive know-how for the next generation of organisms in their population, primarily by transmitting the R_i that they currently carry to their offspring via one or more inheritance systems. They may also pay for the initial assembly of their offspring's bodies with some directly transmitted energy and matter (R_p) resources.

This introduces the second classical component of fitness: reproduction. To reproduce, individual organisms require additional R_p and R_i, beyond what they need for their own survival. They also require some extra phenotypic properties beyond their minimal Maxwell demon–type properties. These additional properties are required both to produce new offspring and, quite often, to supply their offspring with initial energy and matter resources to increase their offspring's initial viability. They may also include the construction of a benign developmental nursery environment. They need all these extra properties to enable the next generation of organisms in evolving populations to be informed indirectly by natural selection. If organisms did not reproduce as well as survive, then evolution would stop. In the long run, enough organisms in each generation must reproduce to ensure the ongoing evolution of their populations.

Physically, individual organisms in populations cannot survive for long. Sequoia trees are among the longest-lived plants on Earth. They can live for up to 3,000 years, which is impressive, but they are still mortal. Animals are less impressive. Probably some species of turtle are among the longest-lived animals. They can live in excess of 200 years. That's also impressive by human standards, but turtles too are mortal. The aging process has multiple causes, but a major reason why all individual organisms are mortal is because, by their active second-law-opposing work, they are bound to generate entropy (or disorder) within their own bodies over time, as well as the entropy that they must export back to their external environments. For example, some genes (called "vacuole-protein genes") are essential in humans because they remove toxins in cells, but the same genes are not essential in mice. It would seem that recent increases in the longevity of humans have rendered these genes crucial, as toxins can now build up for longer periods, to the point at which they are lethal if not removed.[29] The phenotypes of organisms inevitably accrue scars by the wear and tear of their existence. They must eventually age, dissipate, and die. When they die, the residue of the energy and matter (or R_p) resources carried in their bodies is returned to their environment in the form of another kind of detritus: dead organic matter (or DOM, as ecologists call it).

In contrast, the adaptive know-how (or R_i) carried by individual organisms need not dissipate in the same way. If individual organisms manage to reproduce before they die, then they may be able to transmit and spread copies of their R_i, such as genetic information, to multiple offspring and potentially to unlimited numbers of descendants in their populations. It will still cost organisms energy and matter resources to acquire, process, and transmit their adaptive know-how to their offspring, and it will cost more energy and matter resources to produce the offspring themselves. Also, the adaptive know-how may not always be faithfully copied when it is transmitted to the next and succeeding generations. Copying mistakes, such as mutations, can occur. For all these reasons, the R_i resources of organisms may eventually dissipate too. But if organisms in each generation can pay the cost of the evolutionary processes that make their adaptive know-how, and if the transmission of their R_i is faithful, then the adaptive know-how, carried by individual organisms, may last for millions or even billions of years of evolution. Relative to R_p, R_i approaches immortality.

This asymmetry between the physical energy and matter (R_p) resources of mortal individual organisms in their bodies, and the near-immortality of the adaptive know-how (R_i) that they also carry is significant. It is reminiscent of the way in which the potter in my earlier example could transmit her adaptive know-how about how to make coffee mugs to others, both in her own generation and, in principle, also to succeeding generations. If this asymmetry between the R_p and R_i requirements of organisms did not exist, it is hard to see how the evolution of life would be possible.

In spite of this asymmetry, it is of interest that organisms may sometimes be able to transmit energy and matter resources across multiple generations too. They can do so via the ecological inheritance that they bequeath to their descendants as a consequence of their niche-constructing activities.[30] A large-scale example is the oxygen in Earth's atmosphere. The origins of the atmosphere is a complicated story involving both abiotic and biotic sources of oxygen, but crucially, it initially involved the production of oxygen as a waste product of photosynthetic reactions in cyanobacteria. This probably began more than three billion years ago. Cyanobacteria are still with us today, and they are still pumping out oxygen as a by-product of photosynthesis. In NCT, this is called "by-product niche construction." In this case, it is a by-product that has changing natural selection pressures in the selective environments of countless species of organisms, and over eons of time.

JOHN VON NEUMANN'S NATURAL AUTOMATA

Let's consider what extra properties individual organisms need, beyond their minimal Maxwell's demon–type properties, to reproduce as well as to survive. Another physicist, John von Neumann, aided by his colleagues, notably Stanislaw Ulam,[31] was the first to ask that question. He was interested in the minimal properties that any physical system needs to reproduce. His approach was similar to Maxwell's, in that von Neumann developed his theory of natural automata in an attempt to answer his own question.[32] His theory later morphed into cellular automata theory,[33] but I will nonetheless focus on natural automata theory because von Neumann was the first to raise some basic philosophical as well as logical points about reproduction.

Natural automata theory attempted to establish the irreducible minimal properties that any physical system requires to reproduce by making a copy of itself. It combines the properties of a "universal constructor" with the properties of a "universal controller." The theory depends on four interacting components:

- The *universal constructor*
- The constructor's *universal controller*
- The instructions that inform the controller with meaningful information, in the form of appropriate adaptive know-how specifying how to make the constructed entity
- A *copier* that can copy the instructions, and insert a copy of them into whatever entity the constructor constructs[34]

In principle, the universal constructor is reducible to any sufficiently informed and energized system capable of doing the work that von Neumann's theory assigned to it. In his original version, the universal constructor was equipped with a "long arm" that could construct any entity. To do so, it needed a favorable external environment that could provide it with enough energy and material parts (R_p), to allow it to do its constructive work. The universal constructor could therefore construct any entity, including a copy of the natural automata itself, provided that the appropriate know-how was encoded in the instructions and could be copied.[35]

However, the natural automata's universal constructor could not evade the second law of thermodynamics. To do its second-law-opposing work, it needed to be fueled by energy and matter (R_p) resources that it had to take from its external environment. Implicitly, the constructor must have been equipped with Maxwell's demon–type properties, as well as its own natural automata–type properties, to do its work. Originally, von Neumann appeared to take this for granted. He didn't pay much attention to the natural automata's environment either, beyond assuming that it could supply the constructor with everything that it needed.

The universal controller is equivalent to a universal computer (i.e., a computer capable of calculating anything that can be computed, bearing in mind that some things are not computable).[36] Today, a universal computer is usually called a "Turing machine" after Alan Turing, a pioneering computer scientist. The computer is reducible to a finite set of parts and states.

The computer "reads" the instructions, which are reducible to a potentially infinite string of binary digits (0s and 1s) on a tape. The instructions carry "knowledge" (R_i) that is "read" by the control unit and then controls the constructor's work. In principle, the information could cause the constructor to construct anything, but what is relevant here is that it could contain instructions to make another copy of the natural automata itself. The decisive step occurs when the universal constructor constructs another universal constructor, another universal controller, another set of instructions, and another copier and inserts them all into whatever entity it is constructing.[37] That step permits a "parent" natural automata to make an "offspring" natural automata; that is, a copy of itself. Thus, it allows the natural automata to reproduce.

The original natural automata by von Neumann showed that even imperfect parts can lead to a sustained process of replication if there is sufficient redundancy in the instructions or their implementation for error correction to occur. The natural automata's controller is responsible for quality checks, but it is not always the controller that does the checking since it can be done by external factors too. This is significant here, as it implies that if living creatures have natural automata–type properties, quality control could arise through natural selection pressures in their external environments. In addition, von Neumann established that the natural automata is logically capable of constructing automata that are more complicated than itself.[38] In the case of reproducing organisms, that is what allows more complex systems to evolve.

Natural automata have a lot in common with living organisms. There are many similarities, but also some significant differences. Let's consider the similarities first, by translating natural automata into a representative individual organism. Here, we will illustrate the connection by assuming that the only significant inheritance system between generations of organisms in evolution is genetic inheritance. That is oversimple, but it was probably close to what von Neumann was assuming anyway, since that was the orthodox understanding of evolution at the time. The connections between natural automata and individual organisms are then as follows: The phenotypes of organisms correspond to the natural automata's universal constructor. Phenotypes cannot claim to be universal constructors, but they can produce a lot of diversity among their offspring. In this simplified conception, the universal controller of an organism equates to the individual organism's genome, and the instructions correspond to the way in which adaptive know-how is encoded in the strings of nucleotide bases that exist in the DNA that organisms inherit from their parents (i.e., genetic information). The capacity of DNA to replicate copies of itself is also equivalent to the capacity of the natural automata's copier to replicate copies of its own instructions. Moreover, thanks to Watson and Crick (1953) and others, we now know how DNA does replicate. These similarities between natural automata and organisms are remarkable. They are almost self-evident to anyone who knows how organisms reproduce. That was probably no accident. It was, after all, what von Neumann was originally trying to capture.

But there are also some significant differences between the natural automata and organisms. One is that natural automata reproduce asexually, whereas most organisms, including microorganisms, need to combine with other organisms to reproduce. Sexual reproduction complicates reproduction, but it doesn't fundamentally change the natural automata–type properties that lie at the heart of the reproductive process. Once begun, the natural automata's capacity for reproduction can continue indefinitely, generation after generation. It can thereby also mimic the capacity of the evolution of life to continue indefinitely.

The most important difference between natural automata and living organisms is the same as the fundamental difference between Maxwell's demon and living organisms. The natural automata was given its adaptive know-how (R_i) a priori by its parent-scientist (in this case, by von Neumann). But the reproductive activities of organisms must actually be informed by the natural processes of evolution, not by scientists acting as proxies for a divine Creator. It follows that organisms must be instructed to reproduce by the adaptive know-how (R_i) that they inherit from their ancestors via their parents. Living organisms have to be active, purposeful, goal-seeking agents in their own evolution, both to survive as a function of their Maxwell's demon–like properties and to reproduce as a function of their natural automata–like properties.

CONCLUSIONS

Now let's return to Schrödinger's subordinate question: *What else is needed beyond the known laws of physics and chemistry to understand life?* It is clear, thanks to Maxwell, von Neumann, and many others, that living organisms do not evade the laws of physics and chemistry. Rather, they owe many of their vital properties to the underlying lawfulness of the universe. But there are at least two properties of living organisms that appear to be beyond physics. The first is the active, purposeful agency of fitness-seeking organisms. Physics deals with purposeless physical systems on all scales. But so far it has never had to deal with or explain purposeful systems. At present, the concept of purposeful systems is beyond orthodox evolutionary theory too.

Neo-Darwinism assesses the fitness of individual organisms in populations in terms of their capacity to survive and reproduce, but without attributing these capacities to the active, purposeful agency of organisms. In this book, I am going to argue that the purposeful agency of individual organisms plays a central role in evolution.

The second phenomenon of life, which the known laws of physics and chemistry cannot account for, is how the evolutionary process creates the "knowledge" or adaptive know-how that organisms need to live. Knowledge-making processes lie beyond the laws of physics and chemistry. Biologists have to study these processes and attempt to understand them. However, neo-Darwinism, the current orthodox theory of evolution, does not describe evolution as a knowledge-making process, but only as an adaptation-supplying process that occurs thanks to natural selection. This is so, even though adaptations depend on the acquisition of structural and functional adaptive know-how (R_i) by evolving populations of organisms. The adaptations of organisms are contingent on the capacity of evolution to "make" the knowledge that ultimately informs them. Furthermore, the natural selection of genetic variation, while primary, is only one of several knowledge-gaining processes that can occur in evolution (see chapter 7). Here too, orthodox evolutionary theory falls short. It doesn't yet recognize supplementary information-gaining processes in evolution. We will be returning to these issues in later chapters.

THE STRUCTURE OF THE BOOK

Part I of this book attempts to understand the fundamental properties of life anywhere in the universe where life may exist. The remaining chapters of part I concentrate on Schrödinger's subordinate question: *What else is needed beyond the known laws of physics and chemistry to understand life?* Chapter 2 considers what is meant by the term "information" in evolutionary biology. There are many conceptions of information used by scientists and engineers. Chapter 3 explores what is meant by the biological concept of adaptation in the context of adaptive interactive niche relationships between organisms and their local external environments. This will include the use of Ross

Ashby's control theory to specify how the variable properties of organisms relate to the variable properties of their environments. In chapter 4, I will consider meaningful information or adaptive know-how, rather than just bits, relative to the adaptations of organisms. Chapter 5 considers how the bioenergetics and the bioinformatics requirements of life, interact with each other, potentially on any planet where life may exist.

Part II of this book applies the points raised in part I to life on Earth. Chapter 6 addresses the specific challenge of explaining the origin of life on Earth. This chapter owes a great deal to the support of Bruce Damer and David Deamer, the authors of one of the two principal contemporary hypotheses about the origin of life on Earth. Chapter 7 concentrates on the evolution of supplementary information-gaining processes, including developmental processes in individual organisms and social and cultural processes in social animals, including humans. Chapter 8 concerns the origin and subsequent evolution of ecosystems on Earth as a function of the coevolution of niche-constructing organisms with their environments. Chapter 9 summarizes why, in light of all the preceding chapters, it may be necessary to extend the theory of evolution beyond the modern synthesis and neo-Darwinism. Chapter 10 applies all the ideas discussed in the preceding chapters to the question of how humans are currently contributing to the evolution of life on Earth.

2 INFORMATION

In chapter 1, we established that all living organisms have to oppose the second law of thermodynamics without violating it, and that to do so, they must be informed. In such cases, to understand life, we must focus on the natural processes that supply organisms with their adaptive know-how, or R_i. What are these processes, and how do they work?

To answer these questions, we need to go back to the work of Claude Shannon, often called "the father figure of information theory," even though he didn't much like that accolade. Shannon was a mathematician and engineer working at Bell Labs in New Jersey, which is renowned for pioneering research into telecommunications. In the late 1940s, he completed his most celebrated work, establishing the foundations of communication theory—which is what he preferred to be remembered for—and thereby ushered in the new digital age. Not everyone realized it at the time, but the value of Shannon's work was immediately recognized by his fellow mathematicians, engineers, and scientists, and by his colleagues at Bell Labs. One consequence of Shannon's immense prestige was that his employers had the wit to give him a free hand to work on any problem he liked or to fool around in any way he wished. (Among other quirks, he rode down the corridors of Bell Labs on a monocycle while juggling!)[1]

Shannon nonetheless delivered some truly memorable and pioneering work. One project was initiated by his wife, Betty, a fellow mathematician at Bell Labs. She gave Shannon a present, the biggest and most expensive construction kit she could find. It was designed for kids, which raised a few eyebrows, but Betty Shannon knew her husband, and he loved his present. He immediately started using it to construct all manner of things in the basement of their house. Before long, Shannon, with Betty's help, was constructing robots. The first one was called a "turtle," and it did little more than explore a room by traveling around the room until it hit a wall, and then it turned around to travel further and eventually discover another wall by hitting that one too. Not too impressive by modern standards.

Shannon's second robot, however, which he named Theseus, became internationally famous. Theseus was a small wooden mouse, equipped with three wheels, some copper whiskers, and a magnet embedded in its body. He also had an external brain, comprised of seventy-five post office relay switches under the maze. Theseus's job was to negotiate a maze and learn how to find a piece of metallic cheese as a reward. His ability to move depended on interactions between the magnet embedded in its body and the magnet, or possibly magnets, embedded in the maze. It was Theseus's ability to learn the maze by running through it that so intrigued and startled everybody. Remember, this was 1948. Theseus took some time to learn how to negotiate the maze, but once he did, he subsequently went straight to the metallic cheese reward, without making any further errors. In this, the robot was similar to the rats that psychologists were teaching to learn comparable mazes at that time. Shannon's own provocative comment about Theseus's achievement was that the mouse did not solve the maze, "the maze was solving the mouse."[2] That was a joke. But what did he mean by it? And how did Shannon's interest in a robot that was capable of learning relate to his work on communication theory?

Shannon's first insight was that you had to forget about the meaning of a message if you want to understand how a message is communicated. There is a difference between the act of communication and the meanings that are communicated, and the latter is irrelevant to the former. Shannon then established the basic components and mechanisms of communication. A message must have a source and a transmitter. Typically, the transmitter has to encode the message into symbols of some kind and then transmit the message in symbolic form, down a communication channel, connecting the transmitter to a receiver. The channel carries the signal between the transmitter and the receiver. The channel's capacity to transmit signals is limited in various ways. For instance, the speed with which it transmits the message may be limited by power or bandwidth, or the clarity with which it transmits may be limited by the amount of background "noise" that the signal has to compete within the channel. A crucial variable is the signal-to-noise ratio. Crudely stated, if the noise gets too great relative to the power of the signal, the receiver will not get a clear message. The receiver has to decode the message and send it to its destination.

Major questions that Shannon was trying to solve included: How much information was lost during its transmission through the channel? What were the causes of this loss? How could the losses be minimized? These problems were assigned to him by Bell Labs. After all, Bell Labs were owned by the biggest telephone company in the US at the time, AT&T, and these were the kinds of practical problems for which they wanted answers.

To solve such problems, Shannon had to find some way of measuring the amount of information carried by whatever message was being transmitted down the channel from the transmitter to the receiver. The unit of measurement he came up with was a binary digit, or bit. A bit equals the amount of information in, or the amount of uncertainty reduced by, one answer to one "yes" or "no" question. A bit can therefore be represented as either 1 or 0, and any message of any length can be represented by a string of 1s and 0s. Shannon also indicated how measuring information in bits was independent of the meaning of any message. For example, Morse code comprises strings of dots and dashes, equivalent to 1s and 0s. You can send a meaningful message to another person, such as your daughter, like "Your dinner is in the oven," by selecting the appropriate string of dots and dashes in Morse code. But you could also send your daughter a completely meaningless message in the form of a random string of dots and dashes of exactly the same length and with the same number of bits. This example makes the point that, to communicate, both the transmitter and the receiver must share the same symbol system. In this instance, both you and your daughter must know Morse code in order to communicate.

Shannon used bits to measure the efficiency of the transmission of information through a channel, including the loss of information between a transmitter and a receiver. To everyone's surprise, he came up with an equation that captured the efficiency of information transmission, including the loss of information, and was very similar to the equation that measures the loss of order, or the increase in entropy, in a physical system as a function of the second law of thermodynamics. There was one difference, though. The second law of thermodynamics includes a measure of temperature, because entropy is related to temperature, but Shannon's equation did not. His measure of information was dimensionless and had no need to refer to temperature.

Shannon talked to John von Neumann, who was the author of the equation that described entropy in physical systems, about what he should call his finding. In response, von Neumann rather flippantly suggested that he should call it "entropy" because no one understands entropy anyway, and it would make everyone interested. (That may not have been what von Neumann actually said, but it has been repeated in textbooks ever since.) Equipped with his equation, Shannon was then able to tell his employers at Bell Labs a lot about the most efficient way of sending signals between transmitters and receivers along communication channels such as telephone lines. For example, to send messages faster, you need more power. That was obvious. Or you could strip out as much redundancy from the message as possible, so long as the message still made sense. To do this, you need to reduce the number of symbols that you use to carry a message to a minimum. This idea was less obvious. However, to reduce interference due to noise in a channel, you had to do the opposite. You had to add redundancy to make sure that your message still gets through, in spite of the background noise. A simple example might be to repeat the message, such as what occurs at airports where airlines make repeated announcements about gates for specific flights. There is always a balance between reducing redundancy to increase the speed of transmission and increasing redundancy to combat noise in a communication channel. That was a useful result.

Now we can go back to Shannon's mouse, Theseus. How can communication theory be applied to the learning process in an individual mouse, real or robotic? Does it even make sense to ask this question, given that Shannon's communication theory ignores the meaning of messages, whereas learning is always about the acquisition of meaningful information? Shannon presumably thought it did. But how could his communication theory lead him to building a robot mouse that was capable of learning to negotiate a maze?

It is possible to represent any ordered physical system or structure—in this case, the maze—in terms of bits. The mouse is capable of detecting these bits, and therefore acquiring information in this still meaningless but quantifiable sense. This information was derived from the design and structure of the maze that the mouse was running through. The Princeton physicist, John Wheeler, once described the order inherent in organized physical entities as "its" and the information associated with it as "bits," and argued that it is possible to extract "bits from its," and vice versa.[3] For Theseus, the source of its information (i.e., his bits) came from the design, order, and structure of the maze (his "ITs"), as sensed by his copper whiskers. Theseus's task was to translate the information that he sensed via his whiskers into meaningful, internalized "knowledge" in his robotic "brain" (i.e., about the structure of the maze, where the walls and arms are, and where the cheese reward is) and then into adaptive know-how, or Ri (i.e., concerning which way to run and which turns to take to find the cheese). Hence, I am defining three grades of information here. Henceforth, I will restrict the term "information" to meaningless bits, but also talk about meaningful (i.e., declarative) information as "knowledge," and meaningful functional (i.e., procedural) knowledge as "adaptive know-how." Ri refers to the latter.

Shannon's mouse also had to be capable of a minimal degree of niche construction—namely, to make choices about which arm of the maze to enter, to run down the chosen arm, and to register in some kind of memory the outcomes of his actions in the maze. When psychologists teach rats to do much the same thing in mazes, they refer to a rat's activity as an instrumental response or action. Can the learning process in the rats, and in Theseus, be redescribed in terms of the components of Shannon's communication theory? If so, was this what Shannon was indicating about the mouse's learning process when he said, "The maze solved the mouse"? Does the maze "teach" the mouse how to find its "cheese"?

I want to explore the possibility that Shannon's communication theory applies to all knowledge-gaining processes in nature. I will begin with learning, but then consider evolution. There is a preliminary problem, however. We usually think of communication in terms of two people, often nicknamed Alice and Bob by the professionals in the communication business. They could be talking to each other over a phone line. But are Alice and Bob having a two-way conversation, or are they having only a one-way conversation? Is Alice talking to a silent Bob, or is Bob talking back to Alice? If the latter, then Alice and Bob will both be transmitting and receiving messages. In which case, who is communicating to whom (or what is communicating with what?) when an individual learner is learning something about its environment, as Theseus was in the maze, and when no other organism is present? Who or what is the transmitter and who or what is the receiver in these circumstances?

The naive learner, which I'll call the "a priori" learner, transmits a message after it has learned something to a receiver, which I'll call the "a posteriori" learner. For a rat in a maze, the a posteriori learner is just a later version of itself. The a posteriori learner then switches to become the a priori learner again (in Theseus's case, at the start of his next run through the maze). The second time around, however, the a priori learner is no longer naive. It is now equipped with a memory of the outcome of its previous run through the maze, which it communicates to a later version of itself. It is as if an earlier Alice is communicating what she has learned, in the form of memories about the outcomes of previous learning trials, to a later version of herself. The learning process, in any individual organism or in Shannon's robot, is a time-based process. It requires time and communication between the before-and-after states of the individual learner. During this time, the learner is communicating to itself after each of its successive learning experiences.

Now we can consider the learning process by an individual learner step by step, as if it were a form of communication that can be described by Shannon's communication theory. It begins with the source of the information (e.g., the bits extracted from the maze). The mouse acquires this information by interacting with the maze. The outcomes of each run must be remembered, registered, and encoded in the brain of the mouse in the form of knowledge about the maze. In this way, Theseus converts bits into knowledge, a process that relies on the seventy-five post office relay switches that served as Theseus's brain. The a posteriori mouse must then have communicated the revised state of its brain, in the form of the revised configuration of its switches, back to its later self. Then the a posteriori mouse would have become an a priori mouse again at the start of its next run through the maze. However, it would no longer have been so naive. The mouse would now be carrying memories of its past interactions with the maze from its previous runs through it, in the form of the revised configuration of its switches. Functionally and implicitly, the mouse would also have had to assume that the maze had not changed and therefore its memories of its past runs were still relevant. The mouse could then use its memories to guide its next run through the maze. In this way, the outcomes of Theseus's previous experiences were remembered and communicated to the later version of himself.

Eventually, after a number of learning trials, the mouse learned to run directly to its cheese reward without making any mistakes. In this way, useful "knowledge" about the maze was apparently translated into adaptive know-how, or Ri, that specified how Theseus could solve his task. More generally, the same translation of information first into knowledge and then into adaptive know-how should be true not just of Shannon's robot, but of all animals that are capable of learning.

But what motivated Theseus, or for that matter, what motivates any learner, animate or inanimate, and causes it to want to learn? Shannon must have turned his mouse into a "goal-seeking" system that was motivated to find his "cheese" reward. In turning his mouse into a purposive, goal-seeking system, Shannon was doing the same as Maxwell did for his demon in his thought experiment many years earlier. Shannon was giving Theseus the properties that a real mouse would need to stay alive. Among other things, a real mouse must find food and consume it at regular intervals to survive.

SOCIAL LEARNING

Individual learning, however, is only one kind of animal learning. Many species of animals are also capable of social learning: they can learn by imitating or copying each other. Typically, in a social learning situation, one or possibly several animals already possess the knowledge about their world and the adaptive know-how that they need to operate in it. Inadvertently or deliberately, they may then transmit some of their adaptive know-how to a naive animal, who learns from them by copying.[4] Social learning is perhaps more overtly compatible with Shannon's communication theory than individual learning. In a social learning situation, one knowledgeable animal may serve as the transmitter of the message, while another naive animal may serve as the receiver. In every other respect, the communication between them should be the same as the communication between an individual learner communicating its memories of its past experiences to a later version of itself. Learning is clearly a natural information-gaining and knowledge-making process in animals. The knowledge acquired is likely to be meaningful, and therefore probably adaptive, for the animals that gain it, regardless of whether it is cognitive or conscious.

Now we can summarize the chain of relationships between information, meaning, purpose, knowledge, and adaptive know-how (R_i), relative to at least one information-gaining process in nature: learning. Information, measurable in bits, can be acquired by a learner from the causal texture of its external environment via its senses as it interacts with it. At this stage, the information is neither meaningful nor meaningless. It is just another aspect of physical reality.[5]

Meaning is assigned to this information by purposeful organisms when they interact with their external environments. All meaning in the environment is relative to the purposive systems that experience it. There is no meaning without purpose. All living organisms are purposive systems. Some artifacts constructed by organisms, especially humans, may also qualify as purposive systems. They will do so if they are carrying out any purpose on behalf of the organisms that construct them. Shannon's Theseus was one such artifact. The walls of the maze were given meaning by Theseus, either presenting barriers or facilitating his purpose of finding the cheese. If Theseus has not been given this purpose, no adaptive knowledge of how to run the maze could have accrued.

The purposes of organisms in their environments ultimately stem from the universal requirements of all living organisms to oppose the second law of thermodynamics without violating it. This fundamental need arises from the improbability of the internal environments of organisms, relative to their external environments, and their far-from-thermodynamic-equilibrium status. A fundamental need of all organisms, therefore, is to control a flow of energy and matter between themselves and their external environments that opposes the flux of energy and matter favored by the second law. If they fail, the flux would destroy them. Organisms must also avoid or mitigate other threats to their existence, stemming from abiotic sources or from other organisms, such as predators, in their environments. They must also avoid destroying their own habitats by harvesting excessive resources from them or dumping excessive detritus in them. But organisms cannot oppose the second law by chance. Survival and reproduction require organisms to acquire sufficient adaptive know-how (R_i) about their local external environments to persist, using whatever knowledge-making processes are available to them, including the learning process in animals.

Meaning converts bits into knowledge on the basis of its perceived usefulness or biological utility. As it learns, an animal will communicate its earlier memories to its later self—in fact, that is what memory is. Successive learning experiences will accumulate as memories in the nervous systems or brains of the animal. A subset of this knowledge, when put to work by the learner, will become operational knowledge in the form of adaptive know-how, or R_i. A learner may then modify its subsequent behaviors in its environment. It may also modify its environment by its niche-constructing activities in light of what it has learned.

THE ASSIGNMENT OF MEANING

The assignment of meaning to information emanating from the external environment raises a problem not just for learning animals, but for all organisms, at all times. The issue is that the biological needs[6] of organisms fluctuate continuously throughout their lifetimes—food is no longer required once they have eaten, and rest is not a priority after they have slept. These fluctuations are partly due to the fluctuating demands of organisms' own internal environments, such as their state of hydration, but also partly due to the heterogeneity of the external environments that they encounter. As a result, the relationships between individual organisms and their external environments are always changing, on all possible scales. But if the needs of organisms fluctuate relative to their external environments, so do their purposes and motives at different times and places. The (perhaps surprising) conclusion is that the meaning that individual organisms assign to the incoming information that they receive via their senses must fluctuate too.

Superficially, some of the needs of organisms, such as their basic fitness needs, may appear to be stable throughout their lives. For example, all organisms need to survive at all times during their lives. However, in reality, this permanent need to survive decomposes into hierarchies of shorter-term contributing needs, such as for food or shelter, with different priorities at different times and places. These subordinate survival needs may sometimes also compete with each other for priority. To a lesser extent, the same is true of the needs of organisms to reproduce and rear their offspring.

This means that the meaning that organisms assign to the apparently objective information that they receive from their external environments will always be heavily dependent on the subjective states of the organisms themselves. There is no objective description of meaning associated with an environment—it varies from one individual to the next and from one time to the next, which makes characterizing meaning complex. In fact, the resulting complexity is so great that this may have been one reason why Shannon originally ignored the meaning of messages being sent by transmitters to receivers, in order to enable him to concentrate on the process of communication itself. So how can this complexity be handled?

Richard Lewontin was one of the first evolutionary biologists to draw attention to the different meanings, or significances, of the same events in the external environments for different organisms. He did so, not so much in terms of the fluctuating needs of individual organisms during their lifetimes, but more due to the variable needs of different species that experience the same environmental events in shared ecosystems. For example, the same environmental change, say a slight change of temperature, might be insignificant for many organisms but important for others. In that case, the same change in an environmental factor or condition would have different meanings for different species in the same environment.

Other investigators have concentrated more on the fluctuating needs, purposes, and motives of individual organisms during their lives. One approach was initiated by the psychologist James Gibson,[7] writing in the 1950s and 1960s about the perceptions of both animals and humans of events in their external environments. He realized the degree to which the perceptions of organisms combine objective realities with subjective needs. He came up with the concept of "affordances," which characterize the perceived properties of physical objects. Depending on its current needs, purposes, and motives, an animal would assign different affordances to the same external environmental events and to the same external stimuli, at different times and places. The same event might be assigned strong positive affordances and be highly meaningful to an organism at one time but then be meaningless and have zero affordances at another time. For example, a hungry animal may assign high affordance to any cue indicating how it could find food, whereas a satiated animal might not even notice the same cue.[8]

All these approaches have their merits, but none really solve the problem of how the meaningfulness of events in the external environments of organisms could be measured or studied empirically.[9]

EVOLUTION

Can these relationships between information, knowledge, and adaptive know-how, discussed in the context of learning, be generalized to other knowledge-gaining processes in nature? What about the primary knowledge-gaining process of population-genetic evolution? Can evolution also be described in terms of Shannon's communication theory? And is anything gained by doing so? Might the application of Shannon's communication theory to evolution increase our understanding of the bio-informatics of evolution? I suggest that the answer to these questions is "Yes."

Let's go back to evolutionary theory. According to traditional neo-Darwinism, populations of evolving organisms are informed by autonomous natural selection pressures arising from the causal textures and dynamics of their external environments. From this traditional standpoint, individuals in populations are informed by the genes that they inherit from their ancestors about the environmental factors that their ancestors encountered in the past. This information may or may not be meaningful for individual organisms in the current generation relative to their contemporary environments. Whether the genetic information they carry is valid partly depends on whether the prior natural selection pressures, experienced by ancestral organisms, are the same as or similar to those being experienced by their descendants in the current generation. It also depends on the fluctuating fitness goals and needs of individual organisms in the current generation.

Potentially, the biological concept of fitness provides a way of measuring the meaningfulness of incoming information relative to individual organisms in evolving populations. This is where Lewontin's point, that the same physical events in environments may have different meanings for different species of organisms, is relevant.

Now we can apply Shannon's communication theory to the evolutionary process, as described by neo-Darwinism. When we do, the ancestral populations of organisms become the transmitters of potentially meaningful messages to descendant organisms in each succeeding generation of their population, including the current generation, which become the receiver of the messages.

The source of the information carried by these messages is the set of natural selection pressures that the organisms in ancestral populations encountered in the past. The channels of communication between ancestral organisms and descendant organisms in populations are the genetic inheritances that connect them. These genetic channels may be more or less noisy depending on the frequency of random genetic mutations transmitted between generations, most of which will be deleterious. Further noise may be added by the evolutionary process of random genetic drift. Through the process of natural selection in ancestral populations, environmental information (which could in principle be measured in bits) was converted by the ancestors into adaptive know-how (concerning how to survive and reproduce in their environments), and this former know-how is passed on to descendants through genetic inheritance. The know-how that each generation receives from its ancestors is spread across all the individual organisms in the current generation via their individual genetic inheritances. Whether this information becomes the source of meaningful adaptive know-how (R_i) for contemporary individual organisms, then, depends on the two provisos that we have just encountered.

If the natural selection pressures encountered by contemporary organisms in their environments are the same as or similar to those encountered by their ancestors, then the messages that they receive from their ancestors will probably remain adaptive. If the fluctuating needs, fitness goals, and purposes of contemporary organisms are also similar to those of their ancestors, then that too will make it more likely that the messages that they receive from their ancestors will be adaptive. But if either of these provisos are not realized, the former know-how will no longer be adaptive.

Now let's consider the evolution of a population of organisms, all of whom are capable of learning. In practice, the points that I make about evolution in a population of learners apply more broadly to any organism that possesses an exploratory and selective knowledge-gaining process, as I discuss later in this chapter. What difference does this make, relative to a population of nonlearning organisms? The first difference is that the individual animals in the population of learners will no longer be exclusively dependent on the primary evolutionary process of population genetics for their adaptive know-how. They can gain additional adaptive know-how by informing themselves about their own individual local environments through their learning.

The second difference is that because the sources of the information detected by these two knowledge-gaining processes are different, the knowledge that individual organisms gain from them will be different too. The set of past interactions among diverse individual organisms in ancestral populations and their environments will span a very wide region of space and time. Individual organisms in each generation will be informed by their individual genetic inheritances as a consequence of their membership of an evolving population.

In contrast, in the case of individual learners, the samples of past organism-environment interactions that inform their learning will be far less rich. They will be drawn exclusively from the individual learner's past encounters with its own local environment. In the case of social learning, the source of the information transmitted to an individual learner will be the sample of collective experiences of members of the individual's social group. All knowledge-gaining processes are historical; they supply messages about the past to the present. But they refer to different histories, of different individuals or populations, and hence with different content. These varying histories then supply diverse meaningful messages to individual organisms. If we compare individual learning with the evolutionary process, the past interactions of individual learners with their local environments will be more recent than the interaction of their ancestors with their ancestral environments. They will also be based on the learner's own within-lifetime encounters with its local environment. Learning, therefore, may supply more up-to-date and relevant information to an individual than its inherited genes can.

More generally, different types of knowledge stemming from alternative knowledge-gaining processes may have different values for individuals, at different times and places. If a learner lives in an environment that has hardly changed from the environments encountered by its ancestors, the meaningful genetic information that it inherits may be the most reliable source of adaptive know-how for that animal. However, if the learner's own local developmental environment is very different from the environments experienced by its ancestors, perhaps because of changes that have occurred since it was born, then the information gained from its own learning may be a better source of adaptive know-how than its genetic inheritance. This advantage, conferred by learning, may be one reason why the supplementary knowledge-gaining process of learning evolved in the first place (see chapter 7).

The application of Shannon's communication theory to both the evolutionary and learning processes highlights their similarities and differences. The principal similarity stems from the point that knowledge-gaining processes always seem to involve some kind of communication between the past and the present. They therefore always depend on some kind of history. The logic describing the communication of ancestral messages by both the evolutionary process in populations and the learning process in individual animals is also the same, as was established by Shannon's communication theory and more recently argued by Watson and Szathmáry (2016).[10] There always has to be a source of the information, a transmitter to transmit the knowledge extracted from that information, a channel of communication between transmitter and receiver, a receiver of the knowledge, and a final destination for the message.[11] In practice, when considering evolution and learning, where the receiver and destination are the same population or same individual, then the distinction between receiver and destination is blurred.

The principal differences between the evolutionary and learning processes are that information is derived from different histories and the physical mechanisms and components involved in the communication of messages by the two processes are completely different. For example, in evolution, the genetically encoded memories are transmitted by ancestral populations of organisms to a later generation of the same population via genetic inheritance. In contrast, when learning, an individual transmits what it learns about its environment to a later version of itself, utilizing the communication channel of its nervous system or brain.

Given their similarities and differences, how do these two knowledge-gaining processes relate to each other? Learning allows individual animals to acquire modified phenotypic traits in the form of modified behaviors and niche-constructing activities, which may then affect their future behaviors. These phenotypic modifications amount to within-lifetime acquired characteristics. But according to neo-Darwinism, none of the characteristics that individual organisms acquire during their lifetimes can affect the genes that they subsequently pass to their descendants, or the evolution of their populations. That would amount to Lamarckism, which neo-Darwinism rejects. However, the application of Shannon's communication theory to both the evolutionary and learning processes raises doubts about whether neo-Darwinism is right to reject Lamarckism so emphatically.

Shannon's communication theory indicates that the relationship between evolution and learning may be a reciprocal one, in which both processes affect each other. The capacity of an animal for learning is likely to depend on its membership of a particular population. That capacity is itself a genetic given, an evolved capability. But the same is not true of what they learn. What individual animals learn during their lifetimes may be biased by the genes that they inherit from their ancestors, but the details of what they learn cannot be determined in advance. For instance, if the information that individual learners gain from their ancestors via their genetic inheritance is now out of date, relative to their contemporary environments, a learner may be able to adapt to its environment by updating its out-of-date genetic information via learning.

It may help to illustrate these abstractions with a few examples. Imagine that you are a psychologist interested in animal behavior. You are trying to teach a rat in an artificial environment, called a Skinner box, how to press the left lever but not the right lever in the box to gain a food reward. But no rat in the wild could ever have experienced such an artificial environment. It is therefore not plausible that what an animal learns in a Skinner box could be exclusively a consequence of the genes that it inherits from the evolving population of animals to which it belongs. Nevertheless, rats, as well as many other species of animals, easily learn such tasks. This point is brought home by experimental studies showing that through copying, rats will learn to eat low-nutrition or even toxic foods, and birds can acquire fears of even arbitrary objects, such as kitchen utensils.[12] What the genome encodes is the means to learn, not the outcome of the learning.

The same is true of learned niche-constructing behaviors, particularly in the case of opportunistic niche construction. One of my favorite examples is the Galapagos woodpecker finch. This bird learns to exploit a woodpecker niche without being previously equipped by evolution with the typical phenotypic traits of more familiar woodpeckers, such as a sharp bill, long tongue, and reinforced skull morphology. Rather, it learns to use a stick or a cactus spine tool to poke out grubs from the barks of trees.[13] On different islands in the Galapagos archipelago, related species have learned different niche-constructing tricks. On one island, the birds discovered how to become vampires as such, by pecking at large booby birds until they bled and then drinking their blood. On another island, the finches discovered how to eat eggs that were too hard for them to break open with their beaks. If they found an unguarded egg that they couldn't break open, some birds learned how to lie on their backs and push the egg over the edge of a cliff with their feet, causing it to break on the rocks below. After that, it was easy for the bird to fly down and eat the egg's contents.[14]

But if prior evolution cannot account for everything that animals learn during their lifetime, including their niche-constructing activities, what they learn may subsequently affect the evolution of their populations. Probably it will not do so directly, but rather indirectly. This is because what an individual animal learns about the external environment is likely to affect its own fitness. Smart learners may be fitter than dumb learners. They may increase the probability of their own survival through their learning, such as by learning how to forage for food better than their conspecifics. Or they may learn how to modify their own environments by their niche-constructing activities better than their rivals. If they do so, it will be likely to affect their reproductive success. For instance, it may bias which individuals in the current generation of a population make the most contribution to the genetic inheritances that are transmitted to their descendants in the next generation. Or they may learn to consume a novel food, and thereby generate selection for gene variants that improve the metabolism of that dietary item, as has seemingly happened repeatedly in the course of recent human evolution.[15] In these ways, the learned and acquired characteristics of individual organisms may affect the subsequent evolution of their populations.

The application of Shannon's communication theory to both the primary evolutionary process and supplementary process such as learning, therefore, challenges neo-Darwinism's assumption that acquired characteristics of individual organisms cannot affect the subsequent evolution of their populations. Shannon's theory may be able to throw further light on the bioinformatics of evolution.

In much of the rest of this book, we are going to be concerned with how the information accrued by individual organisms in evolving populations through natural selection is translated into adaptive know-how (R_i). Initially, we will focus primarily on population-genetic processes and genetic inheritance systems (see chapters 3 and 4) but later, we will consider other supplementary, evolutionary, and developmental processes, including learning (see chapter 7). In the remainder of this chapter, however, we will be concerned only with the several different ways in which information is described in evolutionary biology and how each of them relates to "meaning."

INFORMATION AND ITS MEANINGS

Different sciences use the word "information" in different ways. It will be sufficient to consider just four of them here. The four alternatives are bits, qubits, semantic information, and algorithmic information. Do any of these four ways of describing information help us to understand the kind of adaptive know-how that is responsible for the adaptations of organisms? I'll elaborate on each of them in turn.

Binary Digits (Bits)

We have already encountered bits while discussing Shannon's communication theory and his robotic mouse, Theseus. This measure of information has turned out to be of only limited use in evolutionary biology. That is because Shannon's bits deliberately ignored meaning, but the information accrued by natural selection in evolving populations must be translated into meaningful knowledge relevant to the adaptive demands placed on organisms in evolving populations by their environments. Fundamentally, the information has to be about the far-from-thermodynamic-equilibrium relationships between the internal environments of organisms and their external environments. Bits have nothing to say about meaning for fitness-seeking, purposive organisms in their environments. They do not specify fitness or adaptive know-how. They only measure the quantity of information in a physical system.[16]

The apparent parallel between information in a communication channel, as specified by Shannon, and order and disorder in a physical system, as characterized by thermodynamics, can also be misleading. As we saw previously, Shannon's measure of "entropy" in his communication theory does not relate to temperature in the same way as entropy does in physical systems. This means that there is no easy way of connecting the physical theory of entropy with Shannon's measure of bits, in spite of their provocative similarities.

Another difficulty is that the amount of information, measured in bits, depends on what you are trying to measure. If you are trying to measure the amount of information carried on a computer chip in a conventional computer, you might come up with a figure of 10^{12} bits, while if you wanted to measure how many bits could be carried by subatomic particles in the same chip, you might come up with a figure in excess of 10^{23} bits.

Initial oversimple attempts to apply Shannon's bits in evolutionary biology did not work either. In the 1970s, it was assumed that humans, being the most complex of all organisms, must carry the most genes and the most DNA, and also the longest strings of base sequences as measured in bits. At that time, people thought that humans must carry about 100,000 genes. In fact, it turned out that several other organisms, such as salamanders, have far more genes and more DNA than humans. Humans are now thought to possess approximately 22,000 genes, a number that is regarded by some as astonishingly small. Strikingly, humans have a similar number of genes to the nematode worm *Caenorhabditis elegans*, which comprises only 1,000 cells and has no brain, heart, or complex organs.

It is also the case that humans, along with many other organisms, do not depend exclusively on genetic inheritances, communicated between successive generations for their adaptive know-how. They depend on other inheritance systems as well, such as learning and sociocultural processes. Human complexity is not just a function of the number of inherited genes; it depends on other processes too. Contemporary geneticists also point out that it is not the number of genes, but rather the details of how they interact with each other, that best explain complex metazoan organisms, such as humans.

In general, there is no simple way of relating fitness to bits, and for these and other reasons, evolutionary biologists have largely lost interest in Shannon's binary digits. With respect to the question of what is meant by adaptive know-how, or R_j, in evolution, bits are not the answer.

Qubits

In the context of quantum mechanics, physicists talk about quantum bits, or qubits, rather than bits. This is another way in which information is described in the sciences.[17] Quantum physics makes any reasonable person dizzy. But I'll have to risk a little dizziness to introduce the topic of qubits. Many years before he wrote *What Is Life?* Erwin Schrödinger made two significant contributions to quantum physics. One was an equation that has been known as "Schrödinger's equation" ever since. The other was his cat.

You will probably have heard of Schrödinger's cat. It is an astonishing creature. Schrödinger used the idea of a strange cat to translate one of the weirdest concepts in quantum mechanics into the world of classical mechanics and the physics of everyday life. His cat was neither alive nor dead, but both at once. It was illustrating the quantum property of "superposition." The cat's life and death were in superposition. You can't ask a simple "yes" or "no" question about whether Schrödinger's cat is alive or dead. Asking that question would not yield one bit of information. You simply can't use bits to measure information in a quantum system. Quantum physicists had to develop the concept of qubits to enable them to account for quantum phenomena.

Quantum physics is concerned with the properties of subatomic particles, such as electrons or photons. It does not immediately refer to macrophysical objects in everyday life. The key to entering the quantum world is another of nature's constants, called "Planck's constant," after the physicist Max Planck, who discovered it at the beginning of the twentieth century. Most physicists assume that the strange world of quantum physics is more fundamental than the familiar world of classical physics. The latter is derived from the former, not vice versa. Quantum physics does not refer directly to the world of cats.

So why did Schrödinger challenge us with his cat? Apparently, he was uneasy about the directions in which quantum physics was heading in the first decades of the last century. He used his cat to satirize some of its apparent absurdities. But why a cat that was both alive and dead? Schrödinger's daughter pointed out that her father's cat need not have been both alive and dead. He could have made the same point, less dramatically, with a cat that was both asleep and awake. But his daughter also tells us that Schrödinger did not like cats.[18] Perhaps he was using his cat to get his revenge on the whole feline species.

Schrödinger's other major contribution, however, undermined his own satire. His equation describes the dynamics of a fundamental phenomenon in quantum physics, known as the "wave function," which refers to the amount of energy in whatever quantum system is under consideration. As I understand it, Schrödinger's equation was originally intended to describe the dynamics of a wave function relative to a single elementary particle, such as an electron. However, it turned out to apply universally.[19]

That meant that Schrödinger's equation could be applied to the whole of the quantum world. And it made quantum physics operational. Apparently, Schrödinger remained none too pleased with his own contribution to quantum physics.

Schrödinger was not the only eminent physicist to feel uneasy about quantum physics at the time. Another was Einstein. Einstein introduced a second weird concept into quantum physics, known as "entanglement," which refers to action at a distance between particles. For instance, if a physicist measures one of the quantum properties of an electron in Oxford (e.g., its rotation or spin), and if the Oxford electron is entangled with another electron in New York, then by measuring the spin of the Oxford electron, the physicist will also fix a complementary quantum property of the electron in New York. It would do the same relative to any other particle with which it was entangled, anywhere in the universe. All of this left Einstein very uneasy, and he worked on it for many years. He described it as "spooky physics." He also did not like the fundamental probabilistic, as opposed to deterministic, nature of quantum physics. His celebrated comment about that was "God doesn't play dice."

There is far more to quantum physics than this, and many physicists are devoting their careers to trying to understand it better.[20] But we can leave it there. The relevant question for us is: Do biologists need to know anything about quantum physics to understand their own subject? Conversely, do quantum physicists need to know anything about biology to understand theirs? The answer to both these questions appears to be "No."[21] It may be that qubits can help quantum physicists understand their subject, but that is not going to help us understand the kind of meaningful information that underpins adaptive know-how, or R_p, in evolution. In this respect, qubits are of no more use to evolutionary biologists than bits are. A mouse needs to know about cats, but it can't know and doesn't need to know about cats that are both alive and dead.

Semantic Information

Let's turn from measurable information to unmeasurable meaning. Meaningful information, or "knowledge," is often called "semantic information."[22]

Semantic information was shunned by Shannon because of the unmeasurability of meaning. It threatened the clarity and power of his theory of communication. Shannon could describe the amount of information, measured in bits, in messages sent by transmitters to receivers through communication channels. But he did not attempt to describe the meaning of those messages. As we have seen, the meaning of a message communicated between two people requires the introduction of some extra subjective variables to account for it. These variables are usually the province of philosophers, psychologists, artists, poets, and lovers, rather than mathematicians.

Nevertheless, evolutionary biologists have to deal with meaning. They cannot ignore meaning in the same way that Shannon did, since organisms cannot be informed by nonsense. Organisms are active, purposive systems, and their purposes define what information is meaningful for them, relative to their fundamental task of opposing the second law of thermodynamics. Biologists come closest to measuring the meaning of the information acquired by organisms when they use the concept of fitness. While there are many concepts of fitness,[23] the adaptations of organisms typically are assessed in terms of their contributions to survival and reproduction. That is an indirect way of measuring the adaptive know-how (or R_i) carried by organisms. Recall that R_i is responsible for the expression of the structural and functional adaptations of organisms. But does reproductive success capture all the adaptive know-how carried by organisms?

At present, commonly deployed fitness concepts in evolutionary biology suffer from at least one shortcoming—namely, they do not yet refer to thermodynamics. These concepts are not derived from the thermodynamic relationships between improbable living organisms and their external environments. And they should be.

Evolutionary biologists have made several adjustments to their fitness concepts over many years. One of the best known is the concept of inclusive fitness, introduced by Bill Hamilton.[24] Hamilton pointed out that to transmit your genes to the next generation, you don't have to reproduce more offspring yourself. You can help your closest genetic relatives to reproduce instead. For example, your brother or sister will carry half of your genes. If he or she can be encouraged to have two offspring, then genetically, that will be equivalent to you having one extra offspring yourself.

The calculation is more complicated in the case of remote relatives, but you get the idea. The theoretical attraction of inclusive fitness is that it can go some way toward accounting for cooperative relationships between organisms that are compatible with neo-Darwinism. It can account for cooperation among close kin. However, inclusive fitness (at least as generally understood) cannot account for cooperation among genetically unrelated organisms.[25] This matters because sometimes organisms can enhance their fitness by cooperating with unrelated conspecifics in their own population (e.g., through mutualistic relationships). However, niche construction theory (NCT) raises a further problem here. It challenges the idea that measuring the fitness of organisms exclusively in terms of the genes that they pass to their descendants is sufficient. NCT proposes that there is a second general inheritance system in evolution, in addition to genetic inheritance, called "ecological inheritance." Niche construction and ecological inheritance also imply an additional form of relatedness, which arises through shared niche-constructed environments.[26]

At the population level, if multiple individual organisms in the same population engage in niche construction in the same or a similar way, then collectively they may cause significant changes in the ecological flows of energy and matter, or R_p resources, in the current populations' environments. They may also affect the energy and matter available to later generations via ecological inheritance.[27] But if niche construction and ecological inheritance are primarily about the modification of energy and matter resources in environments by organisms, then why are they relevant to semantic information? The answer is that when populations of niche-constructing organisms change ecological variables in their environments, they also modify natural selection pressures for their own and each other's populations. Modified natural selection pressures then change the "messages" that are communicated in the form of meaningful adaptive know-how, or R_p, by the current generation of organisms to later generations via genetic inheritance.

In this manner, over time, the two inheritance systems, ecological and genetic, interact with each other. That means that the two fundamental resources that are vital for life: energy and matter resources, or R_p and informational resources, or R_i, also interact with each other over time. However, neither inheritance system fully determines the other. Niche construction is not fully predetermined by genetic inheritance because it is also influenced by other developmental processes, such as learning. Likewise, genetic inheritances are not fully determined by prior niche construction. Genetic inheritances are determined by both those natural selection pressures previously modified by niche-constructing organisms and by autonomous natural selection pressures (and a multitude of other population-genetic processes).[28]

Is there another way to measure fitness that better captures ecological inheritance? Brown et al. refer to two components of fitness that they call "survival" and "production," where production equals growth and reproduction.[29] They are interested in the trade-off between generation times and productive power among all the diverse species on Earth. The reproductive component of fitness is usually measured in the currency of genes; however, Brown et al. prefer to use energy as their currency of fitness. In this way, they find that organisms in diverse species transmit approximately the same energy per gram to their offspring in the next generation, regardless of huge differences in their size and biomass. They argue that their approach supplements and complements genetic fitness. Their point of view is fully compatible with the two inheritance systems, ecological and genetic, proposed by NCT: an energy component and a genetic component. NCT implies just that.

Perhaps a different conception of fitness might better capture the relationship between energy and information in evolution. However, even if that is the case, difficulties remain in relating semantic information to the evolutionary concept of fitness, even though it remains relevant to the evolutionary process, as well as to a subset of adaptations. Semantic information is meaningful information, and thereby it equates to "knowledge," as I have used the term previously. But adaptations do not solely constitute expressions of knowledge about ancestral environments; they also exhibit practical knowledge concerning how to build and operate traits that function effectively in ancestral environments. Hence, semantic information still falls short when it comes to accounting for the adaptive know-how, or R_i, carried by organisms.

Algorithmic Information

One problem with semantic information in the context of evolution is that, while it potentially accounts for the meaningful information or knowledge possessed by organisms about their environment, it is not explicitly operational knowledge. It is not equivalent to the adaptive know-how, or R_i, needed by organisms to act on their world in pursuit of their fitness goals. Hence, it is necessary to supplement semantic information with another kind of information, algorithmic information, to fully capture concepts such as adaptation and fitness. An organism has to know about the consequences of its own actions in its world to operate adaptively. For example, a foraging bird may be sensitive to environmental cues, which indicate the presence of a worm in the soil. That is semantic information, or knowledge about the environment. But the bird also needs to know how to catch the worm. That demands algorithmic information. The bird has to know how to peck the soil in the right place and at the right moment in order to extract the worm from it. The bird needs both semantic and algorithmic information to catch its prey. Both are required for the bird to survive and reproduce.

There is another property of algorithmic information that is relevant to evolution. Meaningful information may have a measurable algorithmic value. It is potentially quantifiable.[30] Algorithmic information is defined by asking what size of computer program is needed to generate some data, where the minimum number of bits is "the algorithmic information content" of that data.[31] Counterintuitively, the less "algorithmic content" there is, the more "meaningful" the algorithm is likely to be.[32] For instance, only a little algorithmic information is needed to generate the value of the number π to a million decimal places, while much more algorithmic content is needed to generate a string of a million random numbers, and "the shortest program for outputting it will be about as long as the number itself."[33] The key variable is the "compressibility" of data. To quote Chaitin, "A useful theory is a compression of the data: comprehension is compression. You compress things into computer programs, into concise algorithmic descriptions. The simpler the theory, the better you understand something."[34]

This compressibility is likely to be relevant to evolution to the extent that natural selection can be regarded as an optimizing process that seeks efficiency gains in information acquisition and use. Natural selection should favor the most efficient way of acquiring adaptive know-how in organisms. Organisms have to be able to go from knowledge about the world to adaptive know-how, but these are complementary forms of information. The better an organism understands something, the more likely it is that it will possess the adaptive know-how that it needs to predict the outcome of its own actions in its world and to respond to the causal texture of its world adaptively. The greater the compression of the data, the more efficient the strategy employed by the organism is likely to be. The kind of information that is relevant in evolution (R_i) is a combination of meaningful semantic and algorithmic information, where in each case the meaning is derived from the fitness purposes of organisms.

GENERALIZING FROM SHANNON

So what have we learned about the bioinformatics of life from applying Shannon's communication theory to both evolution and learning? Have we learned anything more about how organisms acquire the adaptive know-how (R_i) that they need to oppose the second law of thermodynamics without violating it? Perhaps quite a lot. The fact that it is possible to describe evolution and learning in terms of Shannon's communication theory encourages the hypothesis that it may be possible to do the same for all the information-gaining processes that inform organisms in nature. These processes range from the vertebrate immune system, to the insect tracheal system, to sociocultural processes in humans, up to and including the processes of science.[35] They all have a great deal in common.[36] However, there are clearly significant differences between them too. If I am correct, then we ought to find comparable similarities as well as comparable differences to those that I described between evolution and learning. These should reoccur in all knowledge-gaining processes in nature.

Let's start with the similarities. There is one provisional point. It is necessary to distinguish between the physical systems in nature that are doing the actual knowledge gaining and the organisms themselves. It is a subtle but important distinction. An animal is an knowledge-gaining system, relative to its own capacity for learning. But it is not a knowledge-gaining system, relative to the genetic information that it inherits from its ancestors, as a function of its membership of its population. All knowledge-gaining systems depend on organisms. However, some depend on sets or groups or communities of organisms, whereas others depend on individual organisms only. The population is what acquires genetic knowledge, although individuals may subsequently inherit that information. Darwin had to introduce population biology to explain how natural selection pressures could inform organisms indirectly. Following Darwin's lead, it's always necessary to identify the nature of the information-gaining system that is responsible for each different information-gaining process in nature. I will concentrate primarily on the different knowledge-gaining systems in nature rather than on the organisms that they inform.

The first general point is that all knowledge-gaining systems in nature are localized in space and time. They may be able to travel in any direction in all three dimensions in space, as well as in time. They can also stand still in space, for a time. They can even return to a place that they visited before, at some earlier time. Seasonal migrating animals demonstrate this point. But in time, they are much more restricted. They cannot stand still in time, nor can they go back in time. Whatever their nature, organisms can only travel from the past to the present.[37] This localization of knowledge-gaining systems in space and time has two consequences. One refers to their capacity to gain knowledge during time, and the other to gaining knowledge from different places in space.

All knowledge-gaining systems are time-based processes. They can only inform organisms about the outcomes of previous interactions between organisms and their environments in the past. Knowledge-gaining systems can only transmit messages from the past to the present. The past can refer to ancestral populations communicating to the current generation, or, in the case of learning, to an organism's younger self in the past communicating what it has learned to its later self in the present. Either way, all messages transmitted by any knowledge-gaining system can be described by Shannon's communication theory.

The principal consequence of the localization of the knowledge-gaining systems in space is that such systems can gain knowledge only from places that they previously encountered. As we have seen, how much of its environment any knowledge-gaining system can sample in space will partly depend on the nature of its own distribution across space during time. Populations of organisms are likely to be widely distributed across environmental space and time and can sample many different places in their environments, both simultaneously and successively. By comparison, the capacity of individual learning animals to sample their environments will be much more limited.

Knowledge-gaining systems in biology must be able to sense the actual causal texture and dynamics of the environments that they previously experienced in both space and time. They must also have at least a rudimentary capacity to register or remember the outcomes of their previous interactions with their environments. Over time, they have to accumulate a remembered history of these interactions. The histories accumulated by physically different knowledge-gaining systems then provide the data that informs organisms, with each history from each knowledge-gaining system containing a different data set.

At this point, we have to go beyond Shannon because the messages sent from the past to the present must be meaningful. This requires the existence of biological mechanisms that are capable of demarcating between meaningful adaptive know-how and meaningless or maladaptive information. Ultimately, this is the knowledge that organisms need to oppose the second law of thermodynamics. The knowledge is evaluated by the subsequent fates of the organisms that are carrying it. Each different knowledge-gaining process can then communicate meaningful information about what worked before, or was adaptive in the past, with these distinct histories collectively providing the knowledge that allows individual organisms to survive and reproduce. That accounts for how meaningful information can be deduced from data provided by one or more histories of the past.

But this deductive logic does not account for how organisms can use this knowledge to move into their futures. This requires another kind of logic, an inductive as opposed to a deductive step. Constrained by whatever meaningful information they have acquired from the past, organisms can then gamble that their future interactions with their environments will be the same or similar to their interactions in the past. This is the inductive gamble that was first described in chapter 1. What I'm now suggesting is that this same gamble reoccurs in every knowledge-gaining process in nature. It is the only way for any organism to prepare for its not-yet-experienced future. This was first pointed out by the philosopher David Hume in connection with human knowledge.[38] It is never possible to guarantee that that any inductive gamble is going to work. But in practice, the evolution of life on Earth over the last four billion years tells us that it seems to work better than chance. God may or may not play dice, but life must. However, life can load the dice in its own favor, more or less heavily, by knowing something about the past. Life needs messages from the past and needs to act on those messages to constrain its bets about its future.

In spite of these similarities shared among all knowledge-gaining processes in nature, there should also be some significant differences between them. Most obviously, the component mechanisms, responsible for both communicating and evaluating messages, should be different in every knowledge-gaining system. There should also be another kind of difference, which we have not yet discussed. It occurs within, rather than between, knowledge-gaining systems, and it depends on whether the system is communicating to a later version of itself or to some other similar system.

I'll illustrate this and other points by returning to Alice and Bob. Suppose, at the level of individual learning, Alice has been informed by her senses that her plants need watering. She then uses her memory of this event to transmit a message, via a channel of communication, to her own nervous system, telling her later self to water her plants. The transmitter and the receiver, as well as the channel of communication, all reside within Alice. The later Alice then waters her plants.

Alternatively, Alice sends a message to Bob, telling him that her plants need watering and asking him to water them. Now, the transmitter and the receiver are two different people: the transmitter is Alice and the receiver is Bob. The channel of communication between them is also different. Alice's message has to be encoded in the symbolism of words, whether spoken or written. Subsequently, these words have to be decoded by Bob. Bob then acts on the meaningful message that he has received from Alice, watering her plants.

At the population level, the same logic reoccurs, but among completely different transmitters and receivers and channels of communication. Let us suppose that an ancestral population of niche-constructing organisms (population A) modifies selection pressure for a later version of itself, the current generation of the same population. Here, the source of the message is the natural selection pressure modified by the ancestral population. This message is communicated by ecological inheritance, now serving as another channel of communication between ancestral population A and the current generation of the same population. The population may subsequently respond with further evolutionary changes, thereby sending modified genetic messages via genetic inheritances to the next generation of itself. In this case, both ecological inheritances and genetic inheritances are in play as communication channels within a single population.

Alternatively, suppose that ancestral organisms in population A modify natural selection pressures for a second population B, where population B could be a different species from population A. This time, population B may respond with genetic changes, which may cause population B to niche-construct in a different way. B's niche-constructing activities may then feed forward to modify natural selection pressures for population A via an ecological inheritance channel. This two-way communication between populations A and B, via both ecological and genetic inheritance channels. could then initiate their subsequent coevolution in the context of an ecosystem. The general point is that Shannon's communication theory applies to meaningful messages that can be sent by transmitters, in knowledge-gaining systems, to receivers, which are either in later versions of themselves or in other different knowledge-gaining systems.

CONCLUDING REMARKS

I have argued that it doesn't make sense to reduce all knowledge-gaining processes in nature to just the primary process of population genetics, as neo-Darwinism advocates. Unless a population of organisms really is informed only by the primary process of population genetics, this reductionism will distort our understanding of the bio-informatics of evolution. The more pressing problem is how to explain the evolutionary origin of supplementary knowledge-gaining processes in nature. In what circumstances, and relative to what kind of population, is the primary evolutionary process likely to evolve one or more supplementary knowledge-gaining processes? A second problem occurs in populations of organisms that are supplied with adaptive know-how (R_i) by more than one knowledge-gaining process. How do they relate to each other? What happens if knowledge from plural processes is not well integrated? Might this become another source of maladaptation in organisms in some populations? We will return to these problems in later chapters.

3 ADAPTATION

Let's forget science for a moment and revert to common sense. The adaptations of organisms demonstrate "design" to their environments. The giraffe's long neck, for instance, appears designed for browsing leaves from tall trees. To some people, that semblance of design strongly implies an "intelligent designer." As no human designer has the capacity to design even the simplest organism, this might imply that all organisms on Earth were individually planned by an omniscient creator, or God. That inference was famously drawn by the eighteenth-century English savant William Paley. His argument went like this:

> In crossing a heath, suppose I pitched my foot against a stone, and were asked how the stone came to be there, I might possibly answer that, for anything I knew to the contrary, it had lain there for ever. . . . But suppose I found a watch upon the ground and it should be enquired how the watch happened to be in that place, I should hardly think of the answer which I had before given.

Instead, Paley writes he would be forced to make the inevitable inference "that the watch must have had a maker: that there must have existed, at some time, and at some place or other, an artificer or artificers who formed it for the purpose which we find it actually to answer; who comprehended its construction, and designed its use." As "every manifestation of design that existed in the watch also exists in the works of nature," but with nature far more wonderful and complex, Paley concluded that a divine creator or God must exist.[1] Creationists still use a closely related argument to Paley's, even today.

Now let's return to science. Enter Darwin, who introduced the profoundly anti-commonsense idea of design without a designer. With an answer to Paley that is still shocking to many, Darwin described how entirely natural processes could generate the semblance of design. Darwin's response also demonstrates the limitations of common sense.[2] Common sense can't imagine a million years, let alone billions of years. Darwin was lucky to be one of the first scientists to know that the world was far older than anyone had previously thought, an insight that he owed to reading Charles Lyell's *The Principles of Geology*.[3] Darwin still fell far short of realizing the true age of the Earth, but he knew enough to realize that design, caused by natural selection instead of an intelligent designer, was possible given sufficient time.

In this chapter and the next, I'm going to consider how organisms can adapt to their external environments in the absence of an intelligent designer. So far, we have established that the fundamental reason why organisms have to be adapted to their environments stems from the thermodynamic relationship between living organisms and their external environments. The internal environments of organisms are very far from thermodynamic equilibrium, relative to their external environments. This relationship is unstable. If organisms did nothing, they would rapidly dissipate. They would die and return to thermodynamic equilibrium, in obedience to the second law of thermodynamics. Although nothing can violate the second law, in order to survive, living organisms are compelled to resist it for a time by doing active purposeful, fuel-consuming, physical work.

This work involves actively importing physical resources (or R_p) that are relatively high in free energy from their external environments, and exporting resources that are lower in free energy back to their environments, an imperative demanding that organisms continuously interact with their environments throughout their lives. However, organisms cannot interact with their environments productively unless they are adapted to those environments, which requires them to be informed by adaptive know-how (or R_i), acquired through the processes of evolution.

All scientific theories of evolution accept that the acquisition of adaptations by organisms primarily depends on the natural selection of variant phenotypes in populations, combined with the transmission of genetic inheritances between generations. But that still leaves room for different accounts of how organisms acquire their adaptations. I'll concentrate on the differences between the orthodox standard evolutionary theory (SET) and niche construction theory (NCT).

SET accounts for the adaptations of organisms solely in terms of natural selection sorting among phenotypes of varying fitness in the current generations of populations. It further assumes that the phenotypes in these populations are primarily determined by previously selected genes inherited from their ancestors, as well as by chance mutations. SET thus accounts for the adaptations of organisms in terms of a one-way-street or linear causal relationship between organisms in populations and autonomous natural selection pressures in their environments. The feedback recognized in the study of sexual selection, coevolution, or social evolution does not really change this linear characterization of causation: it simply means that the source of selection is another evolving organism. The causal model remains linear: selective environments bring about changes in organisms. SET acknowledges that organisms modify environmental states, but it does not recognize niche construction as a cause of evolution.[4]

In contrast, NCT claims that while the processes that SET describes are necessary components of evolutionary theory, they are not sufficient. While SET fails to accept that the active agency of purposive, fitness-seeking organisms plays a causal role in evolution, NCT emphasizes that the adaptations of organisms depend on both natural selection and the niche-constructing activities of individual organisms. This changes evolutionary theory in at least three important ways.

First, according to NCT, organisms in populations codirect adaptive evolution by modifying some of the natural selection pressures in their environments through their niche-constructing activities. They do this either by physically perturbing their environments or by actively relocating in their environments in ways that transform the target of selection.[5] For instance, by digging a burrow to escape extreme temperatures, burrowing animals reduce the intensity of selection for the ability to cope physiologically with high or low temperatures, and generate selection on burrow-digging capability. Likewise, by overwintering in warmer climes rich with food, migrating animals reduce the intensity of selection on starvation tolerance and create selection for migrating capabilities. Hence, for NCT, the natural selection pressures that organisms encounter in their environments are no longer exclusively due to autonomous environmental events. Many are due to the prior niche-constructing activities of organisms.

Second, insofar as niche-constructing organisms modify the environments of their descendants (e.g., by building nests in which their young are reared), they also modify selection pressures in their descendants' environments. NCT differs from SET in characterizing the transmission of modified selection pressures between generations as a second general inheritance system in evolution, called "ecological inheritance." Offspring receive both genetic and ecological inheritances from their parents, and these two inheritance systems interact with each other in descendant populations. Sometimes the ecological inheritance can persist for multiple generations.

Third, as modified environments can persist across multiple generations or be continuously and repeatedly transformed, components of the environment will change in concert as the niche-constructing outputs of organisms evolve. According to NCT, but not SET, evolutionary theory should not solely be about the evolution of organisms. Rather, it should be about organism-environment coevolution. This evolving organism-environment complex encompasses coevolving abiota in the environments of populations, as well as other coevolving populations, including other species.[6]

In each of these three ways, niche-constructing organisms introduce feedback into evolution—feedback that is not yet fully recognized by SET. The feedback converts the evolutionary process into a nonlinear causal process. In NCT, the adaptations of organisms no longer depend exclusively on autonomous natural selection pressures and genetic inheritances. They depend on natural selection and niche construction, acting reciprocally. They also depend on at least two inheritance systems, genetic inheritance and ecological inheritance, interacting with each other in descendant populations.[7] These nonlinear causal evolutionary relationships between organisms and their environments can be initiated in two ways: either by the natural selection of novel genetic changes in organisms or by the construction of novel changes to environments, and hence to selection pressures. Where the changes are initiated by niche-constructing organisms, these activities may not always be genetically determined and can sometimes be caused by other supplementary processes that inform plastic phenotypes or organisms in evolution, such as epigenetic processes, learning in animals and sociocultural processes in humans.[8]

How do these feedbacks affect organisms' adaptations? This is a difficult question because there is more than one concept of adaptation in biology. Often, the adaptations of organisms are explained in terms of their historical and evolutionary origins. This first concept of adaptation directly addresses the question of how organisms acquire their adaptive designs, relative to their environments, in the absence of an intelligent designer. Traits are recognized as adaptations if they evolved by past natural selection for a particular role.[9] I'll call this first concept of adaptation "historical adaptation."

However, adaptation also commonly refers to the immediate relationships between individual organisms and their local external environments in the current generation. Traits are described as adaptations if they increase the fitness of individuals.[10] As the individual organism-environment relationships are constantly changing, this conception of adaptation varies throughout the lives of individual organisms. I'll call this second concept of adaptation "contemporary adaptation," as it refers to the contemporary match between an organism's traits and the environment it encounters, in the present.

While rarely conceived of in these terms, we might also envisage a conception of adaptation rooted in the future. Adaptations might be construed as those capabilities that allow organisms to resist the second law of thermodynamics for long enough to achieve their fitness goals of survival, growth, and reproduction. Schrödinger's book *What Is Life?* tells us that, in contrast to purposeless nonliving abiota, organisms are purposeful systems *by definition*. This suggests that the origin of life was also the origin of purpose, in the form of purposeful living organisms (see chapter 6). It also implies that the adaptations of organisms refer primarily to whatever traits in organisms enable them to anticipate their future interactions with their environments most effectively. Such "anticipatory adaptations" include the niche-constructing activities of organisms. Organisms anticipate their futures on many scales, ranging from the microscale of a few seconds to the duration of their entire lives. But organisms cannot achieve their purposes unless they are adapted to their external environments at each place and moment during their lives.

Hence, there are past (or historical), present (or contemporary), and future (or anticipatory) concepts of adaptation, and the three forms of adaptation interact in interesting ways. From here on, I am going to focus primarily on the contemporary adaptations of organisms, which are the immediate adaptations upon which the lives of individual organisms depend. However, it is not possible to understand the adaptations of organisms except in the light of their historical evolutionary origins and their anticipated fitness goals.

It is not often acknowledged, but the adaptations that organisms inherit from their ancestors have the logical status of inductive gambles relative to their future interactions with their environments. Their gambles are based on the deductive information (R_i), primarily encoded in genes, about what was adaptive for their ancestors in the past. On the basis of this inherited information, individual organisms in the current generation make inductive bets that whatever phenotypic traits were adaptive for their ancestors in the past are likely to be adaptive for them again in their own individual futures. These inductive gambles refer to both the structural and functional adaptations of organisms.

Why can't organisms do better than bet on their futures? The problem is that the adaptations of organisms depend on the acquisition of adaptive know-how, or R_i, as a function of the evolutionary process (as described in chapter 4). However, it is impossible for any organism to know with certainty what is coming next. The philosopher David Hume may have been the first person to make this point clear, in 1739.[11] Even the laws of physics and chemistry are ultimately stochastic rather than deterministic laws. For instance, none of us can know with absolute certainty that the universe will not evaporate in the next three seconds. The probability of it doing so is very low. Low probabilities can be written as a decimal point followed by lots of zeros and then a 1—however, it might take more zeros to describe this probability than I can fit on this page. Nevertheless, absolute certainty is denied to us and to all other organisms. All organisms have to rely on inductive gambles that the future will resemble the past to have any chance of resisting the second law for long enough to achieve their fitness goals.

Thus the problem of adaptation for individual organisms is how to connect their inherited deductions about what was adaptive for their ancestors in the past to their inductive gambles about what is likely to be adaptive for them in their futures, while always being sufficiently well adapted to their immediate environments in the present. It is still possible for organisms to make better or worse bets about what will be adaptive for them in their futures. For example, insofar as individual organisms are betting on diurnal, seasonal, or tidal changes being the same as they were in their ancestors' pasts, they will be likely to win their bets. Conversely, if organisms in the current generation are betting that the capricious behaviors of other organisms, in their own or other species, will always be the same, they will be more likely to lose their bets.

How does SET on the one hand, and NCT on the other, handle the interactions between these past, present, and future concepts of adaptation? First, let's consider the past. SET and NCT describe the historical origins of the adaptations of individual organisms in broadly the same way. They both seek to understand and explain the origins of the different adaptations of organisms as evolutionary histories arising from natural selection in particular ecological contexts. All evolutionary biologists pay a lot of attention to the origin of such phenomena as photosynthesis in bacteria and plants, eyes in animals or insects, flight in birds, and the origin of bipedalism in humans through natural selection. On a smaller scale, biologists are also interested in the origin of peacocks' tails through sexual selection and the eye-spots on butterfly wings through mimicry.

The principal difference between SET and NCT concerns the source of the natural selection experienced by successive generations of populations. With caveats, SET assumes that natural selection arises from autonomous events in the environments of organisms. The organisms themselves, or other species, may be the source of selection, such as what occurs during sexual selection or coevolution. However, in such circumstances, other organisms are simply treated as an aspect of the external environment. Living organisms are not generally viewed as *part-determining*, or *codirecting*, the natural selection that ensues. In contrast, NCT claims that some of the natural selection pressures that act on individual organisms in the current generations of populations are likely to have been modified by the niche-constructing activities of their ancestors. NCT also expects some novel evolutionary episodes in the histories of evolving populations to have been initiated by the niche-constructing activities of organisms. Natural selection doesn't just happen to organisms; organisms themselves construct selection pressures, and they frequently do so in reliable and consistent ways so as to impose a bias on selection.

What demands are made on the adaptations of organisms by the environments that they currently encounter? Here again, SET and NCT have a lot in common. For instance, both SET and NCT recognize that the contemporary adaptations of organisms are multivariate and dynamic. Organisms always express multiple adaptations, simultaneously as well as successively, relative to the plural selection pressures that they are encountering in their ever-changing environments. They do so at each place and moment during their lives. But there are also some differences. For instance, NCT emphasizes how organisms possess knowledge-gaining capabilities that allow them to improve their fit to the environment. For instance, animals can learn to respond to novel predators (see chapter 4).

What about the future? In addition to possessing sufficient adaptations to cope with their immediate environments, organisms need to carry sufficient general-purpose adaptability to prepare them, in advance, for what is coming next. For instance, a rabbit needs to be prepared in advance, by the past evolutionary history of its population, to respond appropriately to foxes and dogs. It needs to know a priori how to recognize them, hide from them, or run away from them. A rabbit cannot gain these adaptations a posteriori, after it has already been eaten by a fox. But here again, we will see that there are also differences between SET and NCT. For instance, the latter emphasizes how the anticipated future might lead to the expression of niche construction now to change the anticipated future. The rabbit might prepare a new bolt hole in anticipation of encountering a fox in the future.

ADAPTIVE NICHE MANAGEMENT

The multivariate and dynamic nature of the relationships between living organisms and their selective environments turns the problem of adaptation for contemporary living organisms into a complicated niche-management problem. It requires an understanding of how living organisms deploy the adaptations that they possess in their immediate and changing environments. This is true of all organisms in all species. It doesn't matter whether we are talking about bacteria or elephants, or about organisms that live on land, in the sea, or in the air. These same issues arise for both structural and functional adaptations.

As this niche management problem is common to all organisms in all environments, I want to focus on the abstract logic of the problem. I'm going to use game theory to help me do so, but the particular game I want to use may not be familiar to you. I am going to call it the "Adaptive Niche Management Game," as it can be applied to understand the inductive gambles of organisms, although it was originally described by Lawrence Slobodkin and Anatol Rapoport in 1974.[12] The game is played between two entities, any living organism or population of organisms (O) and its local external or selective environment (E). To keep things simple, I'll initially assume that E is exclusively an abiotic environment for O. I'll bring in the other organisms in O's environment later. I'll also restrict the game to one played between an individual organism and its particular local selective environment, although the same logic applies when the game is between evolving populations and their environments.

We can now consider the nature of O's adaptive niche management in more detail.[13] Following Bock (1980), I previously described this game as a "feature-factor matching game" between O and its environment E.[14] When the game is played between an individual organism (O) and its local environment (E), then O's features become a set of phenotypic traits relative to the set of selective factors that it is encountering. Likewise, the environment can be characterized as a set of environmental factors. A match, known as "synerg," occurs whenever a phenotypic trait is well suited to an environmental factor. A mismatch occurs whenever a feature expressed by O is maladaptive relative to a factor in its E, in which case there is no synerg. The implication is that organisms must be able to express sufficient contemporary adaptations, both simultaneously and successively, relative to whatever natural selection pressures they are encountering in their environments, to survive at each place and moment during their lives. Any vital mismatch between the set of features (or adaptations) that an organism is expressing and the set of factors that it is currently encountering in its environment could be fatal. It could terminate the game. If it does, O will lose the game and E will win it.

There are some important asymmetries between the two players in this game. I'll list the main ones. First, assuming that E is exclusively abiotic, E is indifferent to O's fate, but O cannot be indifferent to E's fate. An organism's life depends on its environment, so O cannot afford to be indifferent to what happens to E, or to whatever changes occur in E, because of the requirement that O must be continuously adapted to E if it is to live at E's expense.

The second asymmetry between the two players is associated with the first. O has to adapt to E to stay in the game, but E does not have to adapt to O. E will always persist in some form, regardless of what happens to O.

A third asymmetry between O and E is that O is a purposive goal-seeking system but E is not. O must be capable of some degree of active purposeful agency relative to E, but insofar as E is abiotic, it has no fitness goals of its own.

In reality, O's environment never comprises abiota only. Every O's environment is bound to include some other living organisms. Like O, these other organisms will have fitness goals and purposes of their own. The addition of other organisms in O's environment turns E into a mix of purposeless abiota and purposeful biota. The other purposeful organisms in O's environment may compete with, be irrelevant to, or even be beneficial to O, all of which complicates the game and may make it more difficult for O, but it doesn't change the logic fundamentally.

The final asymmetry between O and E is that while O is a finite-state system relative to E, E is effectively an infinite-state system relative to O. That is because there are always a limited number of features, or phenotypic traits, that any O can express in its E. However, O's local E is an open system. On Earth, E is always open to inputs (e.g., from the rest of an ecosystem or the rest of the biosphere), while the Earth's system itself is open to the rest of the universe. This final asymmetry between O and E accounts for the original name of the niche management game, "Gambler's Ruin." Since a finite-state system can never match everything thrown at it by an infinite-state system, the adaptive niche management game is unfair. O can never win. The best that it can achieve is to delay losing. E must eventually win the game, regardless of whatever O it is playing against.[15]

Reproduction, however, allows an individual organism to go some way toward overcoming the limitation of being a finite-state system in an infinite-state environment. When organisms reproduce, they are contributing to another putative infinite-state system—namely, to the almost infinite number of ways that the evolutionary process can generate variant organisms and species. Evolving life in general can stay in its niche management game for a great length of time—approximately 4 billion years on this Earth, so far. It can also cause life to exist across great intervals of space (e.g., the whole biosphere on Earth).

Up to this point, the adaptive niche management game of organisms is broadly the same game, regardless of whether it is described by SET or NCT. There are only two significant differences between the two theories. First, because SET is not based on the laws of thermodynamics, its starting point is the adaptations of organisms rather than their purposes. SET's treatment of the purposes of organisms is dismissive—organisms are treated as *apparently*, but not actually, purposive. Nor does SET consider purposeful organisms to be agents in evolution. In contrast, partly because NCT has always been closely associated with the second law of thermodynamics,[16] NCT's starting point is the active, purposive agency of organisms. In NCT, the purposes of organisms are to resist the second law for long enough to achieve their fitness goals, with the adaptations of organisms evolving to meet these purposes.

The second significant difference between these theories is that NCT assigns evolutionary causal significance to some "moves" in its niche management game that SET does not characterize as causal. These moves include O's environment-modifying, niche-constructing activities, as well as its active responses to natural selection pressures from E. According to NCT, O's capacity for niche construction gives organisms a degree of control over some of the natural selection pressures expressed by E. This makes a difference to how organisms play the game.

ASHBY'S CONTROL THEORY

Let's return to the question of how organisms deploy their adaptations to achieve their fitness goals. Here, I'm going to turn to Ross Ashby's control theory, which may help us to better understand biological adaptation. Ashby was one of the father figures of cybernetics at a time—in the 1950s—when the word "cybernetics" was used primarily to describe the control of dynamic systems. I'll describe the relevant aspects of Ashby's control theory[17] in general terms first, and then apply it to the concept of adaptation in biology.

Ashby's theory refers to the regulation of the relationship between any two or more interacting dynamic systems. Often one system constitutes some kind of environment for another system. One system may be a self-maintaining and self-controlling system that is attempting to regulate its own relationship with some other system, such as its external environment. Alternatively, the first system may be controlled by some other system. The latter is often true in the case of human artifacts, which are ultimately controlled by human designers, programmers, or users, such as cars controlled by their drivers or planes controlled by their pilots.

In Ashby's theory, the fundamental task of the controller is to protect an "essential variable." The essential variable may be multifaceted and include multiple subvariables. It refers to the variations that can occur in the relationships between two or more dynamic varying systems. The essential variable does not refer directly to either the variance in the controlled system or to the variance in the system with which it is interacting. It refers to a third kind of variance—namely, the variations that can occur in the relationship between two or more varying systems, one of which is the controlled system. It is this third kind of variance that the controller must control to protect its essential variable.

The essential variable is defined by the purposes and goals of its controller. If the controlled system is self-controlling, then the essential variable will be defined by its own purposes and goals. In this case, the self-controlling system must be able to control its own variance relative to the variance in whatever other system it is interacting with. In general, the essential variable must be defined and protected by the controller, regardless of whether the control is exerted by a self-controlling system or a third party. For example, in the case of a plane, the essential variable that must be protected by the pilot is likely to include the route, chosen by the pilot, that will bring the plane to its destination. If the plane strays off course, the pilot must correct it. Subessential variables may include maintaining the orientation of the plane (i.e., the right way up), its internal conditions (i.e., suitable cabin pressure), and its physical state (i.e., undamaged). Whatever the essential variable, the controller must never allow it to stray beyond some kind of tolerance limits. For example, the plane shouldn't fly into a severe storm. These tolerance limits are necessary to ensure that the controlled system succeeds in achieving the goal set for it by its controller.

At the heart of Ashby's control theory is his Law of Requisite Variety (LRV), which states that only variety can "destroy" or "drive down" variety.[18] The controller's job is to protect the essential variable, connecting one varying or changing system to another varying or changing system. The controller must do so by causing the system that it is controlling to express some appropriate, novel, or counteractive variance of its own, at each place and moment, to "destroy" or "drive down" the variance that it is encountering in the other system that it is interacting with, at the same places and moments. For instance, as roads can be dark at night, as well as light during the day, then cars must be equipped with lights that can be switched on at night to counter the darkness if they are to be driven safely.

A further implication of LRV is that, to guarantee that the controller will always be able to protect a relevant essential variable, there must be sufficient variance available in the controlled system relative to the variance in the system that it is interacting with. The amount of variance made available by the controlled system should ideally be equal to or greater than the amount of variance in the system that it is interacting with. If it is not, then sooner or later, the controlled system will be destroyed by the variance in the other system. For instance, at some point, a car without lights driven in the dark would crash.

I'll illustrate this abstract logic with another example. Imagine that you are the controller of a small sailboat sailing on the sea and you are attempting to reach a nearby harbor. The system that you have to control is your boat. The boat has two or more sails and a number of ropes, such as the halyards, that control the settings of the sails and allow them to be raised, lowered, or reefed. All these features can be varied and controlled by you. In addition, a wheel, or tiller, controls the rudder, enabling you to steer the boat. There may also be some moveable ballast. If you are in a small sailing dinghy, the ballast could be another member of the crew. The different states of the boat constitute the variance that you control.

Then there is the variance that you are liable to encounter in your boat's marine environment. This includes the strength and direction of the winds, both of which could be steady or highly variable, and the variable sea states and waves through which your boat is sailing. It may also include variable tides and currents and varying depths of water.

Your job is to protect an essential variable by controlling the variance offered by your boat's equipment, relative to the variance that your boat is encountering in its environment, in such a way that you can reach your destination safely. If things go badly wrong, if the weather suddenly becomes very violent and the sea very rough or if you hit a rock, then you might never get there. You could lose control, and your boat might capsize or sink, having been destroyed by an excess of environmental variance relative to the variance made available by your boat. You will have failed to keep the essential variable within its tolerance limits.

My boat example is intended to mimic the adaptive niche-management problem between organisms and their environments. Unlike the artifacts that we have just been considering, individual organisms are self-controlling systems. A closer contemporary mimic of self-controlling organisms would be a plane when switched to autopilot or the Mars Rover.

What can Ashby's concept of an "essential variable" and his LRV tell us about biological adaptation? He was not greatly concerned with the historical origins of the adaptations of organisms, nor with the problem of their design in the absence of an intelligent designer. Most of the systems to which Ashby's theory has been applied are human artifacts, which are obviously designed by intelligent human designers. However, Ashby's theory does have something to say about the match between organism and environment, as well as the purposes of living organisms.

APPLYING ASHBY'S CONTROL THEORY TO INDIVIDUAL ORGANISMS

Recall that the fundamental purposes of living organisms are to resist the second law for long enough to survive, grow, and reproduce. These fitness purposes define their goals and subgoals, which in turn define the essential variables that individual organisms have to protect in order to achieve their goals. However, as well as having to be sufficiently adapted to their immediate environments here and now, because the fitness goals of organisms lie in their futures, organisms need a minimal capacity to anticipate or predict their future interactions with their environments. This capacity may or may not involve cognition. Usually it does not.

All organisms have to be equipped with adaptations that allow them to perceive or sense their environments. They also have to be able to interpret the meaning of the information that they receive from their senses relative to their purposes in ways that are likely to motivate them to achieve their fitness goals. But, because their fitness goals lie in their futures, organisms also need some capacity to anticipate or "predict" their future interactions with their environments. It follows that organisms can protect their essential variables only by using combinations of adaptive gambles and chance-based mutations.

In spite of all these provisos, the essential variables that living organisms have to protect correspond to the essential variables in Ashby's control theory. Adaptation is not about variant phenotypic traits, or variant organisms, in evolving populations, nor is it about varying natural selection pressures in the environments of organisms. Adaptation is about the third kind of variance, captured by Ashby's essential variable, which concerns the varying *relationships* between two varying systems. Adaptation is about how varying individual organisms, equipped with variant phenotypic traits, manage or fail to manage to protect their essential variables relative to varying natural selection pressures in their environments. Well-adapted organisms are those individual organisms in populations that are equipped by their prior evolutionary histories, with sufficient adaptations and sufficient capacity to control them. Such organisms may then be able to resist the second law for long enough to achieve their fitness goals.

One major difference between SET and NCT concerns the extra causal significance that NCT attributes to individual organisms in evolution arising from their niche-constructing activities. When applied to adaptation in biology, Ashby's LRV concerns how two kinds of variety, environmental variety (V_E) and organism variety (V_O), interact with each other. V_E refers to the variant natural selection pressures that organisms encounter in their environments, while V_O refers to any variant adaptations or variant phenotypic traits that organisms express at different times and places.

Ashby described how variety destroys variety. In the niche management game, the word "destroy" takes on two different meanings depending on what is destroying what. The destruction of V_O by V_E may be literal. Any organism expressing a maladapted phenotypic trait may be killed by environmental components for which it is ill prepared. Alternatively, an animal might learn not to perform a particular behavior in a particular environment, which constitutes another way by which V_O is reduced.

What happens the other way round, when V_O destroys V_E? This occurs when a phenotypic trait expressed by an organism is well-adapted to a natural selection pressure in its external environment. In this case, V_O destroys V_E metaphorically by preventing V_E from killing or harming the organism that is expressing the phenotypic variant. Possibly the organism may do better than that, and the trait may allow the organism to exploit its environment in a way that increases its fitness, such as to harvest additional energy and matter from its environment or dump more of its detritus. The same phenotypic trait may be able to drive down multiple selection pressures in their environments, such as when a turtle's shell or hedgehog's bristles ward off multiple different predators.

In practice, all organisms have to be adapted to multiple selection pressures simultaneously at each place and moment during their lives. If organisms are to protect their essential variables and live, they must destroy all the

V_E that they encounter in the "here and now," and do the same again at each successive place and moment "elsewhere and later" throughout their lives. A single vital mismatch between a phenotypic trait and its environment could kill the organism, no matter how well adapted all its other phenotypic traits are relative to other selection pressures.[19]

How do niche-constructing organisms utilize their capacities to modify natural selection pressures to adapt to their environments in ways not fully recognized by SET? A useful way of answering this question is by considering the maladaptive mistakes that organisms can make, rather than their well-adapted successes during their niche management games with their environments. Logically, there are two ways in which organisms can fail to adapt to their external environments, which I'll call "limited regulator failures" and "fallible regulator failures." A limited regulator failure occurs when an organism does not possess any adaptation that can destroy an environmental selection pressure that it is currently encountering. For instance, an animal that escaped prey by running away might be destroyed by the appearance of a new predator far faster than itself. In such circumstances, V_E will destroy V_O, and the organism may go extinct. In contrast, a fallible regulator failure occurs when O possesses an appropriate adaptive adaptation to cope with the environmental selection pressures that it is encountering, but it fails to express the correct adaptation at the right time and place. A rabbit that decided to fight a fox might be doomed.

One way of illustrating the logic behind these two kinds of failures is provided by another human artifact, a radiator, which is connected to a thermostat and a boiler. The radiator may be required to turn up the heat to a level that it cannot manage because it is out of its range. Here, its capacity to regulate temperature is too limited. Conversely, the radiator might be required to turn up the heat to a level that it could manage in principle, but it fails because it is not controlled correctly due to a faulty thermostat. That would be a fallible regulator failure.

Limited regulator mistakes can be readily understood in the context of both SET and NCT, as organisms either do or don't have a capacity to avoid them, depending on whether they already possess appropriate adaptations. But if an organism makes a limited regulator mistake, what can it do about it? Living organisms needn't passively submit to the V_E that they encounter. Organisms can deploy their niche-constructing adaptations to generate additional V_O to destroy more of the V_E in their naturally selecting environments. For instance, a niche-constructing organism may actively relocate to somewhere else in its environment that is better suited to the tactical adaptations or V_O that it already possesses. Algae respond to diurnal changes in their environments by adjusting their place in the water column, such as typically ascending during the day and descending at night. Similarly, diatoms change their positions on mud banks in estuaries according to the state of the tides. Many animals invest in larger-scale relocations. For instance, swifts, swallows, and terns (among many other birds) migrate annually between northern Europe and southern Africa. They avoid winters by doing so.

Niche-constructing organisms can also change the selection pressures that they are encountering by physically buffering them out. This often involves organisms physically changing their environments through "perturbational niche construction,"[20] in ways that suit the organism itself, to the point where the V_E that it is encountering in its environment can be destroyed by the V_O that it is already expressing. For instance, some animals buffer out winter by hibernating rather than relocating. That may require the animals to store extra fat in their bodies, or possibly to cache extra food in their immediate environments before they hibernate. It also requires them to actively dig burrows or choose places where they can hibernate safely, possibly for months.[21]

Both SET and NCT recognize these possibilities, but the two theories differ in the significance that they attribute to them. For SET, relocating in space (e.g., through migration), or modifying the environment, (e.g., through nest building), are of no great evolutionary significance. Nests can be viewed as "extended phenotypes," which according to Dawkins evolve through natural selection in essentially the same manner as any phenotype.[22] For NCT, on the other hand, these moves are recognized to be important causes of evolutionary change. Niche construction is not simply a product of evolution—it is a central aspect of the process of adaptation. That is because niche-constructing organisms are compelled by their purposes to niche-construct in nonrandom or directional ways, generating order in the external environment, and statistical biases in the selection emanating from that niche-constructed environment.[23] For illustration, studies have shown that where organisms buffer V_E through counteractive niche construction, such as building a nest to damp out variations in temperature,[24] they reduce the strength and variability of the natural selection that they experience relative to selection emanating from components of the environment that they have not built and cannot control.[25] This consistent biasing of natural selection, recognized by NCT but not SET, is what makes niche construction a causal factor in evolution.

Now let's consider the second kind of maladaptive failures, those due to control errors. Organisms can make control errors for quite simple reasons that can be equally well understood by SET and NCT. For example, an organism's senses may misinform it about the state of its external environment, or an organism may misinterpret the meaning of the messages that it is receiving from its senses. A consequence of both these eventualities is that an organism may express the wrong structural or functional adaptation relative to its environment or at the wrong time and place.

To avoid making maladaptive control errors, individual organisms must control their adaptations in ways that cause them to express the right adaptations at the right times and places. For SET, the capacity of individual organisms to control their adaptations is exclusively determined by the genes that they inherit from their ancestors. But this way in which individual organisms can control the expression of their adaptations has its limitations. For instance, in novel or rapidly changing environments, the sample of deductive know-how, or R_i, that organisms inherit from their ancestors via their genetics may be out of date, and as a consequence, they may give individual organisms incorrect information about what adaptations they should express (see chapter 1). Even if it was adaptive at the beginning of the organism's life, years later it may no longer be so. Or organisms may encounter a novel environmental event, such as the arrival of a new virus that has never existed before and that its ancestors never experienced. In the absence of appropriate recognition and analysis of the threat, this could lead to an inappropriate utilization of existing adaptations (i.e., the virus might end up being lethal).

In addition, individual organisms can inherit only a sample of the gene variants carried by the population to which they belong, which means that individual organisms can inherit only a sample of the total deductive know-how, or R_p, acquired by their evolving population. Moreover, that genetically encoded know-how is constantly being shuffled by recombination in a way that will sometimes generate suboptimal combinations, expressed via poor control of adaptations.

Populations of organisms that regenerate very rapidly and in large numbers, such as bacteria, might be able to track any changes in their environments on the basis of chance mutations only. Alternatively, populations of organisms that live in environments that don't change, or change only very slowly, might also be able to adapt to their environments. In such instances, appropriate control mechanisms for adaptations might evolve. However, other organisms would be far less likely to adapt to their environments if they lived longer, regenerated more slowly, or lived in rapidly changing environments, including environments that they themselves were changing by niche construction. SET implicitly assumes that such organisms would sometimes make control errors, risking extinction.

NCT, in contrast, escapes SET's limitation that organisms can inherit their deductive know-how only from their genes by stressing the evolutionary significance of additional information-generating mechanisms and additional sources of control. NCT claims that individual organisms must be self-controlling purposive systems. To put that more accurately, NCT claims that natural selection will favor individual organisms that inherit at least a minimal capacity for the self-control of their own inductive gambles. That self-control is what provides organisms with the capacity to impose direction on evolution through their niche construction.

But if individual organisms are self-controlling systems, and if Ashby's control theory is correct, self-controlling systems have to define and protect their own essential variables. This means that living organisms must evolve some autonomy from genetic control of adaptations, including through the ability to acquire new know-how, or R_p, during their lives, which is capable of updating the control of their adaptations. By enabling individual organisms to be self-controlling, evolving populations should increase their amount of variance that is available to natural selection. Evolving populations should thereby increase their own evolvability, and perhaps the likelihood that their lineage will persist. In addition, if organisms are self-controlling systems, they can no longer be accurately characterized as passive vehicles for genes but rather must be construed as active agents that partly determine the strength and direction of evolution.

The adaptive know-how, or R_i, that they inherit must equip individual organisms with sufficient autonomous decision-making capacity to control their own adaptations during their idiosyncratic niche management games with their environments. However, the particular subgoals that each individual organism has to satisfy if it is to achieve its fitness goals of survival and reproduction are likely to be unique to that organism, at least in detail, rather than being generic to the population. Its subgoals will include having to overcome very particular obstacles, as well as having chances to exploit unanticipated particular resources in its environment. It is these kinds of idiosyncratic fitness subgoals that NCT claims vary between individual organisms in populations. These fitness goals can be met only because individuals in populations are equipped with the same capabilities for supplementary knowledge gain, allowing each individual to tailor its control of adaptation to the idiosyncrasies of its own personal experiences. The reliance on these same capabilities means that (at least sometimes) multiple individuals in the population will respond in similar ways to meet their subgoals, generating a stable modification in selection pressures.

Hence, NCT, but not SET, emphasizes that organisms may switch to controlling their adaptations on the basis of deductive, adaptive know-how (R_i) that is derived from their own individual past experiences and developmental histories instead of being derived from the past histories of their evolving populations. Any population of exclusively genetically determined organisms is vulnerable to invasion by organisms that have at least a modicum of autonomous self-control. Hence, natural selection may also favor supplementary information-gaining processes in individual developing organisms. The evolution of immune systems, epigenetic processes, learning in animals, and eventually human cultural processes allows organisms to go beyond the relatively coarse-grained control of their adaptations granted to genetic predetermined organisms in order to allow additional fine-tuning of this control (see chapter 7).

This means that individual organisms will have the option of controlling the expression of their adaptations by using deductive R_i from more than one source, such as by relying on their own past developmental histories of interacting with their local environments instead of on the past evolutionary histories of their population's interactions with the wider environment. Alternatively, they may control their tactical adaptations by a mixture of more than one kind of adaptive know-how, or R_i, if, for instance, there are genetic constraints on learning or genetic biases in the generation of antibodies. Self-controlling, purposive individual organisms will then have the task of integrating all these sources of information. This may be difficult if and when the different sources of information generate contradictory information for the organism, such as what occurs where inherited genetic information "tells" human senses that sugary or fatty foods are good, while our cultural knowledge tells us that they are bad for us.

The deductive R_i acquired by these supplementary developmental processes will be registered in different physical memory systems in different parts of an organism's body. The deductive R_i that organisms inherit from their ancestors is primarily held in the genomes of organisms, but additional supplementary knowledge is stored elsewhere, including in their brains and immune systems. A further implication is that self-controlling, purposeful individual organisms need to be equipped by the prior evolution of their populations with some kind of higher-order autonomous executive capacity to make some decisions about how they choose among or integrate the various sources of adaptive know-how (R_i) that are available to them. For instance, individual organisms may have to make some decisions about how to control their adaptations, which override the naturally selected genetic information that they inherited from their ancestors.

I'll illustrate some of these points by referring to human artifacts again. The *Perseverance* Mars Rover had to be equipped with some autonomous decision-taking capacity by its ground controllers to enable it to avoid obstacles in its path that could not be anticipated or predicted by its ground controllers. The example has strong implications for some of the differences between SET and NCT, which concern the autonomy of supplementary knowledge gaining systems and the ramifications of that knowledge gain. SET recognizes that organisms have evolved supplementary sources of adaptive know-how. For instance, SET recognizes that vertebrates have evolved immune systems and that animals can learn. However, SET views supplementary knowledge gain to be under tight genetic control, and thereby it denies individual organisms any capacity for autonomous decision making. For SET, the genes retain control, micromanaging the acquisition of new knowledge and biasing it toward what was adaptive in the past. In contrast, for NCT, the entire function of the supplementary knowledge-gaining process is to wrestle control away from algorithms that were adaptive in the past and to replace it with control on the basis of algorithms that are adaptive in the present. One important implication of this insight is that organisms can develop genuinely novel phenotypes rather than phenotypes restricted to those prescreened by ancestral selection, increasing V_E. The lesson of the *Perseverance* Mars Rover is that NCT is likely to be correct, as the strict kind of genetic control anticipated by SET would leave organisms vulnerable to unanticipated events. *Perseverance*'s requirement for an autonomous decision-making capability to function strongly implies that the supplementary knowledge-gaining capabilities of living organisms are likely to have evolved some significant degree of autonomy.

SET also denies that any of the adaptive R_p gained from these nongenetic supplementary processes in individual organisms can affect the subsequent evolution of their populations. NCT disagrees. Historically, SET refuted that any of the characteristics acquired by individual organisms during their lives could be transmitted to their descendants via genetic inheritance.[26] More recently, even orthodox evolutionary biologists have been forced to recognize the existence of one or more nongenetic inheritance systems in evolution.[27] Insofar as individual organisms increase their own fitness by using supplementary developmental sources of R_p, NCT also claims that organisms could bias the inheritances of later generations of their populations by contributing more than the genes that they carry to their descendants. Not just genes, but acquired epigenetic marks, hormones, nutrients, antibodies, a host of symbionts (including bacteria, protists, and viruses), as well as learned and cultural knowledge, are passed from one generation to the next. Any or all of these supplementary inheritances can enhance the regulatory control of individual organisms by updating the information underpinning their inductive gambles. In addition, niche-constructing organisms, including being informed by supplementary sources of adaptive R_i based on their own individual past histories, contribute ecological inheritances to their descendants. These ecological inheritances can also help to ensure that the regulatory control of adaptations by individual organisms is adaptive by reliably producing benign environmental states suited to their offspring's genotype. In this manner, ecological inheritance also helps to ensure that the inductive gamble pays off. SET does not deny these legacies, but it does not regard them to be evolutionarily significant either, since control is assumed to reside with the genes.[28] In contrast, NCT regards these supplementary inheritances as playing a crucial role in evolution, by enhancing the predictive power of R_i, particularly over relatively short timescales. In chapter 10, we will see that there is now considerable evidence that inherited cultural knowledge has directed recent human evolution.

APPLYING ASHBY'S CONTROL THEORY TO EVOLVING POPULATIONS

Can Ashby's control theory be applied to evolving populations? I will consider two alternative scenarios, in which the population does and then does not reduce to the sum of its constituent organisms.

First, for simplicity, let's assume that at any given moment, a population comprises no more than the sum of the current generation's constituent individuals. In such a case, the putative adaptations of a population should be reducible to the collection of the actual adaptations of its constituents. That implies that evolving populations are not self-controlling or fitness-seeking purposive systems in the same way that individual organisms are. It also implies that evolving populations cannot define or protect any essential variables of their own, independent of the essential variables defined and protected by their constituent individual organisms. However, if the apparent adaptations of populations really are reducible to the adaptations of their individual organisms, we may be able to apply Ashby's control theory to populations indirectly. In that case, populations should have a similar status to human artifacts that are controlled by their human operators, with the survival and persistence of populations exclusively dependent on the survival of their constituent individuals. Likewise, the dispersal of populations across their environments should depend exclusively on the capacity of diverse individual organisms to disperse, including for active relocation. Also, the regeneration of populations should depend exclusively on the capacity of individual organisms, or male and female pairs of organisms, to reproduce in each generation.

This does not mean that the contributions of individual organisms to the evolution of their populations will be identical. Each individual inherits a distinctive sample of R_i from its population and experiences a unique personal history, as a result of which individuals vary considerably, inclusive of their variant self-controlling capacities. The usual way of thinking about this variance is in terms of variations in the fitness of individual organisms in populations. The fittest variant organisms are those that contribute most R_i to the next generation of variant individual organisms.

Individual organisms do not exist in isolation but rather live in the context of networks of interactive relationships with other organisms, including conspecifics and individuals of other species. In the context of interactions among members of the same species, the fitness of each individual organism is likely to be affected by the fitness-seeking activities of most of or all the other organisms with whom it is interacting within its population. For SET, many of these interactions are typically understood through the logic of inclusive fitness theory, with its focus on the extent to which organisms share genes.[29] For NCT, other forms of "relatedness" can arise through shared ecological inheritances, which can support mutualistic interactions, including among species.[30]

These interactions with conspecifics are typically two-way-street interactions. Each individual organism may have some capacity to influence or determine its relationships with other organisms. For instance, an organism may have some choice over which individuals it interacts with, and those decisions will be shaped by the individual personal histories of the individuals. But the fitness of a focal organism will also depend on how other organisms interact with it. In general, natural selection pressures, arising from the interactions of individual organisms with their conspecifics, should favor organisms best able to satisfy their fitness goals in the context of their networks of interactive relationships. These should be the same organisms that contribute most R_i to the evolution and adaptation of their populations. Thus, even in cases where it is possible to reduce the apparent adaptations of populations to no more than the collective expression of the adaptations of their individual organisms, it is still possible for populations to influence their own evolution indirectly, in a manner not fully recognized by SET.

Now let's consider a second scenario, where it is not possible to reduce the adaptations of populations to the adaptations of their constituent individual organisms. What happens if the whole is more than the sum of its parts? An example cited by Puckett et al. (2018)[31] is the collective sensing of environmental gradients, such as light, by large shoals of fish. Collectively, a shoal of fish is sensitive to gradients of light, but each individual fish in the shoal is not. This appears to be a genuine example of an irreducible adaptive property of a group or subpopulation that cannot be explained in terms of the adaptations of its individual organisms. Some bird and fish migrations may fall into this category too, such as when members of the group pool their knowledge to solve a problem collectively.[32] Other compelling examples are provided by the collective decision making of social insects, including ants, termites, bees, and wasps.[33] Some researchers suggest that it is not possible to explain the adaptedness of a leaf cutter ant colony or a termite nest, including the divisions of labor among plural castes, solely in terms of the adaptations of their individual constituent organisms;[34] the termite nest itself displays emergent properties.

One approach to this phenomenon is to refer to a termite's nest or leaf-cutter-ant colony as a "superorganism."[35] However, in the context of Ashby's control theory, this implies that to qualify as a superorganism, the collective entity must be able to define and protect its own essential variables and pursue its own fitness goals, independent of the essential variables and fitness goals defined by its constituent individual organisms. This means that a superorganism must also be able to protect a third kind of variance—namely, the variant relationship between the variant superorganism and the variant natural selection pressures that it encounters in its external environment.

None of this precludes the essential variables defined by a superorganism or a population from running in parallel with the essential variables and fitness goals defined by their constituent individual organisms. It probably demands it, in fact. In such a case, how is it possible for the superorganism to ensure harmony between its own putative higher-order adaptations, fitness goals, and essential variables with the essential variables and fitness goals of its constituent individual organisms?

Can there be conflicts between the adaptations and fitness goals of higher-order populations that are not in harmony with those of its constituent individual organisms? Human societies suggest that the answer is yes.

THE MOVES OF THE GAME

Let's return to the niche management game and attempt to specify the various moves that are possible. Once again, I'll start by considering any individual organism in a population (O), and its local selective environment (E), although the same logic applies to populations of organisms in their environments (see Odling-Smee 1988 for details). A move occurs whenever there is any change in the relationship between the organism and its environment. Both the selective environments and individual organisms can introduce changes in this relationship, and therefore initiate moves in the game. This is true according to both SET and NCT, but it becomes more prominent in NCT because of NCT's emphasis on the role of active niche-constructing organisms in evolution. The environment will cause a change in the O-E interactive niche relationship whenever a change in its state leads to a modification in one or more selection pressure relative to the organism. Conversely, the organism will cause a change in the O-E interactive niche relationship whenever it exhibits a changed or novel phenotypic trait, relative to its environment.

Individual organisms and their environments can change their interactive niche relationship in innumerable ways, but these are reducible to eight logically distinct categories, or moves. Four of these comprise the fundamentally different ways in which external environments can change the interactive O-E niche relationship by modifying one or more selection pressures, while the other four constitute distinct ways in which any individual organism can change the same O-E relationship by modifying its phenotype. While it is difficult, or perhaps even impossible, to fully describe varying environments,[36] and it is probably equally impossible to fully categorize varying organisms, it is possible to reduce all changes in the organism-environment relationship to one of these eight moves.

This is where we make contact with Ashby's control theory and his LRV again. The eight moves are the source of the third kind of variance—namely, the varying relationship between the varying phenotypic traits and the varying natural selection pressures. Hence, the moves are the source of the essential variables and its component subvariables described by Ashby, and the essential variable refers to the relationships that self-controlling individual organisms O must continuously protect to stay adapted to their changing external environment E, throughout their lives. The eight moves are illustrated in figure 3.1, with each move occupying a single cell.

Cell 1 The environment changes itself (E changes E). The wider ecosystem can always introduce a change into the individual organism's local selective environment, E, by causing a new change in O's environment. This is because each organism's local environment is open to inputs from the ecosystem in which it exists, which is open to the biosphere, which in turn is open to the universe. Hence, O's local E can be fed by unlimited variance, or V_E, from its ecosystem.

Cell 2 The environment changes the organism (E changes O). The local environment not only provides the physical energy and matter resources that O needs to stay alive, but is also a source of information that shapes development. The change in E that causes O to change could be as simple as a seasonal change, an incoming tide, or a change in the weather. Temperature and light, for instance, are important influences on developing organisms, with hormones commonly mediating the effects of environmental stimuli on gene expression.[37] For example, the butterfly *Bicyclus anynana* switches its phenotype dramatically to either produce wings with or without eyespots

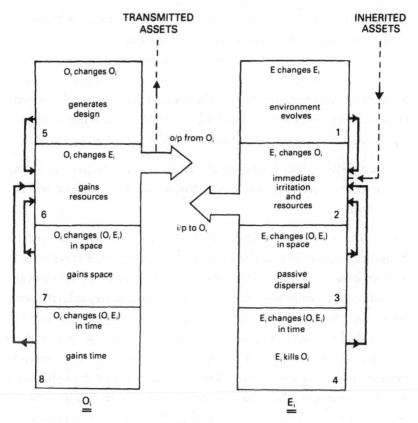

O_i changes O_i generates design 5	E changes E_i environment evolves 1
O_i changes E_i gains resources 6	E_i changes O_i immediate irritation and resources 2
O_i changes $(O_i E_i)$ in space gains space 7	E_i changes $(O_i E_i)$ in space passive dispersal 3
O_i changes $(O_i E_i)$ in time gains time 8	E_i changes $(O_i E_i)$ in time E_i kills O_i 4

o/p from O_i

i/p to O_i

$\underline{\underline{O_i}}$ $\underline{\underline{E_i}}$

Figure 3.1

The moves of the game classified in terms of the set of possible changes in the O-E relationship. Reproduced from figure 5 in Odling-Smee (1988).

depending on the temperature, the switch arising because higher temperatures increase the rate of production of a hormone that affects the expression of the gene determining eyespot size.[38] When O refers to populations, this move represents how selective environments (E) change populations of organisms (O) through the action of natural selection (see Odling-Smee, 1988).

Cell 3   The environment changes the organism's location (E changes O-E in space). The organism's position in space can be changed by passive dispersal mechanisms, such as wind, tides, currents, or waves, or through the activities of other organisms, such as pollinating insects. The environment can also impose isolating barriers on the organism, such as a mountain range or a lake on land, thereby limiting its dispersal across space.

Cell 4 The environment can change the organism-environment relationship in time (E changes O-E in time). The environment, E, can permit the organism, O, to move into its future by allowing it to survive, or alternatively, it can prevent O from moving into its future by killing it. When O refers to populations, this represents the situation in which populations go extinct.[39]

Then there are four fundamental ways in which the organism, O, can change its O-E interactive niche relationship relative to its local selective environment, E:

Cell 5 The organism changes itself (O changes O). The organism can change itself, either in response to, or in anticipation of, a move from E, or in response to internal metabolic changes. For instance, over time, the organism could become hungry, or get tired, or come into reproductive state.

Cell 6 The organism changes the environment (O changes E). For example, an individual organism can take energy and matter resources from its local external environment, dump its detritus in E, or change the environment's physical structure by redirecting a flow of energy or matter through it, such as when earthworms process soil. Organisms may also be able to construct artifacts in their environments, such as nests, burrows, dams, or houses.

Cell 7 The organism changes its own location (O changes O-E in space). An organism can actively relocate in space by choosing to move somewhere else. For instance, a fish may do that by swimming in the sea, or an animal may do so by walking, climbing, or running on land. An organism may also be able to change its own location in space merely by growing larger over time.

Cell 8 The organism can change the organism-environment relationship in time (O changes O-E in time). O can extend its lifetime and thereby move into its future by surviving, or it could end its own life, such as by sacrificing itself to benefit a relative.

This last move involving O moving into its future is the trickiest one to explain, for two reasons. First, it takes us back to the relationship between individual organisms and their evolving populations. Second, it subsumes all eight logically distinct categories of moves in this niche-management game. In particular, it subsumes all four of the logically distinct categories of O's moves because they are all linked to each other by sequences of dependencies, which run as follows: More lifetime for an organism is equivalent to the persistence over time of a very-far-from-thermodynamic equilibrium system. This means that more lifetime demands more free energy to fuel it. Any individual organism requires energy and matter, or R_p resources, to resist the second law and stay alive. An organism can gain energy and matter resources to live only by physically contacting an environment that can supply it with those resources. That environment must also be able to act as a sink for the organism's detritus. While persisting, the organism may grow, shrink, or change in location. Hence, when organisms survive (move 8), they likely also change themselves (move 5), their environment (move 6), and their location (move 7) in the process.

Moves 5–8 constitute all the means available to organisms to protect their essential variables. A change in state of the organism (move 5), such as growing hungry, will trigger appropriate actions, such as foraging behavior, that help to ensure that energy levels remain within suitable bounds. Constructing a nest or digging a burrow (move 6) can allow organisms to escape extremes or temperature, humidity or predation threats. Migrating to warmer climes in the winter (move 7) allows animals to remain within suitable temperatures and levels of food availability. Finally, survival (move 8) is possible only if key essential variables have been protected. Earlier in this chapter, I described how contemporary adaptations can be characterized as a "synerg," or complementary match, between the traits of the organism and the factors in its environment.[40] This conception is evocative of the ecologist Evelyn Hutchinson's (1957) concept of the niche as a multidimensional hypervolume of tolerances relative to a potentially unlimited set of environmental variables. Each dimension refers to an O-E relationship. Hence, we can view moves of the game that preserve the O-E niche relationship and protect essential variables as attempts by the organism to remain in tolerance space regions of its Hutchisonian niche.

Persisting in the game requires both R_i and R_p. Any organism must possess sufficient adaptive know-how (R_i) a priori, either from the prior evolutionary history of its population, or from its supplementary developmental processes, to control its contemporary adaptations in ways that enable it to harvest the energy and matter (R_p) resources that it needs from its environment and to dump its detritus back into it. The immediate point is that these cell 8 requirements that enable O to stay alive in its environment begin immediately, from the first moment when O appears. From the very first moment of its life, an individual organism requires both sufficient energy and matter (R_p) and sufficient adaptive know-how (R_i) to survive, grow, and reproduce and manage its O-E interactive niche relationship adaptively.

How can any new organism get started on its life if it needs both sufficient R_p and R_i resources a priori? The answer is that O's life is dependent on O initially inheriting assets from its parents and ancestors as a function of its membership of an evolving population or species. O must inherit its initial adaptive know-how by inheriting previously naturally selected genes from its parents and ancestors, primarily but not exclusively via genetic inheritance. It must also inherit other resources (namely, energy and matter resources) from its parents and ancestors a priori via other inheritance systems, most prominently via ecological inheritance. Each O must therefore inherit its own initial start-up O-E niche relationship from its parents and ancestors as a function of its membership of its particular evolving population. This includes inherited genes and other sources of inherited knowledge (R_i), and inherited physical resources (R_p), at the initial space-time location that O inherits from its parents and ancestors.[41] For instance, chicks inherit not only genes from their avian parents, but also the hormones and antibodies that are placed in eggs and the nest environment that is temperature-regulated and comparatively safe from predators. Likewise, phytophagous insects not only supply their offspring with genes and all the resources contained in eggs, but also carefully choose host plants on which to lay their eggs that provide their descendants with suitable food sources. In short, each organism must inherit a "start-up niche."[42]

I previously described all these inherited assets as a "free gift" to organisms from their parents.[43] But the gift is not really a gift, it is only a loan, or mortgage.[44] The loan has to be paid off by the organism reproducing, and in its turn transmitting, both R_i and R_p assets by genetic and ecological inheritance to its offspring. The evolutionary trick is that the organism's initially inherited R_i or genetic inheritance almost invariably include the adaptive know-how, plus the instructions, to reproduce as well as to survive. This is how the cost of the evolutionary process itself is paid for by its constituent individual organisms. Hence, all organisms must acquire sufficient resources from their environments, not only to keep themselves alive and to oppose the second law on their own behalf, but also to fuel reproduction of viable offspring.

By producing offspring and contributing to the next generation, individual organisms, O, are not only contributing to the further evolution of their populations, but also permitting natural selection to retest, and hence update, the adaptive know-how (or R_i) that is transmitted between generations in evolving populations.

While I will not dwell on it here, there is a recursion of the eight categories of moves from both O and E in the game, in which O is a population of individual organisms and E has become the selective environment of that population. This happens because any adaptive living system can only rely on the same categories of four fundamental moves to drive down whatever V_E or changing natural selection pressures that it encounters in its interactive O-E niche relationship, now at the population level. At this new higher level, a population may or may not then define and protect an essential variable of its own, over and above the essential variables of its constituent individual organisms. Regardless of whether O is an individual organism or a population, these eight logically distinct moves are universals that apply to the adaptive niche-management problems of all living systems relative to their particular external selective environments.

EMPIRICAL EXAMPLES

After all that abstract logic, we badly need some concrete examples again in order to illustrate the preceding arguments. I have chosen three animal examples, all taken from the work of Scott Turner.[45] Let's begin with termites.

Generally, termites do not have high tolerance for hot and dry conditions. Yet certain termite populations can live in these environments. According to SET, the adaptations that we should expect from termites that do this should be modified physiological adaptations to hot and dry conditions due to the natural selection of genes for high tolerance. This seems to have occurred in the case of *Hodotermes mossambicus*, a southern African harvester termite, which can often be seen foraging on the surface, in full sunlight, in hot and dry conditions. Their physiological adaptations include the development of a highly sclerotized exoskeleton and large size, which help prevent dessication, and retention of eyes by the workers, which is obviously necessary for them to forage in the light. In this case, the natural selection of appropriate phenotype-determining genes, as described by SET, appears to have been paramount. Those termites with physiologies best able to cope with the heat and desiccation were better placed to pass on their genes to descendant populations, allowing the entire population to evolve a suitable physiology. In terms of the aforementioned game, the hot and dry environment (E) has generated a change in a population of termites (O) through natural selection that favored contemporary adaptations that facilitate heat tolerance, such as a highly sclerotized exoskeleton (i.e., move 2). These contemporary adaptations allow the termites to keep an essential variable (their internal temperature) within a suitable range.

However, most termites in similar conditions do not go down this route. Instead, they employ niche construction to modify their selective environments (i.e., move 6). This has changed how they have evolved to cope with harsh, hot environments. "Heuweltjies" are curious landforms common in the winter rainfall regions of South Africa's Nama Karoo.[46] These are low mounds, roughly one-to-two meters tall and around twenty meters in diameter. They are often distributed in regular arrays over the landscape and typically support a distinct plant community that makes them stand out visually from their surroundings. Heuweltjies are the products of a different species of harvester termite, *Microhodotermes viator*. The nests of these termites qualify as constructed niches because the burrowing activities of the termites, which are ancillary to constructing the colony's underground nest and network of foraging tunnels, enhance the infiltration of sparse seasonal rainfalls. Heuweltjie soils are slightly moister and better aerated than surrounding soils, which tend to be compact so that rainfalls are prone to run off instead of infiltrating them. Through niche construction, the termites expose themselves to a suitable temperature and humidity, protecting these essential variables but modifying natural selection in the process. As a direct consequence of their mound-building and other niche-constructing activities, such as mining underground water, termites weaken selection for physiological adaptations that facilitate heat tolerance, such as a highly sclerotized exoskeleton, and strengthen selection for behavioral adaptations, such as mound building. Heuweltjies also qualify as a form of ecological inheritance because the structural modification of the soil by one colony typically persists beyond the colony's lifetime. These modified selective environments are inherited by later generations of termites. Such ecological inheritances can be impressive because some heuweltjies can be several thousand years old.[47]

This example illustrates two alternative means by two different species of termites to the same adaptive end. *H. mossambicus* appears to respond in the manner expected by SET to natural selection pressures with genetically determined phenotypes to these harsh environmental conditions. The second species, *M. viator*, however, adapts in a different way, which is anticipated by NCT. It responds to harsh environmental conditions by niche construction. It changes its environment to suit itself by its activities.[48]

My second example is one studied by Darwin: the earthworm, *Lumbricus terrestris*.[49] Turner[50] points out that earthworms are essentially aquatic oligochaetes, poorly equipped physiologically for life on land. Yet there they are. How can we explain that? We don't know when earthworms first invaded the land, but it is almost certainly over 100 million years ago.[51] We do know an advantage enjoyed by animals that live on land as opposed to in water is that taking oxygen from air is much more efficient than taking it from water. It is easier to do and far less costly in terms of the amount of water versus the amount of air that animals have to pump through their bodies to gain the same amount of oxygen.[52] Given oxygen's considerable contribution to metabolism, that is a major advantage for animals that live on land. This may have been the reason why the ancestors of today's earthworms invaded land in the first place.

But how could earthworms survive on land when they possess a physiology suited to an aquatic existence? Why didn't that physiology stop them unless or until it had been retooled by natural selection, as SET would anticipate? Turner[53] answered this question by focusing primarily on earthworm organs equivalent to kidneys, their nephridia. Turner pointed out something extraordinary. Earthworms clearly possess the "wrong" physiological adaptations for achieving water balance in their terrestrial environments. They live on land, but they have not evolved typical terrestrial "kidneys." They have largely retained the aquatic physiology of their freshwater ancestors.

Turner describes three kinds of physiology needed by different animals in different environments. Animals that live in freshwater are in danger of being flooded by excess water. They need to be capable of excreting large amounts of water, and thus produce lots of dilute urine. Animals in marine environments are in danger of being overwhelmed by too much salt. They have to be capable of excreting very concentrated urine to get rid of those excess salts. Animals living on land are in danger of dehydration. They need to retain as much water as possible. But earthworms have retained their ancestral freshwater physiology even though they now live on land; hence they continue to produce large amounts of urine as if they were freshwater animals. They thereby lose considerable amounts of water, which they would do better to retain. So how have they gotten away with that? The answer is that they have modified their terrestrial environment by diverse niche-constructing activities.

Earthworms don't just live "on" land, they live "in" it. They live in the soil. That complicates matters. Soil scientists divide soils into layers, called "horizons." Three basic horizons reflect the degree to which porous spaces in the soil are occupied by air or water. Proceeding downward from the surface, they are the aerial, edaphic, and aquatic (water table) horizons, respectively. For animals, it is relatively easy to get oxygen from the aerial horizon and easy to get water from the aquatic horizon, but not vice versa. Earthworms are confronted with a dilemma because they need both oxygen and water. They can go up for oxygen, but risk desiccation. They can go down for water, but risk suffocation. So earthworms attempt to get the best of both worlds by living mostly in the edaphic horizon between the aerial and aquatic regions. However, this horizon offers too limited a habitat to permit earthworms to live in it without niche construction. So earthworms enlarge the soil's edaphic horizon by active work, expressing diverse niche-constructing activities in the process.

One major consequence of the worms' niche construction is to make more water available to themselves in enlarged edaphic horizons. The movement of water through soil is determined by the soil's water potential, comprising four principal forces: pressure, gravity, osmotic potential, and the matric potential. The force that is crucially affected by earthworm niche construction is the matric potential. The matric potential arises from interactions between electrical charges on surfaces in interfacial areas. In soils, the matric potential depends primarily on pore size. Left undisturbed, "soil . . . gradually become(s) more and more clay-like in its interactions with water: it will hold lots of water, but will hold it with stronger and stronger matric forces."[54] Earthworms oppose this matric force through niche construction. They do so primarily by aggregating soil particles into larger pieces with increased surface areas. That weakens matric potential and makes it easier for the worms to draw water into their bodies.

In practice, earthworms affect the soil they are living in by several kinds of niche construction. One of them is by opening tunnels in the soil to improve soil drainage and aeration. Another is by passing soil through their guts, which causes the soil to pick up water-absorbing mucus from the worms. Third, the worms' fecal casts glue small soil particles together into larger pieces. Fourth, worms draw in pieces of plant litter from soil surfaces to eat and to line their tunnels. Earthworms also increase the abundance of macropores in the soil. The net effect of all these niche-constructing activities is to expand the soil's edaphic horizon. The earthworms expand their habitats by niche construction, and thus release themselves from "having to evolve" changes to their largely aquatic water balance physiology, in response to natural selection pressures arising from their novel terrestrial environments.

A major consequence of this earthworm niche construction is to "counteract" some of the natural selection pressures that they encounter in their "novel" terrestrial environments, and that they must have encountered when they first invaded the land. Supposing that earthworms had not invested in counteractive niche construction, then natural selection pressures in terrestrial environments would presumably either have prevented earthworms from invading land, or they would have selected for the adaptive "retooling" of earthworm physiology to enable them to live on land. That would be equivalent to move 2 in the game. However, the latter processes would probably have greatly slowed the rate at which earthworms could have adapted to their novel terrestrial environments compared to the rate of adaptation that they seem to have achieved by niche construction. It would also have changed their physiology dramatically, which hasn't happened.

Rather, the transition to land appears to have been reliant on niche construction, as represented by move 6 in the game. That niche construction allowed the earthworms to move onto land while protecting the essential variable of a suitable water balance. In spite of that, natural selection must have played its part in this story too. For example, after invading land, with the aid of counteractive niche construction, it is probable that natural selection would subsequently have selected for additional earthworm niche construction that further enhanced their fitness in their "novel" habitats. Natural selection would very likely have selected for any earthworm niche construction that further expanded edaphic soil horizons. For example, earthworms could not have been selected for dragging leaves into the soil if they had remained aquatic. But once on land, natural selection probably did select for that particular niche-constructing activity in earthworms. Earthworm niche construction did not replace natural selection but merely reset it, transforming the trajectory of these animals.

This earthworm example illustrates another general point. Inceptive (or innovative) and counteractive niche construction can act synergistically with each other. Creating a semiaquatic environment on land is innovative niche construction since it exposes earthworms to a new way of life. Processing soil to reduce matric potentials is counteractive niche construction, as it counteracts the reduced availability of water at higher soil layers. Together, these activities allow for a dramatic change in lifestyle without a fundamental change in physiology. The earthworm example also shows that both inceptive and counteractive niche construction can act synergistically with natural selection. All such interactions are likely to affect the rate of evolution, sometimes speeding it up, and in some circumstances slowing it down.[55]

More generally, inceptive niche construction may allow a species to invade a new habitat. However, the species may subsequently be able to exploit and tolerate its novel habitat only by buffering out some of the novel selection pressures that it then encounters in its new habitat through counteractive niche construction. That in turn may lead to the subsequent natural selection of still more effective counteracting niche-constructing activities. Adaptations depend on the interactions of natural selection and niche construction.

My third example is concerned with the ways in which individual organisms have to control their contemporary adaptations throughout their lives in their niche-management games with their environments.[56] A minority of animals, mostly birds and mammals, are endotherms. They maintain constant body temperatures by diverse internal physiological mechanisms that generate heat internally. But generating heat internally is costly. In our own case, about 90 percent of the food that we eat is needed to maintain our bodies at the homeostatic level of approximately 37°C.

Most animals adopt less costly ways of maintaining their bodies at a more or less constant temperature, ways not determined by internal heat generation. Instead, it is maintained by the clever exploitation of external sources of heat. Creatures that maintain their body temperatures in this way are usually called "cold-blooded," but this is misleading. For example, as Turner[57] points out, several species of lizards can maintain high and stable body temperatures throughout the day. A lizard might spend the early morning sunning itself on a rock to regain heat lost overnight or to the cold morning air. Once it has built up its body temperature sufficiently, it may then make sorties into the colder parts of its environment in search of food. It is then likely to cool down again. Eventually, it will have to stop foraging and return to basking in the Sun to warm up, and so on throughout the day. By shuttling all day between hot and cold spots, lizards can keep their bodies within a narrow range of temperatures. In this way, a lizard can not only turn itself into a warm blooded creature, but it also can be remarkably adept at regulating its temperature at a near-constant level, behaviorally. Its behavioral thermostat then affects almost all aspects of how the lizard fits into its environment adaptively. Its body temperature affects how fast a lizard can move, the rate at which it can digest food, how big its territory is, and its success in foraging. It even affects its capacity for reproduction.

What temperature a lizard maintains is the result of a careful balancing of benefits versus costs. Some are called "opportunity costs." For instance, while basking on a rock, a lizard is giving up the opportunity to search for food. However, good sunbathing spots are often exposed sites, where the risks of predation are higher. What temperature a lizard maintains will therefore depend on a balance sheet of costs and benefits, as well as by immediate and sometimes overriding events in its environment, such as the sudden appearance of a predator.

This third example illustrates the ongoing dynamics associated with the contemporary adaptations of organisms. The lizards' state (i.e., body temperature) constantly varies, relative to their varying environments, on a moment-by-moment basis. They have to balance the costs and payoffs continuously between their varying physiological needs and the varying opportunities and threats offered by their external environment.

This balancing act is equivalent to the lizards attempting to protect an essential variable—its body temperature, as described by Ashby's control theory. This requires a balancing act between three variables: temperature, foraging, and the risk of predation. In reality, and in the case of all organisms, many more variables are likely to be in play in their attempts to protect their essential variables. For instance, another intriguing variable that lizards have to take into account occurs if they become infected by bacteria. Apparently, like endotherms, the lizards have to cause their body temperatures to rise to a temperature that is higher than normal to combat the infection. But they differ from endotherms in having to achieve this febrile state behaviorally, rather than by resetting an internal thermostat. Of course, if a febrile lizard does sun itself more, it raises the risk of predation and loses more opportunities for foraging, which must change its overall balance sheet. As Hutchinson[58] pointed out, in reality all organisms actually live in niches composed of n-dimensional hypervolumes of ecological variables. Organisms don't have to pay attention to all of them, but usually they have to pay attention to many more than three!

CONCLUSION

In this chapter, I've argued that adaptation is not a straightforward unified concept in biology. The adaptations of organisms decompose into three distinct concepts that need to be understood separately. They relate to the past, present, and future. Historical adaptations account for the evolutionary origins of phenotypic design. Contemporary adaptations are concerned with the fit or complementarity between individual organisms and their external environments in the present. Anticipatory adaptations refer to the fitness purposes or goals of organisms in each generation.

After considering each of these concepts separately, it is then possible to recombine them in ways that I have been suggesting could lead to a deeper understanding. It is not possible to understand the repertoires of the contemporary adaptations of individual organisms, nor how contemporary adaptations are controlled by individual organisms, except in the light of the historical evolutionary origins of the populations to which the organisms belong. Nor is it possible to understand how individual organisms control the expression of their contemporary adaptations, except in the light of their fitness goals of resisting the second law of thermodynamics for long enough to survive and reproduce.

Organisms have to constrain their inductive gambles about what adaptations they need to express, where and when in their contemporary environment, by relying on inherited and acquired adaptive know-how, or R_p, about what was adaptive in the past. They also have to constrain their contemporary adaptations by their expected future interactions with their own individual selective environments to achieve their fitness goals. I also discussed how the two theories of evolution that we have been considering here, SET and NCT, differ with respect to how they see the integration of these three components of the concept of adaptation in biology. NCT offers a richer understanding of these complex issues and a stronger interpretation of empirical examples.

This has been a difficult chapter to write, and I fear to read too. In the twentieth century, the developmental geneticist Conrad Waddington once said that how organisms adapt to their unknown, but not wholly unforecastable, futures is the hardest problem in biology. It is this problem with which we have been grappling in this chapter.

4 ADAPTIVE KNOW-HOW IN EVOLUTION

How can the past inform the present about the future? This is a question that any theory of evolution, including both standard evolutionary theory (SET) and niche construction theory (NCT), must address. So how does natural selection inform organisms in evolving populations?

Strictly speaking, that is the wrong question. Biologists usually think of natural selection as a blind and purposeless process.[1] Natural selection alone cannot be a source of "meaningful" information for organisms. By itself, it does not supply the know-how, or R_i, that organisms need to adapt to their environments. For R_i to accumulate, there must also be processes that generate heritable phenotypic variation.

What about the genes that are carried by organisms? Unless they have been subject to prior natural selection, genes cannot inform organisms either. New gene variants typically arise through chance-based mutations. Such mutations cannot inform organisms, except by accident, and then only extremely rarely. Only in combination with natural selection, can the genes that organisms inherit from their ancestors become a source of adaptive know-how, or R_i. For this to be possible, natural selection must be biased toward the differential inheritance of the genetic variants of fit organisms, in each generation, in evolving populations.

The genes that organisms inherit from their ancestors may then be able to supply organisms in the current generation of a population with deductive know-how, or R_i, about which phenotypic traits were adaptive for their ancestors in the past. This R_i is deductive because it is deduced from past interactions of the population with its environment, and it is adaptive because only those interactions that led to successful outcomes resulted in the transmission of genes to the next generation. Ideally, organisms in each generation would inherit the adaptive know-how that maximizes the chances that the inductive gambles that they have to make when they express their contemporary phenotypic traits will prove adaptive again. Similarly, the R_i organisms inherit from their ancestors would ideally maximize the chances that their expectations or predictions about their future interactions with their environment will turn out to be correct.

But why should information about the evolutionary histories of populations, or for that matter the developmental histories of individual organisms, ever have the capacity to tell organisms anything meaningful about their present or future interactions with their environments? To what extent is this inherited adaptive know-how, deduced from the past, relevant to Waddington's problem of how organisms adapt to their unknown but not wholly unforecastable futures? These are the questions that I want to concentrate on in this chapter.

HOW IS IT POSSIBLE FOR ANY ORGANISM TO KNOW ANYTHING?

All discrete systems or objects in the universe, whether nonliving or living, are transitory.[2] They are the temporary manifestations of physical processes. Some (such as clouds) have only a fleeting existence. Others, such as mountains, may exist for millions of years. And planets can exist for billions of years. I'll assume that any discrete system that we are talking about is one that persists with the same identity for a measurable period of time.

All discrete systems are highly localized in space and time. They are also constantly compelled to change their locations in their environments in space and time. Obviously, no discrete system can stop still in time. Each is compelled to move from the past, to the present, and on to the future. The succession of here-and-now moments for all discrete systems becomes a "worldline" in four-dimensional space-time for all nonliving abiotic systems and a lifetime for living organisms in biology.[3]

Superficially, it might seem possible for discrete systems to stop still in space for a time, but that is an illusion. For example, all earthbound systems are traveling in space, as well as in time, if only because the Earth itself is continuously moving in space during time by spinning on its axis and orbiting round the Sun. In spite of continuously having to travel through space and time, every discrete system can interact only with other systems, or with its surroundings, in the specific location where it exists and in the present.

On Earth, discrete abiotic and biotic systems are confronted with varying temperatures, rainfall, wind, gravity, weathering, magnetic fields, light and dark, and so on, but they can interact with any of these phenomena only in the present. In this sense, all discrete systems exist only in the present. Their pasts have ceased to exist, and they cannot now erase whatever interactions may have occurred between them and their environments in bygone days. Similarly, their futures do not yet exist, which means that discrete systems cannot physically interact with their environments ahead of time before they travel into their futures. They are localized systems that exist only in the here and now. In these circumstances, how is it possible for any discrete, transitory system, which is constantly changing its location in space and time, to "know" anything?

Everything that I have been saying until now about localized discrete objects and systems applies equally to both nonliving systems and to living organisms. But this question marks the point at which nonliving and living systems diverge fundamentally. Abiotic systems don't need meaningful information to persist. Abiotic systems are typically probable systems that are not far from thermodynamic equilibrium relative to their surroundings. They do not have to resist the second law of thermodynamics as they travel through space and time. They merely have to obey the laws of nature, including the laws of thermodynamics. They do so by shedding free energy, dissipating by generating disorder and entropy, as they travel toward greater thermodynamic stability. Abiota only have to react to their surroundings. Their reactions may cause further reactions, but nonliving systems are not capable of any autonomous actions. Rare exceptions include radioactive entities, which I'll ignore here.

Unlike abiotic systems, all living systems must actively resist the second law of thermodynamics to stay alive. To do so, they have to continuously import energy and matter, or R_p resources, from their environments and export their detritus back to their environments. Organisms cannot do either of these things without interacting with their environments, and they cannot interact with their environments unless they adapt to them. But organisms cannot be adapted to their environments unless they are informed by appropriate adaptive know-how, or R_i.

The capacity of organisms to resist the second law by establishing a flow of energy and matter (R_p) between themselves and their environments sufficient to counteract entropic processes therefore depends on their acquisition of meaningful information (i.e., R_i). The R_i that organisms need is primarily inherited from their ancestors via their genes and consists primarily of the adaptive know-how that they inherit from their ancestors. However, for the acquisition of adaptive know-how by organisms to be possible, both the organisms themselves and their external environments must satisfy some preliminary requirements.

A completely chaotic universe could never inform organisms. The universal environment must be at least partly lawful, as it is impossible to know anything about randomness. If you happen to know the initial state of a chaotic system, it is just possible to know a little about chaos, but not much.[4] Should the universe ever achieve maximum entropy (i.e., total disorder) then, regardless of everything else, life would be impossible because no organism could know anything.

Fortunately, the universe is at least partly lawful. It expresses the causal laws of physics and chemistry and the putative laws of biology, as well as physical symmetries and some physical constants. However, the laws of nature refer primarily to the dynamics of the universe. They refer to change. In a static universe that was ordered but completely frozen and fixed, there would be no change, and therefore no laws of physics, chemistry, or biology in operation, because there would be no changes to describe or govern. In these circumstances, any imaginary organism that happened to exist would not need to know anything anyway. It would not have to oppose the second law of thermodynamics because the second law would not exist.

Thus, organisms could not "know" anything if the universe were chaotic, and they would not need to know anything if the universe were static. For organisms to know anything, or even need to know anything, their environments must be both lawful and dynamic. In practice, the universe is neither entirely chaotic nor unchanging. As a consequence, organisms must "know" something about how the laws of nature cause changes in the universe.

The hallmark of the laws of nature is that they refer to causal relationships between events, in regularly reoccurring patterns. Causal relationships reoccur in space and time, whereas the hallmark of chance is a lack of repetition. Chance events do not typically reoccur in space and time, and certainly they do not reliably reoccur. They are commonly random events, events of very low probability, or even unique events. The ultimate nature of the relationship between causality and chance, or between regularly repeating invariant event patterns versus nonrepeating variant event patterns, remains unresolved within the field of physics[5] and will not be discussed here.

Before organisms can "know" anything, they must possess some fundamental properties too. They must have at least an elementary capacity to "sense" the present, to "remember" the past, and to "anticipate" the future. Because localized organisms exist only in the present, their senses can only detect the present. They may be able to sense signs of past events in their environments, before the present moment, as is the case when astronomers perceive events that occurred many light-years in the past. But they can detect meaningful information only arising from past events, in the present.

Organisms must also be able to "remember" something about their ancestors' interactions with their past environments. In the case of many animals, they must remember something about their own past individual interactions with their own individual environments too (see chapter 7). Individual plants also register the past through developmental plasticity. However, leaving learning, memory, and developmental plasticity to one side, organisms in evolving populations remember the past primarily in the form of the genes that they inherit from their ancestors. These genes are expressed in phenotypes that proved adaptive in past environments, and in that sense constitute a "memory."

Organisms must also have some capacity to apply whatever know-how they have gained from their remembered pasts (primarily from their inherited genetic "memories") to their own current, and anticipated future, interactions with their environments. The key requirement for all knowledge-gaining organisms is that they must be sensitive to the difference between regularly reoccurring causal relationships between events in their environments and one-off, rare, or random events. Unless they are, organisms will not be able to demarcate between repeating cause-and-effect relationships in their environments and everything else. In such a case, whatever R_i they acquire from the past risks being maladaptive.

How is it possible for any living system to make this vital demarcation? Instead of trying to answer this question by immediately considering evolutionary biology, I'll begin by discussing science.

I'm going to assume that the logic that describes all meaningful information-gaining processes in living organisms is always the same.[6] The logic of knowledge gain is the same, regardless of whether we are talking about how adaptive know-how is gained by evolving populations of organisms, or by learning in individual animals, or by collective sociocultural knowledge–gaining processes in humans, including science.[7] In each case, despite the fact that the mechanisms and processes responsible for the gaining of meaningful information by organisms are quite different, the logic of knowledge gain is the same[8] (for more, see chapter 7).

We can take advantage of this situation by comparing different knowledge-gaining processes with each other. At the moment, we probably know more about how science gains meaningful information about nature than we do about how evolving populations of organisms gain adaptive know-how via the mechanisms of natural selection and population genetics. We may be able to learn more about how evolution works by looking at how science works first.

GAINING KNOWLEDGE IN SCIENCE

Before it was possible to land humans on the Moon, or Rovers on Mars, scientists and engineers had to make correct predictions about how they could travel to these extraterrestrial places, as well as what kind of conditions they were likely to meet when they arrived. The success of these missions proves that it is possible to use information deduced from the past to make correct predictions about the present and the future, although never with absolute certainty.

But how is this possible? The reason stems from the ways in which cause-and-effect relationships work in nature. Causes have effects. Causes precede effects in time. Unless masked by hidden or interfering variables, the same cause always has the same effect. It follows that if scientists observe a particular cause-and-effect relationship and subsequently encounter a repetition of the same cause, they can then, with reasonable confidence, predict the reoccurrence of the same effect.

This, of course, presupposes that the scientist can tell the difference between reoccurring cause-and-effect relationships and chance events in nature. Scientists cannot make correct predictions about the future unless they can demarcate between "true" cause-and-effect relationships and everything else in nature, and do so a priori. To make this vital demarcation, a scientist must be able to "remember" which cause-and-effect relationships regularly reoccurred in the past.

Today, the ability of individual scientists, or scientists working in institutions and wider scientific communities, to make correct predictions about nature is greatly enhanced by modern technology. For example, scientists and engineers have been extending our human senses for centuries. Vision is amplified through the use of telescopes, binoculars, and microscopes, and sound through the use of microphones and amplifiers. Scientists have constructed innumerable artificial senses of enormous sensitivity. We now possess instruments that can monitor previously unknown or unrecognized phenomena on Earth and in the universe, such as radio waves from space. Our technologies have also extended our "memories," which can now be registered in multiple different ways, as well as in human heads. They range from books and journals, to tapes, films, and computers, and to vast stores of digitalized data. Deductions can be drawn from these data about past historical cause-and-effect relationships, by reason, logic, and deductive mathematics, aided by powerful computers that process the data and detect regularities.

Frequently, the consequences of these deductions are the gaining of new scientific knowledge about nature. Initially, this new knowledge is likely to take the form of conjectures and hypotheses. In science, however, it is not enough to sense, observe, and remember putative, reoccurring cause-and-effect relationships. Nor is it enough to form hypotheses about how nature works. In science, hypotheses must be tested empirically. Scientists are constantly doing experiments in attempts to pin down the actual cause or causes of natural phenomena, typically by manipulating one variable at a time and controlling all other potentially interfering or masking variables.

In the context of NCT, our empirical sciences are a form of sociocultural niche construction. In the first instance, the goals of science are typically the gaining of more meaningful information, or R_i, about how nature works. It is seldom about the immediate gaining of more R_p. That usually comes later, for instance through the application of scientific knowledge in technology.

Scientific experiments can greatly speed up the demarcation of repetitive cause-and-effect relationships. Experiments also ensure that previous observations of cause-and-effect relationships were reliable. If scientific hypotheses survive repeated empirical testing, they can eventually be turned into well-established theories. Unless they are destroyed by new observations that they cannot explain, well-established scientific theories are assumed to be reliable (i.e., truthful) statements about the lawfulness of nature. This new scientific knowledge may then be used by scientists and others to constrain our collective human inductive gambles about what adaptations will work for us again with respect to our relationships with our environments in our futures. It might even be used to land humans on Mars one day.

KNOWLEDGE GAIN IN BIOLOGICAL EVOLUTION

Now let's return to evolution. I've already suggested that the gaining of new adaptive know-how, or R_i, by evolving populations of organisms is logically equivalent to the gaining of new knowledge in science, even though the mechanisms involved are typically different. I will focus primarily on the evolutionary processes described by population geneticists.

Organisms in evolving populations are discrete, localized systems continuously traveling in environmental space and time. As they travel, they will encounter regularly reoccurring sequences of events, indicative of cause-and-effect relationships in their environments. For instance, they might learn that lightning precedes thunder, that the sights and sounds of moving water go together, or that earthworms come to the soil surface after rain. They will also encounter chance events, and events or sequences of events of very low probability. These may include a shooting star or a random leaf falling from a tree after a cat walks by. Some regularly reoccurring cause-and-effect relationships will be of particular significance to the organism, as they are indicative of consistently reoccurring natural selection pressures in their environments. The sight of a predator, for instance, might be predictive of danger, while the sound of prey might predict food. To the extent that such regularities affect the organism's fitness, populations can evolve a "memory" of the cause-and-effect relationship through natural selection. Individual organisms possessing genes that predispose them to respond to predator cues by a suitable response to danger, such as fleeing, might have a fitness advantage. As such cause-and-effect relations are reliable and reoccur in time and space, they generate a consistent advantage over those gene variants, and may subsequently spread through the population.

Organisms will also encounter events of very low probability, or irregular sequences of events, indicative of inconsistent natural selection pressures in their environments. Such encounters, because they do not reliably reoccur, will not reliably trigger consistent selection for particular genes. The difference between consistently reoccurring natural selection pressures and inconsistent natural selection pressures should then allow evolving populations of organisms to demarcate between cause-and-effect relationships and all else in nature, leaving a "memory trace" of such relations in their naturally selected genes.

The capacity of populations to make these demarcations is equivalent to evolving populations demarcating between signals (S) and noise (N) in communication theory (see chapter 2). The signals correspond to cause-and-effect relationships in nature, and the noise to everything else. In this connection, it's also important to realize the different roles played by natural selection, relative to evolving populations on the one hand, and individual organisms in those populations on the other.

In populations, the role of natural selection is to sort between fit and unfit individual organisms in each generation. Natural selection thereby biases the adaptive know-how, or R_i, that is transmitted in populations. However, relative to individual organisms in populations, natural selection plays a different role. It is the source of the threats and dangers and opportunities that individual organisms confront in their niche-management games with their local environments throughout their lives. Natural selection pressures are the sources of the selection factors (or V_E) that individual organisms must drive down or destroy by possessing suitable adaptations (V_O) at each place and moment during their lives, as per Ashby's control theory (see chapter 3). Fit organisms manage this by possessing adaptations capable of responding to the environmental conditions encountered. Unfit organisms do not. Natural selection pressures acting on populations then sort between fit and unfit organisms as previously discussed.

Assuming that organisms in evolving populations encounter the same or similar natural selection pressures repeatedly in their environments, the same reoccurring selection pressures should select the same or similar phenotypic traits in successive generations of evolving populations or species. Consistently reoccurring natural selection pressures in the environments of populations should indirectly select for whatever genes are correlated with the reoccurring adaptive phenotypic traits that are expressed by successive generations of organisms in evolving populations. The genes inherited by individual organisms correspond to "memories" of the consistent natural selection pressures that were encountered by their ancestors in the past. Whatever genetic "memories" that individual organisms inherit, serve to anticipate that the phenotypic traits that were adaptive for their ancestors in the past will be adaptive for them again in the present and in their futures.

THE DIVERSITY OF ORGANISMS

Given that the logic that describes how populations of organisms gain their adaptive know-how is the same in all populations, you might be forgiven for expecting that all organisms in all populations could end up with the same R_i and the same adaptations. They don't, for several reasons. First, all evolving populations, and all individual organisms in those populations, have different locations in space and time. The niche relationships between individual organisms and their environments, and between populations and their environments, are always unique, and hence they must differ at least slightly from each other.[9]

In addition, organisms in different populations have very different ecological and evolutionary histories. Their different evolutionary histories promote different adaptations in different populations and species. That is self-evident. You don't have to be a biologist to notice the diversity of the adaptive designs of different species. But you may have to be a biologist to notice some other, less obvious sources of diversity between species. For example, the adaptation that particular species of organisms acquired as a consequence of their past evolutionary histories may lock them into restricted evolutionary futures by causing them to pass points of no return. Insects, for instance, got locked into being small in size by their exoskeletons and their subsequent need to breathe through them via tracheas. Likewise, in bilateral organisms, the two sides of their bodies are symmetrical, or nearly so. It's hard to imagine how any bilaterian species could evolve further in ways that require them to abandon their basic body plan. The same points of no return may not apply to different species with different past evolutionary histories. That will cause different species to diverge from each other irreversibly.

One critical evolutionary breakthrough on Earth was the evolution of the first eukaryotes following at least 2 billion years of simpler prokaryotic cells, including bacteria and archaea. The subsequent evolution of multicellular or metazoan organisms on Earth would not have been possible without this breakthrough, but all metazoa are effectively locked into a eukaryotic cellular structure.

Another source of diversity arises from the fact that different species of organisms may have very different subjective utilities and affordances relative to different components in their environments. This means that the same environmental factors may have very different meanings to different species of organisms. For instance, a 1°C change of temperature in the environment of an elephant would probably be insignificant, but it could be highly significant for a small reptile living in the same location[10] (see chapter 2). Likewise, ocean ecosystems appear to be more sensitive to climate change than terrestrial ecosystems, perhaps because organisms can more easily escape the heat. in the latter than in the former.[11]

THE DIVERSITY OF CAUSAL RELATIONSHIPS

More complex cause-and-effect relationships are harder for the processes of evolution to detect than simpler ones, a point that almost certainly holds true for other information-gaining processes too. "A causes B" is the simplest kind of causal relationship, but it should be more difficult for evolutionary processes to detect more complex cases, such as when $A + B$ causes C, or $A + B + C$ causes D, or when $A + B$, but not C, causes D, or still worse when $A + B + C$ partly causes A again, in a nonlinear causal relationship involving feedback. Examples of relatively simple cause-and-effect relationships in nature are diurnal changes, seasonal changes, and tidal changes. A slightly more complicated cause-and-effect relationship combines changing day lengths with seasonal changes. In this case, A (night or day) + B (seasonal change) is needed to cause C (day length).

No organisms on Earth, apart from humans, can truly know the causes of these phenomena. It took our own species about 300 millennia to realize that the cause of day and night is the Earth spinning on its axis, and that the cause of seasonal changes is a combination of the tilt of the Earth on its axis and its annual orbiting of the Sun. It also took humans hundreds of thousands of years to realize that tides were caused by a combination of the gravitational pull of both the Moon and the Sun on Earth. They pull together and then against each other on a bimonthly basis.

Yet the diurnal and seasonal adaptations of numerous organisms in different species demonstrate that even without knowing the true causes of these phenomena, organisms can still detect all of these repeating cause-and-effect relationships. They must have done so as a consequence of their repeated encounters with the consistent natural selection pressures in their environments that are generated by these phenomena. Probably it will always be more difficult for the evolutionary process to detect and demarcate, on a stochastic basis, more complex cause-and-effect relationships than it is for simpler ones. But the logic should always be the same. It may just take longer for organisms to adapt to more complex causation.

ADDITIONAL SOURCES OF COMPLEX CAUSATION

There are also many other sources of complex causation. One of them is the addition of more background noise (N) to signal-to-noise (S/N) ratios. The signal (S), which here refers to a causal relationship, may be deeply embedded in a lot of background noise in the form of "masking" variables in the environments that organisms have to interact with, which correlate with but do not cause the event. Here again, it may take longer for evolutionary processes to register more deeply embedded causal relationships than superficial ones, but on a stochastic basis, given sufficient time, they should still be able to do so.

Another kind of complexity refers to competing natural selection pressures. For example, if as a consequence of their past histories, organisms in a population end up being selected for trait X, as opposed to trait Y, in some environmental contexts, but trait Y instead of trait X in other contexts, this may make it more difficult for organisms to adapt to their environments. For example, the lizards discussed in chapter 3 had to maintain their body temperatures (i.e., selection for increased basking in the Sun) while still avoiding predators (i.e., selection for reduced basking). This is a familiar problem to most biologists. The end result is usually some kind of trade-off between competing selection pressures and competing adaptations in organisms.

There are also different kinds of natural selection pressures due to their different origins. Three kinds of interactions can occur between localized, discrete systems in ecosystems: (1) interactions between abiota (i.e., between two or more nonliving, purposeless systems); (2) interactions between nonliving abiota and living organisms (i.e., between purposeless and purposeful systems); and (3) interactions between organisms and other organisms (i.e., between two or more, purposeful living systems). This requires a fuller discussion, which follows.

Interactions between Abiota

I'll briefly consider interacting abiota first. Similar to organisms, abiota interact bidirectionally with their abiotic surroundings. Discrete abiotic systems react to inputs from their abiotic surroundings, but they can also cause reactions in their surroundings, including both endergonic (i.e., energy-requiring) reactions and exergonic (i.e., energy-dissipating) reactions. Because abiotic systems are purposeless, their reactions with their surroundings are informed by the laws of physics and chemistry only. Natural selection is not involved. These are the kinds of interactions that occur on "dead" planets (i.e., planets where there is no life). Similar to organisms, however, discrete abiota carry with them the historical consequences of their past interactions with their surroundings as they travel into their futures. Their present here-and-now states may reveal their past histories. For instance, geologists and paleobiologists can often read the past histories of the rocks that they are looking at from the present state of the rocks.

Interactions between Abiota and Organisms

The interactions between organisms and abiotic systems are more complicated than those interactions that solely involve abiota. They comprise two categories of natural selection. The two categories refer to whether the selection stems from unmodified abiota in the environments of populations of organisms, or from abiota that have previously been modified by niche-constructing organisms. If unmodified abiotic sources of natural selection are affecting organisms in evolving populations, then they correspond to autonomous sources of natural selection pressures in the environments of organisms, which are commonly modeled by SET. Conversely, where abiotic sources of natural selection have been modified by niche-constructing organisms, they become a different category of natural selection. The natural selection pressures will no longer stem exclusively from purposeless abiota; rather, they now stem partly from the active purposeful agency of the niche-constructing organisms that previously modified the abiota. In other words, the abiota will bear the signatures of the purposes of the organisms that modified them. This category of natural selection has received much less attention from SET, but it is the focus of NCT. Whether or not the abiota were previously modified by niche-constructing organisms may affect the capacity of organisms in evolving populations to detect the causal relationships that underpin them. It may make it either easier or harder for evolving populations to detect those causal relationships.

While the purposes of niche-constructing organisms vary between species, individuals of the same species generally share similar purposes, which often leads them to interact with their environments, and to modify those environments, in similar ways.[12] These issues are complex and multifaceted, with different forms of niche construction (e.g., inceptive verses counteractive) having different impacts on the intensity and variability of natural selection.[13] However, a recent meta-analysis of selection gradients (which are measures of natural selection in the wild) found that organism-constructed sources of selection that buffer environmental variation (i.e., that result from counteractive niche construction—for example, adaptations to living in nests and burrows) will generate reduced variation in selection in space and time and weaker selection relative to abiotic sources (e.g., adaptations to the climate). For instance, across diverse taxa, nests buffer climatic fluctuations that have been found to be metabolically costly.[14] Likewise, when animals like lizards choose habitats with a suitable temperature, they too have acted in ways that increase the likelihood that the temperatures experienced stay within a tolerable range thereby reducing temperature-related selection.

For illustration, the flow of water in a river could be affected by either rainfall or a dam built by beavers. While the consequences of rainfall for the numerous organisms that live in riparian ecosystems may to some extent be predictable, the precise impact on selection will vary from one year to the next depending on how much rain falls. Selection switches from adaptation to wet or to dry conditions depending on the weather.

In contrast, the immediate consequences of beaver dams on the river, at least in the vicinity of the beavers' lodge, are more predictable. Beavers adjust the height of the dam to control the flow of the river and the depth of the water in the lakes that they create, with families of beavers maintaining and repairing the dam over long periods of time. This allows them to ensure that the lodge is safe from predators, such as bears or wildcats, regardless of how much rainfall there has been. This activity is thought to dampen selection on beaver antipredator adaptations since it reduces the incidence of predation.

The upstream and downstream consequences of beaver dams may be less predictable, but at least with respect to the selection pressures that beaver dam-building directly counteracts, niche construction renders selection more predictable.[15] The differences between these two kinds of natural selection could affect the capacity of organisms in recipient evolving populations to be informed by adaptive know-how, or R_p, about different causal relationships in their environments. These differences help evolutionary biologists to distinguish empirically between these two categories of natural selection in evolution.[16]

By ignoring the purposes of organisms, or treating these as genetically determined, SET does not encourage biologists to distinguish between that natural selection that stems from purposeless abiota and that generated by purposeful organisms. In contrast, NCT assumes that the purposes of organisms are derived primarily from the universal requirement of all life to resist the second law of thermodynamics. Because it is a universal requirement of life, this primary purpose does not involve different natural selection pressures acting on different populations of organisms. However, NCT acknowledges that owing to their past histories, different organisms in different species depend on different adaptations to resist the second law.

Insofar as their adaptations are due to unmodified natural selection pressures stemming from purposeless abiota, the reduction of the purposes of organisms by SET to no more than the consequences of prior natural selection might be justified. But this justification evaporates if these abiotic sources of natural selection have been previously modified by the purposeful niche-constructing activities of their ancestors, or of other organisms, or even of their own prior individual, purposeful niche-constructing activities. In these cases, it will be possible to understand the purposes of organisms only in terms of nonlinear causal relationships based on feedbacks from niche-constructing organisms.[17]

Interactions among Purposeful Living Organisms

Interactions between living organisms involve the second category of natural selection only. Here, the niche-constructing activities of one population can change the natural selection experienced by another population, and vice versa.

Here, I will focus on the three principal classical relationships between different populations of organisms in both evolution and ecology (which are discussed further in chapter 8). These relationships go beyond *trophic* ("who eats whom") and *competitive* interactions, and also encompass engineering interactions (i.e., niche construction).[18]

Competitive relationships between two or more evolving populations occur when the populations are competing for the same environmental resource, or resources. They may either be competing for the same physical locations in their environments or for the same energy and matter (i.e., R_p resources), in those locations. When two populations compete with each other, the organisms in one population impair the fitness of organisms in the other population, and vice versa. In evolution and ecology, this relationship is known as a "minus/minus" relationship. Examples include when plants compete with each other for light and groundwater, or when animals compete for food and territories. Here, the organisms in one population might benefit from transmitting deceptive signals to the other population, making it harder for individuals in the second population to outcompete individuals in the first and undermining the ability of each to detect causal relationships.

Mutualistic relationships occur when one population provides another population with a resource that it needs, and the second reciprocates by providing another resource for the first population. Mutualist relationships are known as "plus/plus" relationships. A classic example is between flowering plants and insects such as bees: the flowers provide the bees with food, and the bees help the plants to reproduce by spreading pollen. In these circumstances, populations have been found to generate particularly consistent natural selection pressures[19] and may transmit "honest" signals to their mutualist partner. This may make it easier for organisms in mutualist populations to detect the nature of the causal relationships between themselves and their mutualist partner, to the benefit of both.

Predator-prey and host-parasite relationships are asymmetric relationships, in which the organisms in one evolving population benefit from their relationship with the organisms in another population, but where the fitness of organisms in the second population is impaired by this same relationship. Asymmetric relationships are known as "plus/minus" (+/−) relationships. There are innumerable examples of both predator-prey and host-parasite relationships in ecosystems: for instance, foxes prey on rabbits, while cuckoos parasitize reed warblers. Here again, there is evidence that organisms in both populations generate deceptive signals to each other—cuckoos have been described as "cheating by nature," for instance.[20] This deception undermines the ability of each to detect causal relationships.

BIODIVERSITY AND COMPLEX ECOSYSTEMS

In simple terms, the more different species there are in an ecosystem, the greater its biodiversity, and the greater the biodiversity, the greater the complexity. Next, I'll discuss two examples of these sources of causal complexity.

First, different organisms in different species may adopt different evolutionary strategies and may affect each other by doing so. A well-known example of contrasting strategies is that between "r-selected" and "K selected" organisms. Typically, r-selected organisms lead relatively simple, short lives and invest little in gaining adaptive know-how, or R_i. Instead, they produce vast numbers of gametes, spores, or seeds, which they may literally throw in the wind to deposit in water or stick haphazardly to the fur of passing animals, relying heavily on numbers and chance events to reproduce. Hence, r-selected organisms produce large numbers of cheap, relatively simple offspring. The inductive gambles made by r-selected organisms are that if they can produce enough offspring, then even by chance, some of them will survive, reproduce, and pass on their genes to the next generation.

K-selected organisms do almost the opposite. They invest far more in the gaining of adaptive know-how, or R_i, and typically, they are relatively more complex, longer-lived organisms. They produce far fewer but relatively more complicated offspring. They also invest, sometimes heavily, in parental care. The inductive gamble made by K-selected organisms is that if they invest heavily in their offspring, inclusive of parental care, that should generate high-quality offspring sufficient to ensure that enough of the organisms will survive and reproduce to create the next generation.

If r-selected organisms interact with K-selected organisms, their greater reliance on chance may generate particularly inconsistent natural selection pressures for the K-selected organisms they are interacting with. For example, bacteria that parasitize vertebrates may generate too many chance mutations for the vertebrates to adapt to, which is probably why vertebrates evolved specialized immune systems to cope (see chapter 7).

A second example of potentially disruptive interactions between different populations of organisms in ecosystems is provided by "invasions." Invasions occur when a species from one community, or ecosystem, invades another community or ecosystem, usually in another part of the world. For instance, presently, Pacific lionfish are spreading along the Eastern coast of the US, while Asian zebra mussels have invaded several North American lakes. Most invasions don't matter much because they don't upset the communities of organisms into which the invaders arrive. For instance, they may not greatly affect the flow of energy and matter between the incumbent populations in the community that they have invaded. However, some invasive species can be very disruptive. If an invader niche constructs in a different way from incumbent species and outcompetes the incumbents by doing so, it may have damaging knock-on consequences for the stability of the community or ecosystem that it has invaded.[21] One example is provided by European earthworms, which have invaded North American hardwood forests and triggered a devastating cascade of consequences for the entire ecosystem.[22]

However, an invasion does not always depend on the arrival of a new species from somewhere else. It may depend only on the introduction of a new kind of niche-constructing activity by an incumbent species in the community. If an incumbent species or population evolves a new kind of niche construction, its novel activity may have similar disruptive consequences for its community as an invading species from elsewhere.[23] For example, novel cultural niche-constructing activities by humans (such as planting crops or building roads) may be equivalent to new invasions in communities and ecosystems each time they occur (see chapter 10).

EVOLUTION AND INFORMATION-SEEKING PROCESSES

Now let's compare knowledge-gaining processes in science to adaptive know-how-gaining processes in evolution again. What more, if anything, can we learn from this comparison? The parallels are illuminating, but counterintuitively, we may learn more about how evolution works from where the processes appear to be most different than from where the parallels are closest. The empirical nature of science is probably the biggest apparent difference between how we think about science versus how we think about evolution. Scientists generate models, hypotheses, and theories about the world; compile data suitable to test their effectiveness; and build on the most useful findings. In this manner, they overtly set out to explore cause-and-effect relations in nature. They are information-seeking. But is this difference real, or do the processes of evolution already include some not yet fully recognized forms of empiricism? I want to explore this possibility.

Empirical science greatly speeds up the detection and demarcation of cause-and-effect relationships in nature, as described previously. Might the processes of evolution sometimes do the same? I suggest the answer may be "yes," albeit on a less overt scale. Instead of relying exclusively on the differences between consistent versus inconsistent natural selection pressures for demarcating cause-and-effect relationships from background noise, organisms in evolving populations may sometimes be selected for anything that enhances their capacity to gain additional adaptive know-how (R_i). This might occur in rapidly changing environments, as natural selection is impotent to detect causal relations that occur on scales shorter than a generation. Supplementary R_i-gaining processes are any processes that speed up the ability of evolving populations to gain additional adaptive know-how about causal relationships. They magnify the capacity of populations to detect causal relations, and they provide shortcuts to new adaptations (see chapter 7).

I'll define such an information-seeking process as any process that, at least in the short term, gives priority to the gaining of more meaningful information, or R_i, over the immediate gaining of more energy and matter resources, or R_p, or other fitness goals. These processes could require a greater expenditure of R_p in the short run in order to gain the additional R_i. The subsequent advantages to organisms could be the more efficient gaining of R_p resources or of mates, or the better avoidance of threats. In general, it could be an enhanced capacity by organisms to resist the second law of thermodynamics for long enough to survive and reproduce.

These information-seeking processes imply the existence of a supplementary fitness goal—namely, the expenditure of energy for the sake of gaining R_i. Any process that enhanced the gaining of adaptive know-how would then be in the service of this supplementary fitness goal. But how and why might organisms ever be naturally selected for a process that increased the adaptive know-how of their descendants? This question becomes more acute when the gaining of adaptive know-how costs more energy and matter resources in the short run. No organism in any evolving population would ever be selected for any process that enhanced the supplementary gaining of R_i unless on average the benefits outweigh the costs.

Perhaps the first and most widespread information-seeking process was the evolution of an increase in the rate of mutations in specific regions of the genome. To the extent that natural selection favors an increase in the mutation rate because this causes parent organisms in one generation to increase the phenotypic variability of their offspring in the next generation, this qualifies as an information-seeking adaptation. The production of more variant offspring is analogous to increasing the variety of trials in a "trial-and-error information gaining process." By analogy to science, it is equivalent to generating more hypotheses that can be subject to testing, while in the context of Ashby's control theory, it is also equivalent to parent organisms increasing the amount of variance, or V_O, among their offspring, relative to the variant natural selection pressures, or V_E, the offspring will encounter in their futures (see chapter 3).

The simplest evolutionary mechanisms for introducing greater phenotypic variability are random genetic mutations. Genetic mutations are transmitted by parent organisms to their offspring in successive generations of populations via genetic inheritance. The additional random mutations could occur in all the genes in the genomes, transmitted between generations. Alternatively, these additional random mutations might occur in only a subset of the genes in the transmitted genomes, possibly as a consequence of different natural selection pressures acting on different populations.

In the short run, parent organisms in one generation may risk reducing their fitness, as a consequence of some of these mutations. The mutations might reduce the fecundity of parent organisms by increasing the number of nonviable offspring that they bequeath to the next generation. But if among the increasingly variable offspring of parent organisms, a few variants were fitter than their competitors, then those fitter variants could become the dominant phenotypic variant in later generations of a population. That might speed up the rate of evolution in a population. In other words, the costs of increased mutation, in terms of wasted R_p might be offset by the benefits, in terms of increased R_p that manifest as greater fitness among descendants.

Sexual reproduction, in conjunction with its associated genetic recombinations, is another source of increasing variability in evolving populations. Through chance recombinations, sexual reproduction is commonly the source of high-fitness phenotypic variants. In sexual populations, both male and female parent organisms must have already survived so far. Perhaps sex provides a means to increase phenotypic variation while biasing it toward previously tested variants. There are many hypotheses for the evolution of sex, many of which stress that it increases phenotypic variability and others that emphasize how it enhances the efficiency of natural selection by making it easier to weed out low-fitness alleles.[24] Bill Hamilton's suggestion that sex may have evolved to combat increased rates of parasitism is one such example. The genetic recombination that accompanies meiosis provides a mechanism for the rapid production of new combinations of resistance factors in the hosts in order to counter the diversity generated in the more rapidly evolving parasites. In terms of Ashby's control theory, sex allows sufficient variability in the organism (V_O) to destroy the variability of the environment (V_E), with parasites as the major axis of environmental variability.

A putative benefit of sex is that it enhances the efficiency of selection. This reduces the energy and matter (R_p) costs associated with an informational (R_i) gain. An alternative mechanism to reducing R_p costs is to produce phenotypic variation that is biased toward high-fitness solutions.[25] Developmental bias occurs when some combinations of traits arise more readily than others.[26] The field of evolutionary developmental biology has identified a large number of examples.[27] The functional integration of organisms arises because developmental systems make the traits of organisms develop in a correlated fashion, such that random mutation has phenotypic effects that are biased toward high-fitness dimensions.[28] Investigations of the size, position, and color of eyespots in butterflies provide an example. A combination of artificial selection and evolutionary developmental biology experimentation has revealed butterfly characters that show clear evidence for developmental bias (e.g., eyespot size and color)[29] with a corresponding impact on evolvability.[30] Developmental bias seemingly allows the generation of more "plausible" phenotypic traits in offspring organisms (i.e., traits that are more likely to be adaptive a priori, relative to natural selection in their environment).[31]

Some organisms have evolved supplementary knowledge-gaining processes, such as adaptive immunity or a capacity to learn.[32] These processes operate in a manner remarkably similar to natural selection. For instance, the vertebrate adaptive immune system operates by generating diverse antibodies, initially at random, and retaining and reproducing the ones that prove most effective, discarding the other variants. This process is costly, as considerable energy and matter (R_p resources) must be invested in variant immune cells, most of which confer little protection. Nonetheless, the system is adaptive because it allows individual organisms to acquire additional R_p in the form of knowledge about the antibodies that are most effective in countering the antigens in their immediate current environment. Again, this makes sense in terms of Ashby's control theory, where adaptive immunity allows each organism to rapidly increase the amount of variance, or V_O, among its antibodies, to counter the variant threats to its health, or V_E, encountered in its environment. This is discussed in more detail in chapters 7 and 9, along with other supplementary processes, such as epigenetic processes, that also add more phenotypic variance among organisms in evolving populations.

If living organisms are sensitive to increases or changes in environmental variability (V_E) and able to adjust by producing compensatory phenotypic variation (V_O), then given that the world is heading toward human-induced climate change, we might anticipate the production of additional genotypic and phenotypic variation in evolving populations. For some populations, this could be the final straw. The additional variants might only have the effect of increasing the number of nonviable organisms in each generation. Most mutations are deleterious. Extra R_p will be expended, but no extra R_i will be derived. However, in other populations, the additional variance might increase their capacity to endure the changing climate by generating additional valuable R_i. Such populations might avoid extinction. However unwelcome the human-induced climate change is, it could offer a test of this reasoning.

HEURISTIC NICHE CONSTRUCTION

Most of the possible ways that we have discussed so far, in which the evolutionary process could become more empirical, are compatible with both SET and NCT. The next alternative is emphasized only by NCT. I'll call it "heuristic niche construction" and define it as the niche-constructing activities of organisms that have the immediate fitness goal of gaining more adaptive know-how (R_i) about causal relationships in their environments, rather than the immediate gaining of more energy and matter resources (R_p). Heuristic niche construction might take the form of exploratory, relocation by organisms in their environments as a means to acquire knowledge about the environment (e.g., foraging ants exploring beyond the immediate vicinity of their nest to identify a new food source). Niche-constructing organisms could also explore the possibility of earning a better living somewhere else in their environments, on a trial-and-error basis. For example, through innovation and social learning, orangutans are known to have invented new food-processing techniques, which allow them to access hitherto-unexploited foods, such as palm hearts.[33] Alternatively, organisms may introduce novel physical perturbations in their environments. Animal artifacts provide many examples. For instance, scent marks confer information about the identity of individuals in a territory, tracks and trails confer information about efficient routes through the environment, and the threads of a spiderweb inform the spider where in the web a fly has gotten stuck, and how big it is.[34] But once again, how could any organism increase its fitness by investing in heuristic niche construction?

Heuristic niche construction must pay for itself by increasing the fitness of organisms in populations. If the immediate gaining of more adaptive know-how, or R_p by heuristic niche construction leads to organisms subsequently increasing their capacity to harvest energy and matter, or R_p resources, from their environments, that could be one way in which heuristic niche construction could pay for itself. Exploration might cost energy in the short run, but ultimately, it increases the fitness of individual organisms in the longer run. It could do so as a consequence of ancestral organisms in populations bequeathing more appropriate ecological inheritances to their descendants relative to their descendants' genetic inheritances. For instance, once superior food sources have been identified, ovipositing insects can lay their eggs on them.[35]

While niche-constructing organisms need to know as much as possible about the cause-and-effect relationships in their environments that are independent of their own actions, they also need to know about the consequences of their own actions. That is because, in their particular environments, they are commonly the instigators of a cause and its effects. Just as they can gain more know-how by being sensitive to the consistency of the natural selection pressures that arise from cause-and-effect relationships that are independent of their own actions, so organisms can gain R_i about the consequences of their own actions by being sensitive to the effects that consistently follow from them.

We have seen that when organisms in evolving populations repeatedly encounter similar natural selection pressures in their environments, this should select for similar phenotypic traits in successive generations of evolving populations. However, that environmental regularity need not arise by chance. It may be self-imposed. Consistent resources and conditions arise in the environments of organisms because organisms consistently and reliably construct them. Nests are regularly found in the environments of birds, webs in the environments of spiders, and dams in the environments of beavers because the animals concerned built those structures. Moreover, each niche-constructing species builds nests, digs burrows, or spins webs or cocoons in a species-typical manner.

This has a number of important consequences.[36] First, it means that the natural selection arising from organism-built aspects of the environment can be unusually consistent. Niche construction sometimes generates atypically strong directional natural selection and other times generates atypically strong stabilizing natural selection or weak directional responses to selection, with these patterns predictable a priori and confirmed in meta-analyses of selection in the wild.[37] Second, this means that closely related species that engage in similar forms of niche construction should generate similar selection pressures and commonly exhibit a similar evolved response to selection. This provides an explanation for the parallel evolution frequently observed, and increasingly reported, in independent populations. Third, niche construction is expected to generate predictable evolutionary trends. For instance, building a nest will generate reliable selection for the nest to be improved, maintained, defended, and repaired; for other species to live in the nest, dump eggs in the nest, or steal nest material; and for chicks and eggs reared in the nest to develop in the predictably encountered conditions of the nest.[38]

The physiologist Scott Turner suggests that many organisms impose regularity on their immediate environments by pushing out the boundaries of their internal environments into their external environments, a phenomenon that he calls "extended physiology."[39] Organisms such as wasps, bees, and ants exert an extended physiological homeostatic control of their nest environments. In many respects, Turner's "extended physiology" is similar to Richard Dawkins's extended phenotypes. However, it differs because Turner is explicit in stressing how extended physiology depends on the active purposive agency of phenotypes, which is consistent with NCT but not SET.

We have already seen some examples of Turner's extended physiology in chapter 3. They include termite mounds, the modification of their soil environments by earthworms, beaver dams, and, for that matter, human houses. All these artifacts increase homeostasis in the immediate environments of different organisms, thereby making them more predictable for those organisms, including as sources of natural selection. For instance, humans often control the internal temperature of their own homes exogenously, almost as much as they control the internal temperatures of their own bodies by endothermic mechanisms. This damps down natural selection for morphological adaptations to the heat or cold.

These observations raise some caveats. First, Turner's extended physiology may be expensive. Extending the domain of homeostasis beyond the organism's body into its external environment inevitably costs organisms additional energy and matter, or R_p resources. Organisms may be limited in how far they can extend a domain of predictable homeostasis by an inability to gain sufficient energy and matter, given their current know-how, or R_i. In simple terms, they may not always be able to pay for it.

Second, by pushing out the boundary of their domain of homeostasis further into their environments, organisms may be taking too big a risk. They may not possess enough adaptive know-how, or R_i, to control the effects generated. Niche-constructing organisms need to be sufficiently informed, not only about the immediate consequences of their extended physiology, but also about its knock-on consequences for their ecosystems, to the extent that these consequences feed back in order to affect the constructor's fitness. If organisms are not sufficiently informed about the consequences of their activities, then they may eventually be overwhelmed by their own attempts to make their unforecastable futures more forecastable. They may do so by inadvertently modifying other natural selection pressures in ways that threaten their own futures (see chapter 10).

Turner's extended physiology provides another means by which organisms can potentially engineer their own futures to some extent. If organisms can demarcate cause-and-effect relationships in their environments, which are independent of their own actions, and they can also demarcate those effects that are contingent on their actions, then they may be able to "anticipate" the likely consequences of their own niche-constructing actions in the context of their niche relationships with their environments. Their anticipations have to be functional, but they need not be cognitive. In most species, they will not be. Either way, niche-constructing organisms may be able to codetermine their own futures by modifying specific natural selection pressures in their environments. They could thereby ensure that their own futures, and possibly the futures of their descendants, were more forecastable. They might do so either for themselves or by bequeathing modified ecological inheritances to their descendants.

Another supplementary fitness goal for niche-constructing organisms could be the achievement of greater thermodynamic efficiency. Greater thermodynamic efficiency implies better or less costly adaptations needed by organisms to resist the second law of thermodynamics. However, more efficient adaptations would need to be better informed by adaptive know-how, or R_i. That might require heuristic niche construction to gain R_i about the interactive niche relationships of organisms with their environment.

Addy Pross (2012) proposes a similar version of this hypothesis in the book that he entitled *What Is Life? How Chemistry Becomes Biology*, in homage to Schrödinger. He starts by talking about an inanimate example of dynamic stability in a flowing river. It's possible to step into a continuously flowing river repeatedly, while at the same time never stepping into exactly the same river twice. This is because the molecules of water flowing past your ankles are always changing. However, inanimate flowing rivers do not have to actively resist the second law of thermodynamics to achieve dynamic stability. The constant renewal of the flowing river depends on repeated rainfall and is ultimately energized by the hydrological cycle, independent of the river itself. The attractor of the flowing river is therefore greater thermodynamic stability, in the same way that it is for all nonliving abiota. Pross argues that in order to resist the second law, all living systems have to travel in the opposite direction, toward a different "attractor." So long as they are alive, organisms must keep heading toward greater "dynamic kinetic stability (DKS)," instead of thermodynamic stability. Greater DKS demands greater thermodynamic efficiency.

Pross describes ordinary chemical reactions that continuously travel toward greater thermodynamic stability or disorder until the point where the reactions stop. He contrasts them with chemical reactions in living systems, which are characterized by self-replication and therefore by what Pross calls "replicative chemical reactions."[40] The problem is that all self-replicating systems have the potential to expand exponentially if supplied with sufficient energy and matter resources, or R_p. This makes living systems inherently unstable.

Stability returns in such populations when the birth of new organisms in a population is balanced by an approximately equal number of deaths. The resulting balance between the birth of new organisms in a population, as a function of reproduction and ultimately of self-replicating chemistry, and the deaths of older organisms is an example of a population in DKS.

This implies an endless turnover of transitory individual mortal organisms in populations, generation after generation, and the dispersal of these organisms across endlessly diverse local environments in space and time.

Pross argues that DKS is a "physical analogue of niche construction" (personal communication). The DKS of all organisms depends on their interactive niche relationships with their environments. The niches of organisms are codetermined by their own niche-constructing activities, and therefore by their own active purposeful agency in resisting the second law. But to resist the second law effectively, the niche-constructing activities of individual organisms depend on them being well informed by meaningful adaptive know-how, or R_i. Organisms gain their R_i primarily via the genes that they inherit from their ancestors. Individual organisms may then gain additional R_i as a consequence of their own niche-constructing activities, including their heuristic niche-constructing activities. Greater DKS must be an attractor for all living systems, including all individual organisms in all evolving populations.

CONCLUSIONS

How far have we got toward understanding how organisms solve the Waddington problem of adapting to their unknown but not wholly unforecastable futures? To what extent can niche-constructing organisms solve the Waddington problem by engineering their own futures?

Organisms cannot engineer their own futures, simply by generating more variety on a trial-and-error basis, but generating variety is often a pathway to acquiring R_i. If organisms are able to gain more R_i, such as by developmental exploratory and selective processes or through heuristic niche construction, and if they can apply their increased know-how to control their own niche-constructing activities and regulate their experienced environment, then organisms may be able to go some way toward determining their own futures. They may be able to make their own futures more predictable and forecastable.

5 LIFE ON ANY PLANET

We have now reached the last chapter of part I of this book. Here, on the basis of the preceding chapters, as well as on examples drawn exclusively from life on Earth, I will claim that everything we have discussed so far should also be relevant to life anywhere in the universe, wherever it exists. That might seem like an overstatement. Shouldn't our understanding of the putative lawfulness of evolutionary biology be restricted to life on Earth?

I think not, for the simple reason that I have described the processes of evolution in a manner that is consistent with the fundamental underlying laws of physics and chemistry, including the laws of thermodynamics, which I maintain will not vary across the universe. To the extent that this assumption holds, it is reasonable to claim that the putative laws of evolutionary biology, which are consistent with the universal laws of physics and chemistry, will be universal too. My claim would fail if the laws of physics and chemistry had not always been the same throughout space and time. For example, it is just possible that these laws were different during an extraordinary period of inflation of the universe shortly after its origin in the Big Bang, 13.8 billion years ago.[1]

Assumptions about the universality of the laws of physics and chemistry can be tested to some extent, however. For instance, astronomers can now look back in space and time to what the universe was like billions of years ago by looking back to the edge of the visible universe. In this manner, astrophysicists and cosmologists have confirmed that the laws of physics and chemistry do not appear to have changed for billions of years. For example, the laws of physics and chemistry that specify how chemical reactions proceed and are closely associated with thermodynamics have apparently been the same for all that time. The same is true of the laws of gravity. It is also possible to check the theoretical models that cosmologists and physicists have developed about the dynamics of the universe. For instance, they can be checked by smashing atoms in the Large Hadron Collider. Once again, the conclusion is that the laws of physics and chemistry are universal.

Recently, it has become possible to start looking for life on other planets in our galaxy. We can do so by assuming that whatever we know about how biological evolution works on Earth should also apply anywhere else in the universe. Niche construction theory (NCT) may have something to offer in this respect. If niche-constructing organisms modify their environments, and if all organisms niche-construct, then the existence of life elsewhere in the universe (e.g., in our own solar system or galaxy) may be detectable from the signatures left behind by the earlier (perhaps very much earlier) niche-constructing activities of organisms. For example, any planet with an atmosphere whose constituent gases are seriously out of thermodynamic equilibrium, as is the case for Earth, would provide an indication of the prior existence of life. The planet's atmosphere would be most unlikely to be in that observed state unless it had been pumped into it by the niche-constructing activities of living organisms. That is what happened, and what is continuing to happen on Earth.

The laws of physics and chemistry alone, however, are not sufficient to understand life. After asking his "What is life?" question, Schrödinger[2] immediately followed it with a supplementary question, which in chapter 1 I paraphrased as: "What else is needed beyond the known laws of physics and chemistry to understand life?" Bear in mind that Schrödinger asked this question more than seventy-five years ago—and since that time physicists, chemists, and cosmologists have learned a lot more about the laws of physics and chemistry[3]—and yet, remarkably, Schrödinger's supplementary question is still relevant. We have already been considering this question in the previous chapters with respect to life on Earth, but I now want to consider it again relative to the possibility of life elsewhere in the universe. Are the same properties of life, beyond the known laws of physics and chemistry, that are required on Earth, also likely to be required by life anywhere else in the universe?

THE PROPERTIES OF LIFE ON EARTH

Let's review the properties of life on Earth that go beyond the known laws of physics and chemistry. There are some initial points that by now may be familiar. First, organisms are very improbable systems. They are very far from thermodynamic equilibrium with their environments. To remain alive, organisms have to resist the second law of thermodynamics while still obeying it, which is possible only for purposeful systems.[4] At some time in the past, purposeless inanimate matter must have become the origin of the first protocells, and subsequently the first purposeful living cells on Earth.[5]

It was because these first cells were not only very improbable, but also purposeful, that they were able to stay alive by resisting the second law while still complying with it. If this is correct, then the emergence of purpose must have been an origin-of-life phenomenon, both on Earth and anywhere else in the universe where life exists. Ever since life first appeared on Earth, the capacity of organisms to stay alive was critically contingent on the active, purposeful agency of organisms. Were organisms not purposeful systems, they would rapidly dissipate, reverting to the status of nonliving, abiotic systems, far closer to being in thermodynamic equilibrium with their environments. The primary purpose of all organisms is simply to stay alive.

Organisms can stay alive only by organizing a flow of energy and matter between themselves and their environments that opposes the flow of energy and matter favored by the second law. That requires organisms to be able to detect events in their environments. It also requires them to innovate by seizing any opportunities to reverse the flow of energy and matter that happened to come their way.[6] But recognizing and seizing opportunities are things that only purposeful systems can do. Opportunities are meaningless relative to purposeless systems. That is another reason why organisms must be purposeful systems to live. It is also the fundamental difference between living organisms and nonliving systems.

Assuming that organisms are purposeful, how do they reverse the flow of energy and matter (or R_p resources) between themselves and their environments through active agency? This returns us to a familiar chain of cause-and-effect relationships. Organisms cannot counteract the flow of energy and matter favored by the second law without interacting with their external environments, and they cannot interact with their external environments unless they are adapted to those environments. Minimally, all organisms must possess semipermeable boundaries between themselves and their environments.

Two-way-street interactions may then allow organisms to import energy and matter resources from their environments, and export detritus back to their environments, through these boundaries, in opposition to the second law. For instance, single-cell organisms interact with their external environments via pores in their membranes that allow atomic and molecular traffic to pass in and out. These two-way-street interactions may then enable organisms to stay alive for a time, at the expense of their external environments.

Most organisms need far more adaptations than just semipermeable boundaries, however. They need appropriate phenotypic traits that are sufficiently well adapted, structurally and functionally, relative to all the natural selection pressures that they are liable to encounter in their external environments during their lives. These selection pressures include both those unaffected by organisms and those previously modified by the niche-constructing activities of organisms. They include natural selection pressures that may threaten the future existence of individual organisms, such as predators or parasites, floods or droughts. However, it is impossible for organisms to adapt to any natural selection pressures in their external environments unless they are informed by meaningful information in the form of adaptive know-how, or R_i. That information primarily stems from their ancestors, as well as from their own prior experiences.

There is also another consequence of organisms living at the expense of their environments. Organisms cannot organize a second law–opposing flow of energy and matter between themselves and their environments without causing changes in their external environments. Those changes are the origin of niche construction, which in turn is the origin of both modified natural selection in the environments of organisms and the ecological inheritances that ancestral organisms bequeath to their descendants. When the evolutionary process comprises both natural selection and niche construction, and both genetic and ecological inheritances, evolution is transformed into a process of organism-environment coevolution.[7]

What are the minimum adaptive properties that organisms must possess to do this active, purposeful, energy-consuming, second law–resisting work on their environments? The minimal properties are those that the physicist James Maxwell gave to his "demon" in the nineteenth century (see chapter 1), but with one significant difference. Unlike organisms, Maxwell's demon was not informed by the processes of evolution, but by Maxwell himself, when he designed it. Maxwell gave his demon its second law–resisting purpose. He also gave it the adaptive know-how (R_i), expressed as the conceptual structural and functional adaptations, that it needed to fulfill its purpose.

Recall that Maxwell was exploring whether it was possible for any discrete system, be it living or nonliving, to oppose the second law of thermodynamics without violating it. He demonstrated that his demon could not oppose the second law so long as it was confined within a closed environment, as the demon required more energy and matter to do its work than it could extract from that environment. However, Maxwell also showed (indirectly) that any discrete system, nonliving or living, anywhere in the universe could potentially oppose the second law of thermodynamics without violating it, provided that the system existed in an open environment that could supply it with the additional energy and matter resources that the demon needed to do its work, and provided that it was equipped with the minimal properties that Maxwell gave his demon. Maxwell thereby revealed the minimal properties that all living organisms need to survive, including some that go beyond the known laws of physics and chemistry.[8]

Maxwell's demon goes some way toward explaining how organisms can resist the second law of thermodynamics, while still complying with it, via their active, purposeful, energy-consuming work. But his demon did not go far enough, primarily because it had to be informed by an intelligent designer: Maxwell himself. In contrast, organisms are informed by the natural processes of evolution.

We have already considered how organisms are informed by evolution (see chapters 3 and 4). Individual organisms in evolving populations are informed primarily by the deductive know-how, or R_i, they inherit from their ancestors via genetic inheritance. The genes that they inherit register memories about what phenotypic traits were adaptive for their ancestors in the past, relative to the natural selection pressures that their ancestors encountered. Individual organisms may also be informed by other supplementary information-gaining processes that cause them to remember their own individual past interactive experiences with their individual environments, and perhaps too by other evolutionary processes such as cultural evolution or the population-level selection and transgenerational inheritance of epigenetic variants. Either way, these memories of the past, whether of their ancestors' past or of their own past developmental histories, can have only a limited capacity to inform organisms about what may be adaptive for them again in their futures.

These limitations stem from three sources. The first problem that must be overcome is that the systems that register memories of the past in organisms (e.g., genomes in all organisms or brains in animals) are imperfect. They may be inaccurate (e.g., because of mutation), or the organism may forget the past. The second problem is that the deductive know-how that is inherited, such as in genes, are nothing more than particular samples of the past and are not comprehensive records of everything that happened. It is therefore possible that some individual organisms will encounter natural selection stemming from past events that were not sampled, and do not provide adaptive information. Third, changes may occur in the environments of organisms that introduce novel selection pressures that were not encountered by ancestral populations. A famous example is the sudden, novel changes that were caused by the asteroid strike that occurred in the Yucatan Peninsula in Mexico about 66 million years ago. It was an abiotic event that triggered so many changes in the environments that it killed off the dinosaurs, and many other species too.

More frequent and less dramatic sources of novel change are the niche-constructing activities of individual organisms in diverse evolving populations. A subcategory of niche construction is *inceptive* (i.e., opportunistic) niche construction,[9] which by definition causes novel changes in environments. While the changes that single microorganisms cause in their microenvironments can be disregarded as insignificant, if populations of microorganisms all niche-construct in the same way because they carry the same R_p, then their collective niche-constructing activities may add up and cause highly significant changes in their environments. For example, thanks to global warming, the permafrost in northern Siberia is melting, causing microorganisms known as methanogens to emit additional methane into the atmosphere. That is significant because methane is a more potent greenhouse gas than carbon dioxide, and the extra methane is accelerating global warming.

Probably the most fundamental source of novel change is the evolutionary process itself, because it is always generating novel variant organisms in populations and novel species in ecosystems. All these changes have the potential to act as the source of novel natural selection pressures for individual organisms in evolving populations. Stuart Kauffman[10] has pointed out that the evolutionary process is "nonergodic," by which he means that it has never realized its full potential and always has the ability to come up with unprecedented novelties, including new species.

In combination, these three limitations ensure that organisms can never adapt to all the natural selection pressures that they may encounter in their unknown but not wholly unforecastable futures. The evolutionary process can never inform organisms with sufficient adaptive know-how, or R_p, a priori to immortalize them as they travel into their futures. All organisms, and all species, are mortal.

REPRODUCTION

The mortality of living beings highlights the second strategic fitness goal of organisms: reproduction. When organisms reproduce, they solve two problems. First, even though all organisms are mortal, they can survive by proxy via the genes (and other forms of inherited knowledge) that they transmit to their offspring and subsequent descendants when they reproduce. They can do this indefinitely for generations. Reproduction is therefore the second way in which mortal organisms can resist the second law of thermodynamics while still complying with it.

However, individual organisms could not resist the second law indefinitely, even by proxy, if they only made exact copies of themselves when they reproduced. If they did just that, their exact copy offspring would probably run into exactly the same adaptive problems as they confront themselves. Fortunately, because of faults in their gene-based memory systems, reproducing organisms can seldom or never reproduce exact clones of themselves. They are almost bound to make copying mistakes and generate variant organisms when they reproduce. The primary sources of these copying mistakes are random genetic mutations. As we saw in chapter 4, natural selection may also favor increased rates of mutations in subsets of the genes in their genomes in some environmental circumstances. Sexual reproduction may further increase the variation among the offspring of male and female parent organisms by the recombination of their genes during meiosis.

All these sources of increased variance among the offspring and subsequent descendants of reproducing organisms then solve a second problem. Organisms in the current generation are bound to supply the next and subsequent generations of their population with new variant organisms when they reproduce. These inputs of variation in each successive generation, when subject to natural selection, may then allow evolving populations to update themselves by tracking any novel changes that they encounter in their environments through changing their phenotypic traits over time. Novel changes in individual variant organisms in evolving populations may introduce novel adaptive solutions to novel problems, generation after generation. Ultimately, such changes may also lead to the introduction of new species in ecosystems.

How do organisms reproduce? Just as it is possible to reduce the properties that organisms need to survive to the minimal properties that Maxwell gave his demon, it is also possible to reduce the properties that organisms need to reproduce to the minimal properties that John von Neumann gave his natural automaton (see chapter 1). Like Maxwell and his demon, von Neumann informed his natural automaton, in advance, by giving it all the adaptive know-how, or R_i, that it needed to reproduce.

Recall that the minimal properties that von Neumann gave his natural automaton were as follows: First, it had to be equipped with a "universal computer," otherwise known as a "Turing machine." Second, it also had to be equipped with a "universal constructor" that was able to construct anything, provided that it had sufficient energy and matter, as well as appropriate instructions. Hence, third, von Neumann had to give his natural automaton suitable instructions, including the information, or R_p, that his natural automaton needed to reproduce. Fourth, the automaton needed a universal copier. Fifth, the copier must copy the natural automaton's own instructions, inserting these into any "offspring." Since von Neumann's day, there have been considerable practical and theoretical advances in automata and robotics,[11] but von Neumann's key points are the most relevant here.

Maxwell and von Neumann were both physicists who presumably assumed that the laws of physics were universal. Maxwell's demon and von Neumann's natural automata both go beyond the known laws of physics to Schrödinger's supplementary question about what else is needed beyond the known laws of physics and chemistry to understand life.

Maxwell's demon describes the minimal properties that organisms need to survive by resisting the second law of thermodynamics without violating it. Also, von Neumann's natural automata describes the minimal properties that organisms need to reproduce. Given that in combination, these are the minimal properties that all organisms need to survive and reproduce on Earth, and given that these properties have been derived from the universal laws of physics by Maxwell and von Neumann, the same minimal properties are likely to be the minimal properties of life, wherever it may exist in the universe. It may therefore be possible to build a universal theory of evolution.[12]

A universal theory of evolution does not mean that life anywhere else in the universe need be the same as life on Earth. Life elsewhere in the universe might have to comply with the same lawfulness of the evolutionary processes as life on Earth, but that still leaves a lot of scope for life to be very different from life on Earth. Imagine a different planet, hundreds of times more benign to life than Earth. Organisms on excessively benign planets could be flooded with energy and matter resources by their environments without them having to do much or to know much to stay alive. In such worlds, individual organisms in evolving populations might approach immortality by living for extremely long lifetimes. Or imagine the opposite—imagine an extremely impoverished world in which life forms may nevertheless exist.

Such environments do exist on Earth, and we do find forms of life there that are somehow adapted to their very harsh conditions. These organisms are known as "extremophiles." One example is the bacteria that live in hot volcanic springs in Yellowstone National Park in the western US. They can tolerate temperatures in excess of 90°C.[13] Other examples of extremophiles are the bacteria that live in sediments, and even in rocks far below the ocean floors on Earth. Some extremophile microorganisms stay alive by practically shutting down their metabolism. They seem to be able to reduce their metabolic activities until they are roughly equivalent to a single animal heartbeat in a year.

Another imaginary organism that might exist somewhere else in the universe and yet still be compatible with a universal theory of evolution could be an organism that carried so much redundancy that it could adapt to almost any natural selection pressure that it encountered.

Science fiction writers invent other exotic life forms, such as organisms on wheels;[14] however, to exist in the actual, rather than the imaginary, universe, organisms would have to comply with the same universal theory of evolution.

To what extent might any of our current theories provide a foundation for a universal theory of evolution? And to what extent do current theories of evolution fall short? I'll only consider the two approaches to evolution that we have been discussing here, SET and NCT. Could either of these contemporary approaches to evolution become a parent theory for a future universal theory of evolution? Before trying to answer that question, let's consider how far we have already traveled, thanks to Maxwell and von Neumann pushing the known laws of physics of their day to their limits.

Maxwell and von Neumann's conceptual artifacts demonstrated that the minimal properties that organisms need to survive and reproduce are potentially universal. That was a good start, but it's not enough. Both Maxwell and von Neumann were the intelligent designers of their conceptual artifacts. That allowed them to take short cuts when it came to supplying their artifacts with the same two fundamental resources that organisms need to live: energy and matter, or R_p, and meaningful information, or R_i.

Maxwell and von Neumann both gave their artifacts sufficient R_i, up front, to allow them to survive and reproduce. They also accounted for the energy and matter resources that their artifacts needed to do their work. Maxwell explored his demon's capacity to acquire sufficient energy and matter from within its closed environment to do its work, but it couldn't. And von Neumann assumed that his natural automaton existed in such a benign environment that it had access to all the R_p it needed, both for itself and to build its copy offspring. However, von Neumann's natural automaton really dealt only with the bioinformatics of reproduction. In reality, to be viable, offspring organisms also need to inherit environments relative to which their inherited instructions have at least some chance of being well adapted. Strictly speaking, organisms need to inherit viable niche relationships, comprising both meaningful information about their environments (i.e., R_i) and environments capable of supplying organisms with the necessary energy and matter that they need to survive and reproduce (i.e., R_p). It is not enough to inherit copies of instructions alone.

Hence, even though Maxwell's demon and von Neumann's natural automata are already modeling some of the universal properties of life, they do not provide a foundation for a universal theory of evolution. Their artifacts fall short because they were the intelligent designers for these artifacts, and also, in the case of von Neumann, because he failed to consider bioenergetics adequately. That is enough to rule out their artifacts from being a satisfactory foundation for a future universal theory of evolution. Nonetheless, it is helpful to understand in what ways they fall short, as this sheds light on the requisite properties of a universal theory of evolution.

In contrast, both SET and NCT are genuine theories of evolution. They do not appeal to intelligent designers, whether supernatural or otherwise, to account for the processes of evolution. Do they suffer from other limitations? I'll consider SET first.

STANDARD EVOLUTIONARY THEORY

What does SET get right, and where does it fall short? It is hard to imagine any universal theory of evolution that is not based on the natural selection of heritable variation in evolving populations, as currently described by SET. To this extent, SET could be a forerunner of a future universal theory of evolution. Darwin's description of evolution as a process of descent with modification is likely to apply to the evolution of life wherever it may exist in the universe. But we have also seen some of the ways in which SET is falling short. Even though Schrödinger wrote his book *What Is Life?* in 1944, orthodox evolutionary theory, initially in the form of the modern synthesis and later in the form of contemporary evolutionary theory, which I call SET, did not, and still does not, fully recognize the relevance of the laws of thermodynamics to biological evolution.

SET also falls short by failing to acknowledge that organisms are purposeful systems. The purposes of organisms do not demand cognition or consciousness, but they always demand the presence of meaningful information, or adaptive know-how, in organisms. SET recognizes that the adaptations of organisms depend on information carried in their genes. But it does not recognize the connection between the adaptive know-how carried by organisms and their active purposeful agency in resisting the second law, which organisms must express to survive and reproduce.

These two shortcomings of SET then have several consequences. SET fails to recognize that organisms must be purposeful systems merely to stay alive. Minimally, to stay alive, active, purposeful organisms have to organize a flow of energy and matter between themselves and their environments that opposes the flow of energy and matter favored by the second law. When organisms do this, they are bound to cause changes in their environments, if only because they have to take physical resources from their environments and dump their detritus back into their environments, to oppose the second law. Although individual organisms may have only a very slight impact on their local environments, collectively populations of organisms are likely to have a measurable and sometimes considerable impact on their own and each other's environments.

As a result, SET fails to emphasize how the changes that second law–opposing populations of organisms cause in their environments will almost inevitably modify some of the natural selection pressures in their environments. That is, SET fails to recognize that niche construction is an obligate feature of evolving organisms. The natural selection pressures that change through niche construction may subsequently feed back to act on the populations of organisms themselves or on other evolving populations in their ecosystems. The purposeful, niche-constructing activities of organisms then introduce the second category of natural selection that we met previously, comprising purposeful natural selection pressures, in contrast to the first category of purposeless natural selection stemming from all other environmental events (see chapter 4). Because organisms are the source of this second category of natural selection, the niche-constructing activities of organisms codirect their own and each other's evolution.

However, because SET does not recognize the relevance of the second law of thermodynamics to evolution, and because it does not recognize that organisms must be purposeful systems, SET does not yet discriminate between these two categories of natural selection. That prevents SET from recognizing niche construction as a codirecting process in biological evolution, and from recognizing that evolution must be a process of organism-environment coevolution rather than just a process of organisms evolving in response to quasi-autonomous[15] natural selection pressures in their environments.[16]

SET is also falling short in the way in which it is currently handling the relationship between the two fundamental resources that organisms need to live: energy and matter resources, or R_p, and adaptive know-how resources, or R_i. SET acknowledges that living organisms need both these kinds of resources to survive and reproduce, but it does not deal with the relationship between them even-handedly. At least since the discovery of DNA in 1953, SET has concentrated primarily on the bioinformatics of evolution, and hence on R_i. It has usually left the bioenergetics of evolution to the ecologists.[17] For instance, SET describes the major transitions of evolution almost exclusively in terms of bioinformatics rather than in terms of bioinformatics and bioenergetics in combination.[18]

For instance, Maynard Smith and Szathmáry[19] do a good job of describing in bioinformational terms each major transition in evolution, but they pay far less attention to the associated bioenergetics that must also have been required. In contrast, Nick Lane and William Martin[20] deal with the missing bioenergetics of one of the greatest transitions of all—namely, the transition from prokaryotes to eukaryotic cells. Lane and Martin point out that this transition could not have happened without the extra energy supplied to novel eukaryotic cells by their endosymbiotic mitochondria.[21] Recently, SET has been paying more attention to the relationship between evolution and ecology, but until niche construction is recognized by evolutionary biology, reconciling evolution with ecosystem level ecology, rather than just population and community ecology, will remain difficult.[22]

There are also some less obvious ways in which SET falls short of providing a solid basis for a future universal theory of evolution. One of them takes us back to the purposes of organisms again. Neither evolution, nor development, nor inheritance would be possible without meaningful information, or R_i. However, meaning is derived from purpose. There can be no meaning without purpose. There can be regularities, or natural laws, without purpose, such as what arises through the laws of physics and chemistry. But that lawfulness is meaningless to purposeless systems. Purposeless abiotic systems comply with the laws of nature, but that is all they can do. They cannot generate meaning. In contrast, purposeful living organisms, which are informed by meaningful information or R_i, about the cause-and-effect relationships that arise from the laws of nature, are sometimes able to harness the laws of nature for their own purposes, provided that they still obey them. In the process, they can generate new meaning. The purposes of organisms may therefore be the origin of much of "what else is needed beyond the known laws of physics and chemistry to understand life."[23]

The problem here is that SET ignores the purposes of organisms. But if organisms are not purposeful systems, how can ancestral organisms transmit meaningful information in the form of adaptive know-how, or R_i, to their descendants, and how can this information accrue and change over time? SET's answer to that question goes something like this: Purposeless natural selection pressures in the external environments of organisms select those variant organisms in each generation of evolving populations that are the fittest organisms with respect to their capacity to survive and reproduce. According to SET, the apparent purposefulness of living organisms is either a mirage or an irrelevancy. From SET's perspective, living organisms have no purpose. They evolve solely because they vary in fitness, and they owe their fitness exclusively to the purposeless natural selection pressures that they encounter in their environments, in combination with the particular genes and chance mutations that they inherit from their ancestors and their parents. For SET, the ultimate source of the meaningful information, or R_i, that organisms need to survive and reproduce is reducible to prior purposeless natural selection pressures that act on their purposeless ancestors, combined with genetic inheritances that must have originally stemmed from chance-based genetic mutation. The problem with this reasoning is that unless their ancestors were purposeful organisms in the past, populations of organisms could not acquire or inherit meaningful information, or R_i, in the present.

SET also suffers from a second, less obvious limitation that we have not yet discussed. That arises from what I have previously called a "reference device problem."[24] I suggest that, currently, SET uses an inappropriate reference device with which to understand the processes of evolution. What does that mean? Well, it is not possible to understand anything relative to nothing. Everything always has to be understood relative to something, and that something is its reference device. This applies not only to biological evolution, but to all sciences, including physics and chemistry.[25] The questions are: What reference device can we best look at to understand the biological processes of evolution? What cause, or causes, are sufficient to explain biological evolution, and what is their source?

Given that SET rules out the purposeful agency of individual organisms as a contributing cause of evolution, and leaving aside random processes such as genetic drift, the theory's only option is to treat natural selection as the sole directing cause of adaptive evolution. That turns the external environments of organisms into the fundamental reference device relative to which SET understands evolution.[26] We can, for instance, understand that giraffes have long necks because trees are tall and their forage source is high, or that cacti have spines rather than leaves because deserts are arid and leaves lose lots of water through transpiration, but spines do not. We would be puzzled, in contrast, if giraffes fed on low bushes or cacti were found in swamps. The adaptations of organisms make sense only relative to the external environment that is the source of the natural selection that favored them.

This approach has been labeled the "externalism" of SET since the properties of organisms are understood relative to the properties of their external environment.[27] The use that SET makes of the external environment as its reference device for describing evolution makes intuitive sense, but it is responsible for reinforcing some of SET's other shortcomings. For instance, SET's externalism is incompatible with the idea that the purposeful, niche-constructing activities of organisms can be a codirecting process in evolution, in conjunction with natural selection. SET struggles to recognize how niche construction can be regarded as a cause of evolution since niche-constructing traits must themselves be understood relative to the external environment that favored their evolution. From this mindset, niche constriction is a proximate rather than an evolutionary cause.[28] SET's externalism is also incompatible with the concept of organism-environment coevolution. Because SET uses the external environment as its reference device to describe evolution, it cannot recognize that populations of niche-constructing organisms have the ability to cause their environments to coevolve with them. An exception is where the relevant source of selection in the environment is another organism, in which case we have organism-organism coevolution rather than organism-environment coevolution. For this reason, SET cannot accept that when purposeful, niche-constructing organisms in populations modify natural selection pressures in their environments, they often cause their environments to coevolve with them.[29]

At this point, the philosophy and history behind these ideas may be of some interest, even though they are seldom considered by evolutionary biologists. This is a brief sketch of these areas. Prescientific explanations of life usually invoked creative deities of some kind. In particular, the three major Abrahamic religions that originated west of the Indus—Judaism, Christianity, and Islam—all explain the adaptive design of organisms relative to their local environments in terms of an external, transcendental, and creative deity. In each case, God is deemed to have created organisms that are well suited to their environments.

Darwin proposed that natural selection was the source of the adaptations of diverse species of organisms relative to their diverse local environments. His idea that natural selection, which stems from the external environments of organisms, is responsible for the adaptation of organisms is minimally logically consistent with the prescientific ideas about an external deity being the source of the design of organisms, regardless of all the other difficulties that Darwin raised for those religions. To borrow a quip from Richard Dawkins,[30] it is as if "The God Delusion" merely had to be replaced by natural selection. But that simple substitution risks creating a new delusion. It is not that natural selection is unimportant—it is *very* important. However, as I've been arguing throughout this book, natural selection is not the sole cause of the adaptations of organisms.

Darwin's thinking was probably also indirectly influenced by the field of physics, which by ill luck and unfortunate timing was dominated at the time by Newtonian mechanics. Newton's second law of motion is $F = ma$, which is to say that an external force (F) acts on objects with mass (m), which react by accelerating (a) in the direction of the force. Newton's theory was very successful because, among other things, it explained gravity. It was, therefore, vastly authoritative in Darwin's day. Translated into biology, it encouraged the idea that a putative external force (i.e., natural selection), acts on objects (here, populations of organisms), that react by moving in the direction dictated by the force (i.e., evolving to be suited to the environment). The same idea discourages recognition of the contribution of the purposeful, fitness-seeking activities of organisms to the evolutionary process through their niche-constructing activities.[31]

When Darwin published *On the Origin of Species* in 1859, a comprehensive theory of thermodynamics did not yet exist. That theory was progressively developed by physicists, initially by Nicolas Carnot and ultimately by Ludwig Boltzmann, during the fifty years following the publication of Darwin's book.[32] When Darwin introduced the idea that natural selection was responsible for the adaptations of organisms, his idea was consistent with the prevailing physics of his time—Newtonian mechanics. Eventually,

Newtonian mechanics was not only superseded by thermodynamics but also subsumed by Einstein's theories of special and general relativity, published in 1905 and 1915, respectively. Post-Einstein, we end up with the much more complicated and fundamentally relativistic idea that "mass acts on space-time, telling it how to curve" and "space-time acts on mass, telling it how to move."[33] It might be too glib to claim that Einstein's theory is an "interactionist theory"[34] involving interactions between space-time and matter, but it is certainly a relativistic theory. The relativism of organism-environment coevolution is what we have been advocating in the previous chapters. That is why I am arguing that SET still falls well short of being a candidate parent theory for a future universal theory of evolution.

NICHE CONSTRUCTION THEORY

NCT avoids many of the limitations of SET. First, NCT recognizes the relevance of the laws of thermodynamics to evolution. Second, it recognizes how the active, purposeful agencies of organisms contribute to their own and each other's evolution through niche construction. And third, NCT recognizes the coevolution of organisms and their environments in the context of their shared ecosystems. That is a good start, but it immediately raises a further problem. We have just noted that both niche construction, when regarded as a codirecting process in evolution, and organism-environment coevolution are incompatible with the externalist reference device used by SET. If everything requires a reference device, then we will need a new reference device for NCT. What should it be?

The external environment is the obvious reference device for understanding natural selection, even though it may have limitations as the foundation for a universal theory of evolution. Similarly, organisms might be thought of as an alternative reference device for understanding niche construction since organisms are the source of that niche construction. Such an account would be labeled "internalist," as the traits of organisms are being understood relative to some internal properties of organisms.[35] Lamarckism has been described as an internalist theory, as the traits of organisms were seen as evolving due to an internal cause—namely, the organism's inner drive for complexity. That is not what NCT is advocating. We cannot use organisms as a reference device for understanding evolution, as that would merely replace an incomplete externalist evolutionary theory with an incomplete internalist evolutionary theory.[36] Instead, "we need to start in the middle," as Santa Fe Institute director David Krakauer once put it.[37] We need a reference device that can catch the relativistic interactions between natural selection pressures in environments and the purposeful niche-constructing activities of populations of organisms. How can we do this? Let's go back to Ashby.

According to Ashby's control theory, organisms have to protect one or more essential variables to be adapted. His Law of Requisite Variety (LRV) tells us that "only variety can 'drive down' or 'destroy' variety." That means that it is only possible for organisms to drive down (i.e., cope with) the variable natural selection pressures, or V_E, that they encounter in their environments by expressing adaptive phenotypic variation, or V_O, that can drive down (i.e., neutralize or exploit) the variable environmental conditions that they encounter. Recall that here, essential variables do not refer directly to variant natural selection pressures in the environments of organisms, nor to the variant structural and functional phenotypic traits of organisms. Rather, essential variables refer to that third kind of variance, the *constantly fluctuating variable relationships* between variant natural selection pressures in the environments of organisms and variant phenotypic traits of organisms, or V_{O-E}. Essential variables therefore refer to the interactive niche relationships between organisms and environments.[38]

According to NCT, these interactive niche relationships can be modified by both changes in the natural selection pressures in the external environments of organisms and changes in the purposeful, niche-constructing activities of organisms as they manage their niche-management games with their local environments. All these sources of change are summarized in the moves shown in figure 3.1 in chapter 3.

Changes in the interactive niche relationships can drive changes in individual organisms during their lives, including the developmental changes that contribute to the variability of organisms in successive generations of evolving populations. The same distinct categories of change can also drive changes in evolving populations of organisms over time, as well as changes in aspects of the environment subject to niche construction, some of which are passed to future generations as an ecological inheritance. Organism-environment coevolution can thus be described relative to the niche-constructing activities of organisms and the natural selection pressures that they encounter. All relevant aspects of change in evolving organism-environment systems can be understood relative to changes in the interactive O-E niche relationship. For all these reasons, I previously proposed that the variable interactive niche relationships between organisms and their environments should replace the externalism of SET as a more appropriate and less restrictive reference device for understanding evolution.[39]

There are at least three nontrivial advantages of this relativistic reference device compared to the externalism of SET. First, while the externalism of SET can only explain the adaptations exhibited by populations of organisms, the relativistic reference device can potentially apply to any individual organism or to any focal population of organisms, in relation to their environment, and therefore to any niche relationship, on any scale. Second, unlike SET, which is biased toward explanations based on natural selection, the relativistic reference device approach is indifferent to whether the niche relationship is changed by natural selection or niche construction. In both cases, it reduces change to the common currency of modifications of the interactive O-E niche relationship (see chapter 3). Third, because it can act as a general-purpose reference device for both the development of individual organisms and the evolution of populations, changes in the O-E niche relationship can be understood as driving the development of individual organisms in their individual niches, just as they also drive the evolutionary changes that occur in the populations of organisms.[40] This brings the twin advantage that it helps set development in an evolutionary context and helps reveal how, contrary to SET, developmental processes can play an evolutionary role.[41] Many of these points were anticipated by Waddington in 1969, as well as by Lewontin in 1983.

The interactive niche relationship also has at least one disadvantage when used as a reference device. That is, it introduces a third kind of variance (V_{O-E}) that is more difficult to think about or comprehend than both V_E and V_O. Sources of natural selection in environments (V_E) are physical, observable, and measurable, and the same is true of the variant phenotypic traits (V_O) expressed by organisms. But the varying relationships between natural selection pressures and varying phenotypic traits (V_{O-E}) are more abstract and harder to comprehend. Nevertheless, I maintain that the adaptations of organisms are not primarily about varying natural selection pressures in their environments, nor are they about the varying phenotypic traits expressed by organisms. They are primarily about the varying relationships between selection and phenotypic traits, which introduces a third dynamical variable (V_{O-E}), determined by the dynamics of the other two (V_E and V_O).

Nevertheless, the interactive niche relationships that encapsulate the varying relationships between varying organisms and their varying environments are empirically tractable. It is possible to investigate whether these relationships are adaptive, and it is possible to identify different sources of adaptive variation, corresponding to different moves in the game, as depicted in chapter 3. For illustration, let's reconsider the termite examples discussed in that chapter. Among some species of termites, variant individuals that differed in the thickness of their outer cuticles (V_O), possessed varying abilities to cope with the arid desert conditions in which they lived (V_{O-E}), which led to natural selection sorting between individuals, and eventually to these populations evolving thick cuticles that limit desiccation (move 2), thereby protecting the essential physiological variable of water balance.[42] In contrast, other species of termite, exposed to desert conditions that varied in their aridity and the availability of subterranean water (V_E), were able to find water deep below the desert surface and bring it back to their nests.[43] As a direct consequence of this niche construction, differences in the thickness of their outer cuticles (V_O) did not lead to varying abilities to cope with the arid external desert conditions (V_{O-E}), because these animals had protected the essential physiological variable of water balance through a different move—namely, niche construction (move 6). In doing so, however, varying abilities to mine for water (V_O) may have led to varying moisture levels within and between nests (V_E), generating fitness differences (V_{O-E}) that favored an enhanced capacity to locate and transport subterranean water (move 2).[44] The example suffices to illustrate that the deployment of a relativistic reference device, as well as the identification of moves in the game, are empirically tractable.

Odling-Smee et al. (2003) cited hundreds of examples, organized into tables, of adaptations that could have been due to the modification of natural selection pressures by niche construction. These tables of examples were dismissed as "factoids" by some of NCT's critics.[45] However, all these examples are open to empirical investigation, and many have been with a role for niche construction confirmed. That could turn these factoids into newly established facts about the causes of adaptations, while at the same time exposing others as erroneous hypotheses. In 1957, the ecologist Evelyn Hutchinson pointed out that the niche relationships and their environments break down into n-dimension hypervolumes of variables, all of which, in principle, are open to empirical investigation. In practice, it is possible to investigate only a few of them at a time. Following Hutchinson's lead, my collaborators and I advocated investigating only the most important niche variables relative to whatever research questions were being investigated at the time.[46]

Used in this way, the advantages of the V_{O-E} interactive niche reference device for understanding evolution far outweigh any disadvantages. Unlike the use of the external environment as a reference device for evolution by SET, which reinforces its shortcomings, the interactive niche reference device strengthens NCT. But NCT, at least as it was laid out prior to this publication, is still an unfinished theory of evolution. That is because NCT still fell short in the way that it handled the relationship between energy and matter (or R_p) and meaningful information (or R_i). Organisms cannot gain R_p from their environments unless they are informed by adaptive know-how, or R_i, a priori. But the opposite is also true. Evolving populations of organisms cannot gain adaptive know-how, or R_i, without paying for it a priori with energy and matter, or R_p. For example, the major extra energy and matter costs involved when organisms reproduce have to be paid for in advance by parent organisms in the preceding generation. In many animal species, these costs include the considerable costs of parental care.

Dilemmas in the R_p–R_i relationship have occurred and reoccurred repeatedly throughout the evolution of life on Earth. The existence of these dilemmas implies that the processes of evolution must have had a capacity to resolve them in different evolving populations, and in different environments. Thus far, neither SET nor NCT has yet been able to fully explain how

R_p–R_i dilemmas are resolved by evolution. However, for the reasons given in this chapter, NCT provides a much more promising framework on which to build a universal theory of evolution. It may only be possible to understand how evolution has resolved R_p–R_i dilemmas in the past by working out the relationship between the bioenergetics and bioinformatics of evolution in more detail, building on the secure foundations of NCT. One highly relevant domain in which R_p–R_i dilemmas frequently play out is major evolutionary transitions. Evolutionary biologists do consider the major transitions in the bioinformatics of evolution, but typically without closely integrating the associated bioenergetics. In the remainder of this chapter, I examine whether it might be possible to extend NCT in ways that potentially will allow R_p–R_i dilemmas to be better understood.

NICHE CONSTRUCTION THEORY AND THE R_P–R_I RELATIONSHIP

Let's go through some of the basic points about the R_p–R_i relationship again. If I give you an energy and matter resource (say, an apple), I lose the apple and you gain it. But if I tell you about an apple tree that is currently bearing fruit, I give you some meaningful information about something that you did not know before, and yet I still retain the knowledge myself. Thus there is a fundamental asymmetry between energy and matter resources and information resources.

In spite of this asymmetry, both R_p and R_i can affect the fitness of organisms. For example, if I give you an apple, I may be reducing my own fitness while increasing yours. If, on the other hand, you are one of my offspring, or another close relative, I could increase my inclusive fitness by giving you an apple. If you are a friend, or a member of my social group, I could be increasing my social status by my generosity, which might confer fitness benefits in the longer term. But if you are a stranger, someone whom I would not see again, I would be reducing my fitness by increasing yours.

Contrast that to what happens if I give you the knowledge of where to find that apple tree. I still know where to find the tree myself, which changes the fitness payoffs slightly. By sharing this knowledge with a relative, I can increase my inclusive fitness. I might also increase my social status and fitness indirectly by sharing my knowledge within my social group. But if I tell a stranger where to find an apple tree, it may have little impact on my fitness—unless, that is, apples are a limiting resource, in which case I might reduce my fitness by giving adaptive know-how to a competitor. The next time I go to the apple tree, there may be no apples left. Every time there is a change in the R_p and R_i relationship in evolving populations, these problems are likely to reoccur.

Let's now consider how these differences play out in major evolutionary transitions. I suggest that each of the major transitions in evolution discussed by Maynard Smith and Szathmáry in 1995 not only represent a new breakthrough in the bioinformatics of evolution, but also represent a new breakthrough in our understanding of the relationship between bioenergetics and bioinformatics. This point is illustrated by Lane and Martin's work on the prokaryote-eukaryote transitions in evolution,[47] which I discussed earlier in this chapter, but it should apply more generally. That is because organisms cannot gain R_i without paying for it in advance with R_p, and neither can they gain R_p from their environments unless they are informed in advance with R_i.

Is there anything in the bioenergetics of evolution equivalent to the major transitions in bioinformatics? In a remarkable paper published in 2017,[48] Olivia Judson suggested that there is. Leaning heavily on earlier work by Timothy Lenton and colleagues,[49] Judson identified five energetic epochs in the history of evolution on Earth. First, there was an epoch labeled "geochemical energy," when only geochemical resources were available to support life. A second epoch subsequently arose, called "sunshine," when some bacteria acquired the adaptive know-how for photosynthesis, allowing them to utilize the radiant energy of the Sun directly. The third epoch, which Judson called "oxygen," refers to the time following the so-called great oxidization event, when multiple organisms acquired the capacity to exploit oxygen. The fourth epoch, labeled "flesh," refers to the capacity of organisms to feed off other organisms. And the final epoch, which Judson called "fire," refers to the ability of our own hominin ancestors to control fire as a way of increasing our human capacity to harvest energy and matter resources from our environments.

To be consistent with the concept of the major bioinformatic transitions in evolution, I'll call Judson's five successive energetic epochs the major bioenergetic transitions in evolution. These two kinds of transitions, bioenergetics and bioinformatics, not only interacted with each other in the past, they also changed and triggered transitions in each other. Expansions of the sources of energy available to living organisms led to concomitant increases in the diversity and complexity of both organisms and ecosystems. Apart from the origin of life itself, which probably only depended on geochemical sources of energy and matter, it is likely that all the other epochs in the major bioenergetic transitions depended, at least partly, on the activities of organisms (i.e., on niche construction).

For instance, in the geochemical epoch, Judson describes how early ecosystems may have quickly diversified to resemble a microbial mat, where the waste products of one type of organism fed the metabolism of others. This inference is consistent with a recent experimental evolution investigation in bacteria, which showed that huge biodiversity could emerge in a completely homogeneous environment through niche construction.[50] San Roman and Wagner describe how bacteria create new ecological niches when they excrete waste products that sustain other bacterial populations. Thousands of new niches were created in this manner.

Similarly, with respect to sunshine, oxygen would never have built up in the atmosphere of Earth had cyanobacteria never evolved. Sunlight is available across much of the Earth's surface, which allowed life forms to disperse around the globe. Oxygen is a very efficient source of energy for any organisms that can utilize it.[51] Oxygen was just a waste product of the photosynthesizing chemical reactions of cyanobacteria. Accordingly, NCT describes the production of oxygen by cyanobacteria as "by-product niche construction". Nevertheless, its production subsequently modified natural selection pressures in the environments of countless other organisms.[52] Most obviously, the oxygenation of the atmosphere created an abundance of oxygen-rich niches, which supplemented the existing anoxic niches.

It is also worth noting that the production of oxygen by cyanobacteria must initially have been toxic for many organisms. This is because most organisms at that time were anaerobic. They lacked the adaptive know-how to utilize oxygen. Eventually, atmospheric oxygen became a positive resource for multiple species of organisms after they had evolved the adaptive know-how to use it. For instance, Judson describes how the availability of oxygen also may have permitted the construction of new molecules, such as collagen, which allowed further diversification.[53] This may be an example of organisms initially degrading the environments of many contemporary organisms by their niche construction, but subsequently enriching the environments of multiple other species.

It may be no coincidence that during this epoch, two major evolutionary events took place: the emergence of eukaryotes and the appearance of the lineage that would lead to land plants. Each represents an important shift in the capacity for organisms to transduce energy. Judson describes how the emergence and diversification of eukaryotes provided new niches for prokaryotes to occupy. As Lane and Martin describe, this then fueled the expansion of eukaryotes into occupying a far wider variety of niches.

The "flesh" epoch allowed the acquisition of energy through the hunting and eating of other life forms. This led to the emergence of the classic pyramid of trophic levels in ecology. It also led to innumerable species of organisms interacting with each other in ecosystems, and it introduced another celebrated question: "Who eats whom in ecosystems?" The first trophic level at the base of the pyramid comprises autotrophs that are primary producers of energy and matter, or R_p, in ecosystems, including microorganisms and plants. These organisms gain their energy and matter either from geochemical sources (from epoch 1) or from the Sun (from epoch 2). The autotrophs produce resources for the second trophic level, typically comprising herbivores that eat the plants. The third trophic level comprises smaller carnivores, which get most of their resources from herbivores. The fourth trophic level comprises larger carnivores, which live off the smaller carnivores. In reality, the trophic pyramid is a good deal more complicated than this simple description. For instance, it includes omnivores that feed at more than one trophic level.

We'll be returning to the trophic pyramid in part II. For the moment, it will suffice to make some brief points. First, the entire food web is ultimately reliant on the capacity of autotrophs to extract energy from geochemical sources or from sunlight through simple, but profound forms of niche construction. Second, the advent of flesh eating allowed for increases in organism size, which in turn generated new and more complex kinds of ecosystems. Third, the evolution of predation not only led to diversifications in morphology among predators (e.g., teeth, claws) and prey (e.g., shells, scales, spikes), but also to the creation of refuges (e.g., nests, burrows), and hunting artifacts (e.g., webs, pit traps), which demanded niche construction. Nest building created new resources for inquilines to exploit. Widespread burrowing, in turn, creates a mixing of sediments known as "bioturbation," which redistributes nutrients and aerates sediments and soils. Animals also produce feces, which strongly affect how nutrients are distributed and the rates at which they cycle.

Judson highlighted how Earth is the only planet in our solar system where fire could exist. That is because fire not only depends on sources of ignition, such as lightning strikes, which are common to many planets, but also requires the presence of oxygen and combustible fuel. Both of these are products of prior biotic evolution and are present on our planet only because of the niche construction of ancestral populations. Even on our planet, fire cannot have existed before the invasion of land by life and the growth of vascular plants on land as a source of fuel.

The use of fire depended on the prior acquisition of appropriate adaptive know-how, or R_p, by our ancestors. Judson discusses some of the ways in which the controlled use of fire increased our ancestors' capacity to harvest more energy and matter resources from their environments. For example, our ancestors used fire for cooking. Cooking releases more nutrients from food than can be obtained from raw food. Judson also refers to the use of fire by humans for smelting metals, for making tools for hunting, and later for agriculture. More advanced tools then enabled the agricultural revolution, urbanization, the Industrial Revolution, and our contemporary societies (see chapter 10).

Judson's list of the bioenergetic transitions in evolution also raise some additional questions. One question is: Why don't successive generations of organisms progressively degrade their external environments of energy and matter resources until the evolution of life stops? What seems to have happened on Earth is the reverse. The biosphere of our own planet seems to have got richer rather than more impoverished over time. It has increased rather than decreased potentially available "ecospace" for diverse organisms to exist over time.[54] It is possible that during the first bioenergetic epoch of life on Earth, organisms did degrade their environments. The evolution of life could have stopped. It didn't, though, because some organisms on Earth acquired the necessary adaptive know-how, or R_p, to achieve photosynthesis. That allowed them to utilize the radiant energy of the Sun. More generally, throughout the history of life on Earth, progressive expansions of ecospace took place over time, in many instances following the emergence of new forms of niche construction.

As far as the evolution of life on Earth is concerned, the Sun can be thought of as an infinite source of energy. It is possible that on some planets elsewhere in the universe, life did originate, but then it stopped. Life may have stopped prematurely because organisms on those planets used up too much of their within-planet geochemical energy and matter, or R_p, before photosynthesis evolved and allowed them to harvest radiant energy from their stars. Waddington claimed that once the processes of organism-environment coevolution begin, evolution cannot stop. However, I suggest that it might stop if evolving populations of organisms can no longer tap into sufficient energy and matter, or R_p, to pay for their own future evolution.

Indeed, there have already been some dangerous moments when the future evolution of life on Earth was severely threatened, which correspond to the mass extinctions of life that occurred repeatedly during the last 4 billion years.[55] There are controversies among paleobiologists about how to define mass extinctions and how to count them. However, there is a broad consensus that there have been five mass extinctions since the Cambrian explosion 542 million years ago. In reverse order, the last two mass extinctions were the Cretaceous or Cretaceous Tertiary mass extinction, 66 million years ago, and the Permian mass extinction, about 250 million years ago. They had overlapping but different causes. For example, the Cretaceous-Tertiary mass extinction was due to abiotic environmental events. It was caused by a preliminary period of intense volcanic activity in the region of modern India, followed by the sudden asteroid strike in the Yucatan in Mexico.

The Permian mass extinction, which nearly ended life about 250 million years ago, was due to an "unthinkable volcanic nightmare."[56] It is sometimes called "the worst of times" by paleobiologists.[57] The Permian extinction led to 96 percent of all life in the sea going extinct, which raised the question of why the remaining 4 percent didn't go extinct too. The extinctions on land were more complicated, but also very severe. According to the paleobiologist Doug Erwin,[58] most of the losses in a mass extinction probably don't come from the initial shock, but rather from the secondary cascade of failures that follow. The Cretaceous-Tertiary and Permian mass extinctions may have been due more to the resultant atmospheric dust, which deprived life on Earth from access to a major source of energy, sunlight, which prevented autotrophs from supporting the food webs that underpin most ecosystems.

There were also multiple pre-Cambrian mass extinctions, about which far less is known. An example was the "Snowball Earth" mass extinction, which occurred in the Ediacaran era about 632 million years ago.[59] There were probably other "Snowball Earth" extinctions that had occurred earlier.[60] During a "Snowball Earth" extinction, the Arctic and Antarctic snow ice sheets expand from the north and the south until they almost meet at the equator. It is assumed that in the Ediacaran era, life only existed in the sea, and apparently, most of it went extinct.

What do all these mass extinctions have in common, relative to the bioenergetics-bioinformatics, or $R_p–R_i$ relationship? Can we account for them in the context of evolutionary theory?[61] Let's first consider the extinction of a single species. Why does a species go extinct? It's a safe bet that all species go extinct sooner or later, although how much sooner or later varies considerably with different species, in different environments. To avoid extinction, there must always be enough organisms in a species that are, capable of surviving and reproducing in each generation. That requires an adaptive match or synchrony between the sources of energy and matter, or R_p, that are present in the environments of organisms and the adaptive know-how, or R_i, carried by those organisms.

This adaptive synchrony has two prerequisites. First, the environment must have the capacity to supply the organisms with sufficient energy and matter to enable enough of them to survive and reproduce in each generation, and it must also be able to act as a sink for their detritus. Second, there must always be enough organisms in the species that are sufficiently well informed to harvest the energy and matter afforded to them by their environments. If, for whatever reason, one or both of these preconditions for the future evolution of a species are not satisfied, then the species risks extinction.

There are several reasons why a species-environment relationship could fail to provide it with sufficient R_p resources. For instance, a species could damage its own habitat by overexploiting it or dumping excessive detritus into it. More likely, a focal species could be outcompeted by another rival species that is equipped with superior R_i, for harvesting the same R_p resources in their shared ecosystem. Or another species might destroy the habitat of the focal species. There are other possibilities too, but I will not dwell on them here.[62]

Given these kinds of threats, how do species avoid extinction? There is no general solution to the R_p–R_i dilemmas that at all times and places threaten the ongoing evolution of a species. Each individual species repeatedly has to find its own way into its future. The way in which any particular species does this may largely depend on the capacity of individual organisms in the species to engage in "inceptive" or opportunistic niche construction. To echo Stuart Kauffmann,[63] individual organisms in a species need to be able to seize new opportunities by taking advantage of whatever potential new sources of R_p happen to come their way relative to the adaptive know-how, or R_i, that they already possess.

Individual species can avoid extinction if they can repeatedly restore the synchrony between their own R_i and the R_p afforded to them by their environments. Organisms may do this either by responding to the natural selection pressures that they encounter in their environments without changing them, or by modifying some of the natural selection pressures that they encounter by purposeful active niche construction. These alternatives correspond to the two different categories of natural selection that we met in chapter 4. They are unmodified, purposeless natural selection pressures versus natural selection pressures that have been previously modified by the active, purposeful agency of niche-constructing organisms. If they cannot do either of these things, they may go extinct. If they can, they may evolve further. They may even speciate.

Now we can go back to those mass extinctions again. How can evolutionary theory account for them, if at all? Mass extinctions are logically equivalent to the extinction of individual species, except that multiple different species go extinct almost simultaneously. But here, the word "simultaneous" refers to what paleobiologists call "deep time," rather than ordinary "human time." For instance, simultaneity in deep time could refer to 100,000 or 200,000 years in ordinary human time.[64] With that proviso, we need to account for the near-simultaneity of mass extinctions

Encouraged by SET, paleobiologists usually look for anything that may have caused significant changes in the natural selection pressures acting on many species at approximately the same time. There are several possible abiotic agents with the capacity to change natural selection in the environments of many different species simultaneously. For instance, a mass extinction could be caused by a major change of climate, a major change in an ocean current, a huge volcanic eruption, or an asteroid strike. Any one of these events could be the source of changes in the natural selection pressures encountered by many different species simultaneously.

However, it is also possible that there were some biotic sources of mass extinctions in the past as well, often working in conjunction with abiotic sources. We may have met one of them already, in the oxygen produced as a by-product of the photosynthetic activities of cyanobacteria. Another much more recent example occurred in the Holocene, during the last 11,000 years. It's called "large-scale mega-faunal extinction" rather than mass extinction.[65] These mega-faunal extinctions were partly due to abiotic events, but they were also almost certainly due to the arrival of modern humans, equipped with weapons and hunting skills, in new regions of the Earth. The arrival of humans in these new regions rapidly drove most of the large animals that they encountered to extinction. This process is still continuing today. NCT raises the possibility of biotically induced mass extinction. A mass extinction could be caused either by significant changes in unmodified natural selection pressures, due to abiotic events, the modification of natural selection pressures by the particularly potent niche-constructing activities of organisms (see chapter 4), or both.

One reason why a species could be a particularly potent niche constructor is by it evolving a major breakthrough in either the bioenergetics or the bioinformatics of evolution. A niche-constructing species could thereby disturb the synchrony of the R_p versus R_i relationship for multiple other species, more or less simultaneously. This is always a possibility because of the ways that the bioenergetics and bioinformatics of evolution interact with each other and cause changes in each other, generating chain reactions that ripple through ecosystems. The cyanobacterial production of oxygen, as a by-product of photosynthesis, is a case in point. It had huge consequences. It modified natural selection pressures in the environments of multiple other species, through niche construction, in the process introducing the third bioenergetic epoch, oxygen.

We do not know enough about this major bioenergetic transition in evolution, partly because it happened so long ago, and partly because the oxidization of Earth's atmosphere was a complicated process that took a long time. However, it must have upset the synchrony between the adaptive know-how, or R_i, carried by numerous species of anaerobic microorganisms and this new source of energy and matter resources, or R_p, in their environments. It probably drove many species extinct and drove others into oxygen-free environments. Eventually, it drove some other populations to evolve further in response to this biotically induced change in the natural selection pressures that they were encountering.

My second example of a potent niche-constructing species is closer to home. It refers to our human exploitation of fossil fuels in our environment. Most of these fossil fuels were originally laid down by diverse species of organisms in the Permian era, about 250 to 300 million years ago. For example, when trees in the Permian forests fell and died, they did not rot in the same way as trees in modern forests rot when they fall and die. This was largely because at that time, there were no species of detritivores in existence that carried the appropriate adaptive know-how, or R_i, to harvest or use all the nutrients made available by the fallen trees. In particular, there were no detritivores capable of breaking down the lignin in the rotting trees.

Consequently, over millions of years, the dead trees sank into the ground and gradually transformed into coal.[66] The coal remained an untapped resource in the environments of multiple species of organisms for millions of years. It couldn't be harvested by any organisms. At that time, no species of organism existed that had evolved the appropriate adaptive know-how, or R_i, to exploit this resource, or R_p, in their environment. There was no adaptive synchrony between the R_i carried by any species of organism and the R_p, in their environments in the form of coal—until, that is, the arrival of humans.

Today's humans, in our contemporary societies, are finally equipped, as a consequence of another putative major transition in the bioinformatics of evolution, fire, and its subsequent use in the Industrial Revolution, when humans acquired sufficient adaptive know-how, or R_i, to harvest coal. The crucial putative major bioinformatics transition consists of the extra knowledge-gaining, or R_i-gaining, processes of human sociocultural evolution (see chapter 7). We are now using coal, as well as other fossil fuels, to generate electricity in power stations. Ironically, the detritus produced by our power stations when they burn coal are now threatening to change Earth's atmosphere again, not with oxygen this time, but with carbon dioxide. As a result, we may currently be threatening much of life with another future mass extinction due to global warming.

If we compare the cyanobacteria example with the human fossil fuel example, what do they have in common, and where do they differ? They differ in that cyanobacteria were responsible for a major transition in the bioenergetics of evolution. In contrast, the human capacity to exploit fossil fuels arises from a putative major transition in the bioinformatics of evolution (although, according to Judson, this has triggered a further major transition in bioenergetics). Humanity's ability to harness this source of energy depends on another putative R_i-gaining process in evolution, "science," and its associated technologies (see chapter 10). It may be the case that whenever a major transition in evolution occurs, whether in bioenergetics or in bioinformatics (possibly in the limiting case), initiated by the niche-constructing activities of just a single species, a dangerous moment may follow for many other species in their shared ecosystems. Such moments could be caused by the fracturing of the synchronous adaptive relationships between the energy and matter resources available to species in their environments and the adaptive know-how that they carry, primarily as a consequence of their prior evolution. Sometimes this fracturing may cause mass extinction. Whenever a major transition in the evolution of life occurs, it possibly ushers in associated dangerous moments. In which case, a potential mass extinction would be induced by biotic agents rather than by abiotic events. They could be induced by the processes of evolution themselves, through which species of organisms achieve major breakthroughs in either the bioenergetics or the bioinformatics transitions of life, often through new forms of niche construction. That may or may not be true, but I suggest that it is a hypothesis worth exploring.

CONCLUSIONS

NCT may be a forerunner for a future universal theory of evolution because it recognizes that the second law of thermodynamics is relevant to evolution, that organisms must be purposeful systems to be able to resist the second law, and that to do so, organisms must interact with their environments in ways that modify them through niche construction. Crucially, for NCT, evolution is about organism-environment coevolution, not just about the evolution of organisms. Any future universal theory of evolution will need to be a universal theory of organism-environment coevolution. It will need to understand how evolving organisms solve perpetual R_p–R_i dilemmas.

The same R_p–R_i dilemmas, whereby each depends on the prior acquisition of the other by organisms, raises a well-known problem for how life began in the first place. Which came first, the harvesting of R_p or the acquisition of R_i? How could life have gotten started when each needs the other, a priori? This is the classical "origin of life" problem, which is the focus of chapter 6.

II LIFE ON EARTH

6 THE ORIGIN OF LIFE

If my arguments in part I are correct, the origin of life on Earth must also have been the origin of purposeful systems on Earth. Insofar as purposeful systems are able to make "choices" between alternatives, they are decision-making systems. That need not imply that purposeful systems must be cognitive or conscious, but it does require that they have access to meaningful information. It also implies the existence of at least a crude form of intentionality in purposeful systems. This point is obscured, when either for philosophic or religious reasons, people insist that intentionality is exclusively a human property. To satisfy their fitness goals, organisms must possess some level of intentionality, which establishes that intentionality is not an exclusively human attribute.

Today, there are three sets of fundamentally different kinds of systems on Earth. There are *inanimate, purposeless systems*, which have to comply with the laws of nature, but cannot do much else. Rocks and mountains can change over time (e.g., through erosion), but they do not have any fitness goals that they must meet. That is in marked contrast to living organisms, which are *animate, purposeful systems*. They too must comply with the laws of nature, but if animate purposeful systems can acquire sufficient meaningful information in the form of adaptive know-how, or R_i, about their interactive niche relationships with their environments, they may be able to use some of the laws of nature to fulfill their own purposes of survival and reproduction. Third, there are *inanimate artifacts*, which are constructed by animate purposeful systems. Artifacts have an unusual status with respect to purpose, as they lack the inherent purposefulness of living organisms. They have no fitness goals of their own. They are supplied with artificial purposes by the organisms that constructed them and may help their animate constructors to fulfill their fitness goals. They include birds' nests, animal burrows, beaver dams, and so forth, but currently the most sophisticated artifacts are made by humans, such as computers, iPhones, and robots.

This classification into the three logically distinct types of system found on Earth raises two questions. One applies to some of our most recent human artifacts. Might it be possible in the future for some of our most sophisticated artifacts, which can currently only act on the basis of their human-supplied purposes, to acquire purposes of their own? If so, might such artifacts also be able to pass on their own purposes to "descendant" artifacts? Must our artifacts always be constrained by their present inanimate limitations of having to receive their purposes exclusively from their human constructors? Or will technological advances mean that they eventually bypass these limitations by constructing purposes of their own? Could our artifacts program themselves one day to achieve their own "fitness" goals, as science fiction writers have fantasized? If that happened, it would introduce a fourth category of systems on Earth—namely "purposeful, abiotic life." We might then want to know how it was possible for this kind of abiotic life to have originated from human artifacts. I'm not going to deal with this question here; I will touch on it in chapter 7.

Here, I want to focus on the other question raised by my classification, which concerns the origin of life. How was it possible for a process with a nonliving assemblage of simple chemical compounds and energy to give rise to the first living cells or organisms, approximately 4 billion years ago? How could inanimate, purposeless systems produce purposeful animate systems? This takes us back to where we left off in chapter 5. To exist, life needs two kinds of resources—energy and matter (R_p) and meaningful information (R_i). Living organisms need both of these resources to counteract the flow of energy and matter between themselves and their environments, favored by the second law of thermodynamics. Organisms have to resist the second law continuously by their active, purposeful, fuel-consuming work.

These two requirements immediately raise the R_p versus R_i dilemma again. The acquisition of R_p by organisms requires R_i, a priori, but equally, the acquisition of R_i has to be paid for in advance with energy and matter, or R_p resources. So which came first at the origin of life, R_p or R_i? How did life get started? This is the classical origin-of-life question.[1] It is often phrased as "Which came first: metabolism or reproduction?" The adaptive know-how (R_i) required for reproduction implies the prior existence of parent organisms, which are not available in an origin-of-life scenario. For that reason, I'll stick with my version of this same question: "Which came first at the origin of life, R_p or R_i?" What must we look for when trying to answer this question?

Given that life is cellular, we need to look for the origin of the first living cells. But because living cells can exist only by counteracting the second law of thermodynamics, by taking R_p resources high in free energy from their environments and dumping detritus lower in free energy back to their environments, it raises another origin of life demand. We need to look not only for the origin of the first living cells, but also for the origin of the first interactive niche relationships on Earth, between the first cells and their initial environments. That means that we need to consider the relationships between the first cells, or microorganisms (O), and their first local external environments (E). How and where did the first interactive O-E niche relationships between living cells and their environments originate on Earth?

One hypothesis is that it didn't. According to some authorities, life did not originate on Earth but somewhere else in the universe. It's called the "Panspermia hypothesis," and it's associated with the cosmologist Fred Hoyle,[2] among others. According to this hypothesis, microorganisms from somewhere in outer space were responsible for the origin of life on Earth. The hypothesis gains plausibility from the fact that when the Sun and its surrounding planets first formed, approximately 4.5 billion years ago, space was teeming with dust, meteorites, asteroids, and comets, many of which were known to have bombarded the early Earth and were involved in the creation of Earth itself. Earth is still being bombarded by meteorites, comets, and an occasional asteroid, although these bombardments are far less frequent now. But it is known that many of these foreign bodies are awash with organic compounds, including amino acids, lipids, and sugars. They might therefore have led to the origin of the first microorganisms that seeded Earth with life. The trouble with this hypothesis is that it does not solve the origin of life problem—it merely transfers it from Earth to somewhere else in the universe. For this reason, I'll ignore this hypothesis here and assume that life did originate on Earth.

Given this assumption, let's consider the two leading contemporary hypotheses about how life originated on Earth. The first hypothesis is that life began in deep-sea hydrothermal ocean vents.[3] The second hypothesis is that it originated in fresh water, in small puddles or ponds on land, that were constantly forming and drying out.[4] I will consider both of these hypotheses in turn, primarily from the point of view of niche construction theory (NCT). I want to consider whether NCT is compatible with each of these hypotheses. Before I do that, though, there are two preliminary points to make.

First, I must declare my own limitations. Research into the origin of life demands a detailed knowledge of both chemistry and biochemistry. Since I am neither a chemist nor a biochemist, I can only refer to the work of others who are. In the case of the deep-sea hydrothermal vent hypothesis, regardless of its validity or otherwise, I will do no more than indicate where there might have been a role for niche construction that is consistent with this hypothesis, even though its authors don't mention niche construction. In the case of the hypothesis that life originated in fresh water on land, I will describe the possible part played by niche construction in slightly more detail, and in terms that are already being explored by its authors. However, because of my limitations, I'll keep this chapter short.

My second point is that I want to draw attention to the correspondence between chemical reactions and the laws of thermodynamics, based primarily on personal communications with David Deamer and Bruce Damer, both of whom are authorities on the origin of life at the University of California, Santa Cruz. Deamer points out that chemical kinetics go hand in hand with thermodynamics. The term "chemical kinetics" refers to the rates at which chemical reactions take place. When the chemical reactions that lead to complicated molecules occur more rapidly than the chemical reactions that break them down again, then thermodynamically far-from-equilibrium systems can persist for extended periods of time. Deamer refers to such circumstances as a "kinetic trap."

All life today exists in a "kinetic trap." All organisms spend most of their lives living in a steady state, away from thermodynamic equilibrium. Deamer proposes that the origin of life must have incorporated a kinetic trap. If a chemical reaction exists that rapidly drives chemical complexity uphill away from equilibrium, and if the downhill reaction is slower, more complex molecules can exist in a transient state, far from equilibrium. Deamer illustrates this point with a simple human example. We don't dissolve when we take a shower because the downhill chemical reactions of hydrolysis are so much slower than the uphill chemical reactions that built our bodies in the first place. The concept of a kinetic trap is relevant to both of these hypotheses about the origin of life. It is particularly relevant to Deamer and Damer's hypothesis, discussed in the next section.

THE DEEP-SEA HYDROTHERMAL VENT HYPOTHESIS

The deep-sea hydrothermal vent hypothesis is summarized in Nick Lane's book *The Vital Question*,[5] and more critically by David Deamer in his book *Assembling Life*,[6] although it is derived from the work of other authors dating back to the 1960s, notably Peter Mitchell[7] and Michael Russell.[8] Scientists studying the origin of life have made progress with experiments, designed to re-create as closely as possible the early chemical processes on Earth, via which the antecedents of cells, known as "protocells," could have formed. Protocells consist of a bilayer membrane around an aqueous solution. Lane describes how in recent years, researchers have succeeded in creating self-assembling protocells in an environment similar to that of hydrothermal vents. His hypothesis is based on the relationship between deep-sea alkaline vents and the surrounding slightly acidic sea. The acidic sea is more proton-rich than the alkaline deep-sea hydrothermal vents. This asymmetrical distribution of positively charged proton ions in the vicinity of the vents, compared to the surrounding sea, sets up a natural gradient in which proton ions diffuse from an area of higher to a region of lower concentration. According to Lane's hypothesis, energy and chemical gradients could have been established across the elementary semipermeable membranes, separating the first protocells from their external worlds. This putative relationship is also associated with a process known as "chemiosmosis," which refers to the movement of ions across a membrane-bound structure, down an electrochemical gradient. Lane assumes that this energy gradient was responsible for setting up a potential difference or force between the inner world of the earliest protocells and their external environment. Following Peter Mitchell[9] and others, Lane calls this force a "proton motive force."

The proton motive force is a relatively recent discovery. Before it was discovered, origin-of-life researchers had concentrated on trying to understand the energizing of life almost exclusively in terms of flows of negatively charged electrons across protomembranes, and therefore on redox chemistry. Redox chemistry involves an exchange of negatively charged electrons between electron donors and electron acceptors. Electron donation, for example by molecules, is referred to as "oxidation," while electron acceptance by other molecules is referred to as "reduction"—hence the term "redox." Redox chemistry was long regarded as the most likely source of energy for protocells and the earliest cells. If we ignore fermentation (e.g., in yeasts), redox chemistry underpins respiration and metabolism in all living organisms.

The significance of the proton motive force is that it could have kick-started and then sustained the subsequent redox-based metabolism to which it is now coupled, by supplying protocells with an initial gift of free energy in the form of adenosine triphosphate (ATP). The ATP molecule is the universal energy currency in all organisms because of its ability to release energy, when it breaks down into adenosine diphosphate (ADP). Lane suggests that a flow of positively charged protons, from the external worlds of the first living cells into their internal environments, could have led to the subsequent synthesis of ATP.

According to Lane, this initial gift of free energy could have been a vital first step toward the origin of life. It could have initiated and subsequently sustained metabolic activities in the first living cells. Initially however, it could only have been a brief first step because as soon as the flow of protons across the semipermeable membrane between the first cells and their environments achieved equilibrium, the flow of proton ions would stop. To maintain the proton motive force, and therefore the continuing synthesis of ATP, the cell must keep returning either excess protons or equivalent positively charged ions such as sodium ions from inside the cell to its outside environment. A cell must do this to preserve the energy and chemical gradients between its inner and outer worlds, upon which the synthesis of ATP and its metabolism depends.

Lane proposes that the export of charged particles may have been achieved by an antiporter molecule or protein in the membrane of the first living cells. The job of the antiporter molecule is to restore the electrochemical gradients between the inner and outer worlds of the first living cells through active nonrandom work. Lane loosely refers to this molecule as a "turnstile molecule." Insofar as the antiporter molecule is an origin-of-life adaptation in the first living cells, the cell must be informed by meaningful adaptive know-how, or R_i. But the first living cells can have been informed only by the processes of evolution, and primarily by natural selection. Assuming that Lane's antiporter molecule is an adaptation, how does it relate not only to the origin of life, but also to the origin of the processes of evolution? At the origin of life, natural selection could have acted only on populations of variant individual protocells, and then cells. Natural selection could not have informed individual protocells or cells directly. It could have informed individual protocells or cells only indirectly, as a function of their membership of evolving populations of protocells or cells.

This limitation of natural selection strongly suggests that the origin of life cannot refer exclusively to the origin of the first living cells. It must refer to the origin of the first population of cells, including component adaptations of those cells, in this case the structure and function of the antiporter molecules. The antiporter molecules could then be informed by prior natural selection, acting on previous generations of populations of protocells, and later cells, in the usual way. Specifically, individual cells could be informed by heritable, deductive, adaptive know-how, or R_p, registered in elementary memory systems in populations of variant individual cells and specifying what was adaptive for earlier generations of cells. Their informed properties could then provide individual cells with a basis for making their inductive gambles about what might be adaptive for themselves, as well as for their descendants in future environments (see chapters 3 and 4).

The active export of positively charged ions by the antiporter molecules in the first living cells would also have introduced the first two-way interactive niche relationships between the first cells and their local environments. These first interactive niche relationships must have affected the cells' own internal environments. They must also have caused at least some microchanges in the cells' external environments. It is currently unknown whether these initial interactive niche relationships may also have introduced the second category of natural selection that we discussed earlier. The second category is natural selection pressures previously modified by the purposeful, nonrandom niche-constructing activities of the first cells

The putative niche-constructing activities of the first living cells must have included the expulsion of positively charged ions by the antiporter molecules. These metabolic activities would have included the harvesting of resources from the cells' environments and the return of detritus to those environments. The impact of these putative niche-constructing activities by each individual living cell would have been minuscule. However, the collective activities of large populations of cells might have had a measurable impact on chemical gradients in their immediate external environments, in restricted domains. Origin-of-life populations of variant cells might therefore have started to introduce some modifications to natural selection pressures in their immediate microenvironments, or in the microenvironments of some individual variant cells, by their environment-modifying activities. This is the first way in which niche construction might possibly have played at least a minor role in origin-of-life scenarios, as described by the deep-sea hydrothermal vent hypothesis.

All of this raises the next question: How could populations of living cells be generated by earlier protocells? Presumably this required the reproductive or self-replicative activities of the protocells through proto cell division. These first populations of protocells could then have been evaluated and informed by natural selection, and possibly by niche construction too. That might have been sufficient for adaptive phenotypic traits in individual cells in populations to be informed as a consequence of their membership of their populations.

In this context, the reproductive activities of the first protocells would have depended on what another origin-of-life researcher, the Israeli chemist Addy Pross, calls "replicative chemistry," which eventually leads to the exponential growth of populations of origin-of-life cells. The reproductive activities of early protocells introduce a second way in which niche construction could have played a role in origin-of-life scenarios. I suggest that in all organisms, reproduction is itself an elementary kind of niche construction. This is because when organisms reproduce, if only by cell division in the case of single cells, they immediately introduce additional organisms into their own and each other's environments.

Organisms must thereby modify their environments by their reproductive activities. Reproduction necessarily modifies the selective environments of organisms and must have done so from the outset, even in the case of protocells undergoing a crude form of reproduction. In origin-of-life scenarios, some of the offspring introduced by the reproductive activities of protocells would be potential competitors, while others might be potential cooperators. If and when the first protocells start to cooperate with each other, this introduces the possibility of more advanced or complicated adaptations appearing in origin-of-life scenarios based on interactions between sets or groups of protocells in interacting networks.

The Nobel Prize–winning German chemist Manfred Eigen and his colleagues[10] proposed an early example of cooperative interactions among networks of molecules with his concept of the "hypercycle." A hypercycle is a cycle of connected, self-replicating macromolecules that are linked to each other such that each of them catalyzes the creation of its successor, with the last molecule catalyzing the first one again. In this way, the cycle is self-perpetuating. It has been shown that hypercycles can originate naturally and evolve through natural selection. As a result, not only does the system gain information, but its information content can be improved. In the process, the hypercycle potentially becomes a more complex adaptive system, comprising a set of cooperating interacting units (in this case, genes in a network). That was not enough to account for the origin of life, but it was a step in the right direction because it showed how biological complexity could increase over time through interactive relationships among its composite elements. It was a fascinating idea at the time.

Again, it suggests that life could not have started with the origin of the first protocells in the context of single, interactive niche relationships. Rather, it must have started with networks of variant protocells and later cells, in populations, with multiple diverse niche relationships between individuals and their individual local environments. Pross[11] proposed a similar idea with his concept of "complexification," which involves interactions among networks of similar prebiotic, and later biotic, units. It may demand catalytic relationships among these interacting units, leading to the emergence of more complex autocatalytic adaptive systems. Pross argues that complexification played a crucial part in setting up the origin-of-life scenarios.

The production of the first protocellular populations would still have depended on the capacity of individual protocells to reproduce. Reproduction always demands adaptive know-how or R_i. The first living cells would have needed additional adaptive know-how to reproduce. They would have needed sufficient R_i to inform both their R_p processing metabolisms and their reproductive activities. But in both cases, the first cells would have needed energy to pay for their R_p, possibly taking us back to life by being kick-started by ATP, synthesized by the proton motive force and then to the redox chemistry to which it is coupled.

Might there also be a further role for niche construction, as well as natural selection, in these scenarios? Insofar as the first protocells may have cooperated with each other to generate more complex traits, they may have done so by modifying each other's environments by mutual niche construction. One quite likely form of cooperation was that protocells could have acted as catalysts for each other's chemical reactions. Mutual relationships are best understood in terms of natural selection pressures, previously modified by the niche-constructing activities of their mutualist partners. Mutualist partners typically benefit each other by modifying natural selection pressures in each other's environments, by their niche-constructing activities. It is more difficult to understand mutualist relationships exclusively in terms of unmodified purposeless natural selection pressures. This is a third way in which niche construction might have participated in the scenario described by Lane's deep-sea hydrothermal vent hypothesis.

Like all hypotheses about the origin of life, the deep-sea hydrothermal vent hypothesis must confront the "Which came first?" dilemma: metabolism or reproduction. The origin of life is the first major transition in evolution. It is a transition from nonlife to life. There obviously must have been some kind of solution to this dilemma, and it is difficult to escape the idea that both these fundamental requirements of life must have been satisfied simultaneously.

One possibility, suggested by Lane's hypothesis, is that the "proton-motive" force, based on a flow of positively charged protons across membranes in both directions, might have initiated both metabolism and self-replication, but in different variant cells. The proton-motive force could have granted a free-energy gift to prebiotic entities to kick-start both active metabolism and active self-replication in the first protocells. These variant protocells may then have collaborated in prebiotic mutualistic relationships, combining to form an emergent, more complex adaptive system capable of both active metabolism and self-replication. In combination, they could have accounted for the first true organisms.

The principal strength of this deep-sea hydrothermal vent hypothesis is that, in using the proton motive force, it introduces a hitherto unknown source of activation energy to drive increasingly complex chemical reactions. It is also possible to test some of the ensuing ideas in the lab, although this is by no means easy. Unfortunately, there are also some major challenges for this hypothesis, which to some authorities at least, appear to rule it out.[12] For example, a saltwater environment is extremely hostile to the self-assembly of lipid membranes. But in the absence of membrane-bounded compartments, or barriers, it is difficult if not impossible to see how the necessary concentration of chemicals, required for work to be done through energy gradients to initiate combinatorial reactions, could have occurred.

A second problem is to explain how the products of these complex chemical reactions could have been preserved for long enough in a deep-sea environment to have eventually led to the origin of life. If chemical reactions had built more complex polymers out of simple monomers, then the polymers would almost certainly have been destroyed by subsequent hydrolysis, caused by invading water molecules in an underwater marine environment. The consequences of these reactions could not have been preserved in a kinetic trap underwater. Deamer has argued that chemiosmosis cannot occur in conditions of hydrothermal vents.[13] In addition, to be credible, this hypothesis must be tested in the wild as well as in the lab. For obvious reasons, this is difficult to do, and it has not yet been done. But until it is tested in the wild, the deep-sea hydrothermal vent hypothesis remains conjecture.

DARWIN'S WARM LITTLE POND HYPOTHESIS

The second hypothesis is that life originated on land in fresh water, in small pools or ponds. The principal authors of this idea, David Deamer and Bruce Damer, call it "Darwin's warm little pond hypothesis," in honor of a conjecture that originally came from Darwin, in a letter that he wrote to a friend, the biologist Joseph Hooker.[14] Again, I won't go through this hypothesis in detail. I'll just sketch some of the principal steps toward the origin of life as described by Deamer and Damer.[15] Both these authors believe that life originated in small pools or ponds associated with hot springs and geysers in volcanic regions of Earth. These ponds undergo repeated rounds of hydration and dehydration, known as "wet-dry cycling," as they continuously fill up with water and then later dry out again. This introduces a dynamic process into origin-of-life scenarios.

The cycles can occur on varying time scales. For instance, wet-dry cycles may take seconds to minutes, such as when geysers splash water on hot rocks. Other cycles may take longer, such as when they are caused by the rise and fall of pool levels due to fluctuations in the hot springs. Still other cycles, lasting hours or days, may be caused by the complete evaporation of the ponds and subsequent refilling by rainfall. Warm, small ponds with these properties are quite common in volcanic regions of Earth today, such as Yellowstone National Park in the western US. This is significant because it means that it is possible to test some components of this hypothesis in the field as well as in the lab.

Deamer and Damer's approach to the origin of life circumvents many of the obstacles confronted by the deep-sea hydrothermal vent hypothesis proposed by Lane (2015) and others. The warm little pond (aka "hot spring") hypothesis is described by Deamer in his books *First Life* and *Assembling Life*, and also in Damer and Deamer's articles.[16] I suggest that it's the most plausible idea at the moment about the origin of life.[17] The key reason why it is more likely that life originated in freshwater rather than in the sea is because salty seawater inhibits the self-assembly process and many crucial prebiotic reactions.

I'll use some of the personal communications that I've had with Deamer and Damer to summarize the dynamics of the wet-dry cycles. First, I'll consider the wet phase, when Darwin's warm little pond is full of water and may be overflowing. Under such circumstances, the pond is likely to be connected to networks of streams and rivers and other ponds, ultimately ending up in the sea. This implies that the products of any chemical reactions that occur in the initial pond may subsequently be distributed across a whole drainage system.

Deamer and Damer assume that during the wet phase, when the pond is full, it will contain diverse monomers, simple polymers, amino acids, and nucleotides. Monomers are small molecules that bond together to form more complex structures, such as polymers. Polymers are strings of monomers of varying lengths. The ponds may also contain some amphiphilic compounds, which are molecules such as lipids that are attracted by water at one end and repelled by water at the other end. As a result of this and other properties, amphiphilic molecules spontaneously form membranes and vesicles that can encapsulate other molecular contents. This self-assembling property of membranes by lipids has been referred to by Stuart Kauffman as "order for free."[18]

Now we can turn to the dry phase of a wet-dry cycle. Evaporation during the dry phase concentrates otherwise dilute potential chemical reactants, including monomers. As a pond shrinks during a dry phase, it will often leave a rim of concentrated chemicals at its edges on some kind of surface, perhaps basaltic or silicate minerals. Deamer and Damer draw a loose analogy with the dirty ring left in a bathtub after someone has taken a bath. The difference is that the rim in a dirty bathtub is likely to include some cells and complex biotic molecules bequeathed by whoever last used the bath. Nothing like that could have been left behind originally when Darwin's warm little ponds dried up, but there would still have been a more condensed concentration of chemicals left at the edges of dried-up ponds, particularly on mineral surfaces.

Damer also realized that between the wet and dry phases, there must be a gel-like moist phase where protocells can begin to fuse and exchange useful polymers that have survived earlier wet-dry cycles. This means that the wet-dry cycles are really wet-moist-dry cycles. Deamer and Damer consider aggregated protocells in the gel phase to have been a primitive version of Carl Woese's "progenote,"[19] a hypothetical, communal, preprokaryotic stage in cellular evolution.

Combinatorial chemical reactions inside these vesicular protocells promote condensation reactions during the subsequent dry phase, leading to the polymerization of monomers and to the further buildup of complex chains of polymers. Polymers made in the dry phase would then be captured within the multilayered matrix of amphiphilic compounds. It has been shown that during the next wet phase, the polymers would become encapsulated in membranous compartments that spontaneously break off to form very large populations of microscopic protocells. This budding process creates populations of variant vesicles. Initially, the budding process probably involves only the informing of replicates by binary digits or bits (to be discussed shortly), rather than being informed by meaningful information, or R_i.

The transmission of information by templates is limited but efficient. To give an everyday example of its efficiency, suppose that you want another copy of your front door key, so you take your key to a key-cutter and ask for a copy to be made. They will use the original key as a template. Information in the form of binary digits, or bits, encapsulated in the structure of the original key will be transmitted to the copy of the new key by the key-cutter when the original key is used as a template (see the discussion of Shannon's communication theory in chapter 2). But nobody bothers about binary digits when cutting a new key. Each replicate protocell, whether produced by template information, bits, or meaningful information (R_i), is likely to differ in composition from all the other replicates, at least slightly. That should be enough for protonatural selection processes to start sorting between different variant protocells with different survival capabilities.

The buildup of more complicated chemical reactions should also be aided by the kinetic trap phenomena. Not all polymer variants, assembled during a previous dry phase, will be wiped out by subsequent hydrolysis during the next wet phase. That means that during any subsequent dry phase, the polymerization and the process of building up more complex chemical reactions do not have to start over from scratch. The following dry phase would inherit a residue of chemical products left over from earlier dry phases by a crude form of inheritance over successive generations of wet-moist-dry cycles.

The kinetic trap allows a progressive buildup of more complicated chemical reactions during successive cycles. In combination, the populations of variant protocells, due to the "inheritances" of the variant polymers over successive generations of wet-moist-dry cycles, introduces a "prebiotic evolutionary process." By "prebiotic evolution," I mean a selective process across repeating generations of wet-moist-dry cycles, leading to the buildup of more complex systems. This dynamic process introduces a ratchet effect leading to longer-chained polymers, which may eventually have included both proteins and nucleic acids. Deamer and Damer call this process "combinatorial selection" prior to the start of gene-based "Darwinian selection." A well-developed experimental field known as chemical or molecular evolution has demonstrated the viability of this approach.[20]

This is an attractive feature of Damer and Deamer's hypothesis. It introduces a potentially creative prebiotic evolutionary process into origin-of-life scenarios. The process includes the protonatural selection of heritable variant polymers in populations of protocells. It combines all three of the fundamental processes necessary for evolution to occur[21] at a prebiotic stage—namely, phenotypic variation among polymers, the protonatural selection of that variance, and differential inheritance of the fittest variance.

FIVE POSSIBLE ROLES FOR NICHE CONSTRUCTION AT THE ORIGIN OF LIFE

Where, if anywhere, might there be a role for niche construction in this prebiotic evolutionary process? Let's go back to the cycling scenario. One possible candidate for protoniche construction is the formation of the rims of condensed organic compounds around the edges of ponds during the drying phases of the wet-moist-dry cycles. The succession of wet-moist-dry cycles are exclusively abiotic events that apparently have nothing to do with niche construction. Yet the condensation and the composition of the lipid membranes, polymers, and other solutes, on the rims around the edges of drying ponds, form distinctive structures with properties conducive to key prebiotic chemical reactions. As the polymers formed through dehydration synthesis between drying lipid lamellae accumulate and lengthen, the complexity of the environment grows. When water is reintroduced, these polymers enter vesicular compartments, budding off from the lipid layers to form protocells. Polymers trapped within these compartments can stabilize the membranous boundaries surrounding them. Through an early form of mutualism, the protocells are more likely to survive the aqueous immersion phase and return to a moist aggregate, fusing with neighboring protocells and delivering their cargoes of polymers.

It has been observed, in the lab and in recent field experiments at volcanic hot springs, that the aggregate and its populations of polymers grow through multiple cycles. Could these aggregates and the protocells and polymers which comprise them be regarded as a form of prebiotic "proto niche construction"? Whatever we might call this self-assembling bathtub ring sludge at the edges and bottoms of pools four billion years ago, it may have been an essential medium in the emergence of the first communities of living cells. Until now, NCT has assumed that organisms already exist and niche construction refers to the interactions between organisms and natural selection pressures in their environments. NCT has not previously been applied to origin-of-life scenarios, although it has been applied to the internal environments of living organisms.[22] But could NCT be applied to events within the internal environments of protocells in origin-of-life scenarios, as implied by Damer and Deamer? Possibly yes, even though Damer and Deamer's' suggestion raises some difficulties for NCT. For instance, who or what is the niche constructor within the internal environment of a protocell?

One of the earliest established examples of niche construction is preserved in rock, in the form of laminated structures known as "stromatolites." Stromatolites are widely represented in the fossil record from more than

3.2 billion years ago. These laminated textures were laid down through the niche-constructing activities of microbial mats.[23] Entire colonies of variant bacteria, including cyanobacteria, produce the constructed product (namely, stromatolites). There is therefore a precedent for recognizing that niche construction can be produced by a network of multiple, diverse entities. The primary difference is that in the case of the microbial mats, the constructors are a network of living cells rather than a network of prebiotic protocells, but in other respects, the comparison is striking.

A second possible role for niche construction arises from the following question: What could be the source of an internal natural selection pressure in the protocell, which could be modified by proto niche construction? This question refers to interactions between polymers of different complexities in the internal environments of protocells. Might different polymers aid the extension of other polymers, or possibly do the opposite, by outcompeting them or destroying them? If we cheat a little, by arbitrarily running the evolutionary clock forward by over 2 billion years to the first appearance of multicellular metazoan organisms, it is possible to ask more precise versions of this question. For example, there appear to be intriguing coevolutionary processes going on between the products expressed by the nuclear DNA and the products expressed by the mitochondrial DNA in the eukaryotic cells of metazoan organisms.[24] Might this coevolution between mitochondrial organelles and their host cells involve some kind of internal niche construction as well as internal natural selection? It is known that eukaryote cells formed through endosymbiosis, in which originally prokaryote cells engulfed mitochondrial organelles. Rather than digesting them, the prokaryote cells benefited from energy resources produced by the engulfed mitochondrial organelles, such as from their production of sugar or ATP. At the same time, the host prokaryotic cell constituted a benign environment for the mitochondrial organelles.[25]

A third role for niche construction in origin-of-life scenarios is more definite. It refers to the multiplication of variant vesicles, protocells, or progenotes by their reproductive activities. It may not have mattered much whether the replication of vesicles or protocells initially only involved replication via templates, or whether their reproduction may have involved meaningful information, or R_i later. The multiplication of protocells by some form of replication or reproduction would have had two effects. First, it would have caused replicating protocells to generate populations of variant protocells, thereby allowing protonatural selection pressures to start sorting between them. Second, the replication of protocells would have added protocells to their own and each other's environments. Adding extra protocells would by itself have been enough to modify some of the natural selection pressures in the external environments of the populations of protocells. For this reason, replication through these primitive forms can be viewed as an elementary form of niche construction in prebiotic evolution, as we saw before in Lane's deep-sea hydrothermal vent hypothesis. The multiplication of protocells in populations by replication becomes both the origin of protonatural selection, acting on variant protocells and the origin of protoniche construction, modifying those particular natural selection pressures in prebiotic evolution.

A fourth possible example of protoniche construction in protocells is the active construction by polymers of pores in the membranes that encapsulate the vesicles that contain them. There is some evidence from laboratory experiments that internalized polymers can fix themselves to membranes and create dislocations or pores in the membranes, which create opportunities for small molecules to pass through them. Concentrating mixtures during the dry-down phases would force solutes through these pores, passing close to the polymers that formed them and creating opportunities for chemical reactions. If true, it would be another example of protoniche construction. It would have created the opportunity for the exchange of resources between the vesicle and its external environment.

A fifth candidate for niche construction refers to the potentially active control of the atomic and molecular traffic that passes in and out of protocells via these pores. This example corresponds closely to conventional niche construction. The import of energy and matter, or R_p resources, from the external environments of organisms, and the export of detritus back to their external environments by active work in both cases, are conventional forms of niche construction. Through both the import and export of resources, the metabolic activities of the protocells would be likely to modify some of the natural selection pressures in their microenvironments.

In comparison with prebiotic protocells, the active metabolic work of living cells should have had an even greater effect on both their internal and external environments on either side of their membrane boundary. Some definitions of life list the characteristics that living organisms must possess, one of which is their capacity to actively import resources from and export detritus to their environments. If we accept that definition, then the origin of life could not have begun prior to the origin of these forms of elementary niche construction.

We can now see that these five proposed examples of putative niche construction by protocells, or their constituent polymers, represent increasingly credible candidates for a role, or roles, for niche construction in the origin of life on Earth. But in conjunction with Deamer and Damer's "warm little pond hypothesis," they raise a further question: Was protoniche construction necessary for the origin and subsequent evolution of life on Earth? I suggest that it was.

Protoniche construction by protocells, and later niche construction by living cells, should have introduced the second category of natural selection into evolution (namely, natural selection pressures modified by prior niche construction from the outset). Damer and Deamer's observation that flimsy protocells in isolation cannot undergo complexification suggests that some form of niche construction was necessary for life and evolution to begin. This implies that the first protocells, and later the first living cells, were cocontributors to their own prebiotic and later biotic evolution.

From the origin of life onward, evolution was about the coevolution of protocells, and later of living organisms, with their contemporary environments. I submit that origin-of-life researchers should be looking not only for the origin of the first living cells or organisms, they also should be looking for the origin of the first interactive niche relationships between active and eventually purposeful living organisms and their external environments. This is not just a semantic point; it could have some bearing on how origin-of-life researchers do their work. Some of these points were already anticipated by Deamer and Damer.[26]

Finally, we are still left with the R_p versus R_i dilemma. Given that life required both energy and matter resources and meaningful information resources to resist the second law of thermodynamics, which came first at the origin of life, R_p or R_i? Deamer and Damer's answer to this question is that there is no dilemma. If the abiotic Earth had a mixture of potentially reactive chemical compounds in aqueous solutions, as well as an abundant energy source, there need not have been a "Which came first?" dilemma. Everything would have happened simultaneously. In this respect at least, both the deep-sea hydrothermal vent and Darwin's warm little pond hypotheses agree.

One advantage of Deamer and Damer's hypothesis is that during the last fifteen years, it has been subjected to empirical tests in both the lab and (more to the point) in the field, at hot springs in a number of active volcanic regions of Earth. However, some of the stepping stones toward life may never be accessible to empirical tests. For example, some steps may actually have required millions of years of wet-moist-dry cycles ratcheting up increasingly complex chemical reactions. It is hard to see how any tests in either the lab or the field could mimic all of them adequately. But there are two ways in which this second hypothesis about the origin of life might be tested. If it eventually proves possible to synthesize life artificially in a lab from prebiotic precursors, then the way in which it is achieved will itself be a test of Deamer and Damer's hypothesis. The paths to this synthesis of life will either be consistent or inconsistent with their hypothesis. Likewise, if extraterrestrial life is ever discovered, it will again either be consistent or inconsistent with their arguments. It might also raise hitherto unsuspected surprises. If this reasoning is correct and if, for example life did once exist on Mars, even if it no longer does, then it should have left traces of prior biological niche-constructing activities in the Martian landscape, and some of them may still be detectable. One day, we may find out.

7 SUPPLEMENTARY KNOWLEDGE-GAINING PROCESSES IN EVOLUTION

Max Tegmark's intriguing 2017 book, *Life 3.0: Being Human in the Age of Artificial Intelligence*, talks about three distinct stages of life on Earth. For Tegmark, "Life 1.0" refers to almost all conventional evolutionary biology, while "Life 2.0" refers to human, sociocultural, language-based life. Tegmark's "Life 3.0" refers to a future artificial life based on artificial general intelligence (AGI) and robotics. It assumes that AGI robotic life will, in the not-too-distant future, exceed human intelligence, eventually by orders of magnitude. Life 3.0 doesn't exist yet, but Tegmark wrote his book to encourage his readers to think about where we may or may not be heading as a result of the digital information revolution. He is neither an optimist nor a pessimist; he stresses both the opportunities and dangers that contemporary humans face.

To this end, Tegmark has also been instrumental in cofounding the Future of Life Institute (FLI) at Berkeley, California, and Cambridge, Massachusetts. The relevant point that I wish to draw attention to here is that almost all the people that have assembled under the rubric of the FLI are physicists, cosmologists, mathematicians, economists, neuroscientists, philosophers, and computer scientists. To the best of my knowledge, there isn't a single evolutionary biologist among them. Why not? One reason is that Tegmark and his colleagues assume that biological evolution was a necessary initial evolutionary step toward launching future artificial life. However, they also assume that Life 1.0 is destined to become either irrelevant to Life 3.0 or extinct altogether.

So, why bother about evolutionary biology? In reality Tegmark's classification is a little too simple, as it is now well established that there are other species aside from humans that also have a capacity for cultural evolution, but for the moment, I will ignore this caveat. Life 2.0, which for Tegmark is restricted to human life, may coexist with artificial life for a time, in some kind of hybrid or mutual coexistence. But the implication is that if artificial life really takes off, human life will also drop out of the picture. That leaves postbiological artificial life, or Life 3.0, as a different kind of breakthrough. Given this conceptual framework, it may explain why Tegmark and his colleagues do not seem to think that evolutionary biology has much to offer. For example, they think that neo-Darwinism, or standard evolutionary theory (SET), is not sufficiently relevant to the future of life to be worth including evolutionary biologists in the FLI.

In this chapter, I want to take up this implicit challenge by reconsidering how a more comprehensive evolutionary theory, which includes niche construction theory (NCT) and various other components of the extended evolutionary synthesis,[1] may be more relevant to the future of life on Earth than SET. It could be of greater interest than SET to Tegmark and his colleagues at the FLI. I suggest that there is an evolutionary continuum connecting Life 1.0, Life 2.0, and a future Life 3.0. Of course, there are many possible reasons why evolutionary biologists may not be involved with the FLI initiative. However, plausibly, it is because SET deals with a truncated version of Life 1.0, and not much else, that Tegmark and his colleagues appear to think that it is irrelevant to a possible Life 3.0.

I will start by considering why SET seems to have got stuck with Life 1.0 only, and why it seems to Tegmark to be irrelevant to the future of life on Earth, or anywhere else in the universe where life may exist. The point that Life 1.0 may be irrelevant to life anywhere in the universe was also anticipated by Tegmark, who is a cosmologist as well as a physicist. He wants us to realize that human Life 2.0 could be on the threshold of waking up the universe by introducing superintelligent forms of artificial life that have the potential to disperse beyond Earth. Life 3.0 could do this by colonizing our solar system, and then by colonizing other planets in our galaxy, up to the limits allowed by the laws of physics.

THE LIMITATIONS OF STANDARD EVOLUTIONARY THEORY

We have already considered one of the limitations of SET at some length—namely, its failure to recognize niche construction as a distinct causal process in evolution. SET also fails to recognize the two categories of natural selection: purposeless, natural selection and natural selection previously modified by the active, purposeful agency of niche-constructing organisms (see chapter 4). There is another limitation of SET, however, that may be more relevant here. SET fails to fully recognize the capacity of the primary evolutionary population genetic process to bootstrap.

What is meant by "bootstrapping" in evolution? I mean evolving supplementary mechanisms for adaptive knowledge gain. For SET, in the majority of species, the phenotypic traits of organisms are exclusively informed by population genetic processes. In all such populations, the phenotypic traits of individual organisms are largely, if not exclusively, determined by the naturally selected genes that the individual organisms inherit from their ancestors at the start of their lives. The individual organisms subsequently use the genetic knowledge that they have inherited by expressing it in their structural and functional adaptations. Hence, as organisms cannot gain any additional genetic information for themselves during their lives, there is no capacity for bootstrapping in these populations for SET.[2]

In contrast, I suggest that bootstrapping occurs when the primary population-genetic processes give rise to phenotypic traits in organisms that are capable of acting as supplementary, knowledge-gaining processes in evolution. There is a nested hierarchy of such information-gaining processes, operating at up to three levels. At level 1, at the base of the hierarchy, there is the primary knowledge-gaining process of population genetics, which is common to all evolving populations whether they bootstrap or not.

In bootstrapping populations, there may be up to two supplementary information-gaining levels of knowledge-gaining processes, which are themselves products of genetic evolution at level 1.[3] At level 2, the supplementary information-gaining processes comprise the gaining of additional adaptive know-how, or R_i, by individual organisms during their lives. For instance, later in this chapter, I will discuss how the vertebrate adaptive immune system and animals' capacity for associative learning are level 2 knowledge-gaining processes. Other examples include the vertebrates' nervous system, the vertebrate vascular and muscular systems, the insects' tracheal system, and the microtubular system, all of which operate through the generation of variation, largely at random, and selective retention of functional variants; and hence they are to some degree level 2 knowledge-gaining processes.[4] Individual organisms may gain knowledge as a consequence of their own individual interactions with their local environments. Unlike in level 1, they may gain additional adaptive know-how at any moment during their lives. There may also be another level of supplementary information-gaining process in some species of animals, including humans. Level 3 refers to the information that is gained by the sociocultural group to which individual animals or human beings belong. Like level 2, but unlike level 1, individual organisms may gain additional information from their sociocultural groups at any moment during their lives.

Bootstrapping populations, which possess level 2 knowledge-gaining adaptations, differ from nonbootstrapping populations in two main ways. Most obviously, individual organisms in bootstrapping populations are informed by more than one knowledge-gaining process. Less obviously, organisms in bootstrapping populations are informed by more than one kind of biological system: their population's genetic history (at level 1), their individual experiences (at level 2), and their social group's culture (at level 3). These two differences have a significant evolutionary consequence for organisms in bootstrapping populations. Organisms in bootstrapping populations are likely to gain qualitatively different adaptive know-how from each of the biological systems that are informing them. There are two reasons why. One is that organisms in bootstrapping populations can be informed only by the prior histories of the knowledge-gaining systems that they possess, and different biological systems will have different histories of the past. The second is that each biological system, at each level, has different operating characteristics. They use different biological mechanisms to inform organisms.

I'll consider both of these points, starting with the different histories of each knowledge-gaining system. At level 1, the relevant history of the past is the history of a collective. The collective is the particular evolving population, or species, to which an individual organism belongs. At level 3, the relevant history is that of another kind of collective—namely, the particular sociocultural group to which an individual cultural animal, or human being, belongs.[5]

At level 2, the relevant history is not the history of any group of organisms. Instead, it is the past developmental history of an individual organism.[6] Whatever additional adaptive know-how is gained by individual developing organisms at level 2 contributes to the phenotypic plasticity of individual organisms in bootstrapping populations.

These histories of the past also operate on different time scales. At the primary level of genetic evolution, the relevant time scale ranges from a few generations to millions of years. At level 2, the relevant history of individual organisms could be on a time scale of a few hours, days, or years. At level 3, the history of a sociocultural group is likely to range from days or months to centuries, or even millennia. These radically different timescales are another reason why the different knowledge-gaining systems will be liable to provide individual organisms with qualitatively different adaptive know-how, or R_i.

Now we can turn to the various operating characteristics of each of the different biological systems that inform organisms. First, we need to consider the fundamental requirements that all knowledge-gaining systems must satisfy. What must they all have in common? Then we will consider how they differ.

There are three fundamental requirements that all knowledge-gaining systems in nature must be able to satisfy. First, they have to be sensitive to at least some cause-and-effect relationships in nature. As we have seen, they have to be able to demarcate between signals and noise, where signals refer to reoccurring cause-and-effect relationships in nature and noise refers to everything else (see chapters 3 and 4). Second, knowledge-gaining systems must be able to register information about these cause-and-effect relationships in some kind of memory system. Third, they must be able to express some of the knowledge that they have previously registered in their memory systems via some kind of communication channel, passing this information either to other individuals in the population or to themselves at a later time (see the discussion of Shannon's communication theory in chapter 2).

But these knowledge-gaining systems do not have to fulfill these requirements in the same way. For example, knowledge-gaining systems may not be
sensitive to the same cause-and-effect relationships in nature, nor do they all
have to register the knowledge that they gain in the same kind of memory
systems. There are many kinds of memory systems in different species of
organisms: the genome is one kind, but brains are a second and memories
of effective antibodies are a third. Similarly, the informing biological systems
do not have to communicate the remembered information that they have
previously gained to individual organisms in bootstrapping populations via
the same kind of communication channels. The ways in which they communicate previously gained knowledge need to be consistent with Shannon's
communication theory, but other than that, they can differ.

Now let's consider how each of these biological systems satisfies these
three fundamental requirements, starting with evolving populations at
level 1. Evolving populations can only demarcate between true cause-and-
effect relationships in nature and everything else, as a consequence of the
differential survival and reproduction of organisms in a population. Like
many processes that operate by generating excess variation at random and
selecting functional solutions, this is not an efficient knowledge-gaining
process, as many dysfunctional variants will have been generated, but not
used. Population genetics may be particularly inefficient, given the large
number of selective deaths necessary to acquire adaptive information (see
chapter 3 and 4).

How is information registered by evolving populations at level 1? One
problem is that there is no way that any evolving population can register
or store the totality of the information that it has previously gained at the
population level itself. There is no such thing as a population-level genome.
A population can register information that it has previously gained only by
distributing different samples of the information across the genomes of all
the individual organisms in each successive generation. This distribution of
information occurs when individual organisms in each generation inherit
the particular sample of genetic information from their ancestors, via their
parents, at the start of their lives. After that, individual organisms cannot
gain any additional genetic information at any later stage.[7]

If fit individual organisms survive and reproduce, they may pass on genetic information that has been reevaluated by natural selection to the next generation of their population. This involves a reversal of the flow of information between individual organisms and their populations. After initially receiving their sample of genetically encoded information, fit organisms return their samples of genetic information, after it has been reevaluated by natural selection along with some mutations, and typically with some mixing through recombination in sexual populations, to their populations via their offspring. If enough fit organisms in each generation of a population do this, then that allows successive generations to "communicate" (i.e., pass on) reevaluated and potentially updated information to each other, thereby allowing their populations to continue to evolve.

At level 2, where the knowledge-gaining systems are communicating with themselves at a later time, the biological mechanisms involved are easier to understand, at least conceptually. The capacity of individual organisms to demarcate between reoccurring cause-and-effect relationships in nature and everything else is largely a genetic, predetermined gift to individual organisms. This gift is provided by the prior evolution of their populations at level 1. For example, different species have evolved different senses. They have also evolved different abilities to perceive different kinds of cause-and-effect relationships in nature. These differences largely depend on what organisms in different species need to know about to survive and reproduce in their particular environments. However, level 1 evolution cannot predetermine precisely which cause-and-effect relationships will be encountered by individual organisms during their lives. For instance, genetic evolution has provided animals with a capacity to learn, but it hasn't specified exactly what each individual will learn: the specific idiosyncratic experiences of each individual will determine which cause-and-effect relations they detect. A probable reason for the evolution of the supplementary adaptive know-how, or R_i, gaining processes at level 2 is that they provide additional adaptive flexibility to individual organisms in evolving populations, such as by allowing them to respond to features of their environment that change on shorter than generational timescales. Hence, this adaptive flexibility goes beyond what can be conferred in advance by level 1 evolution.

I'll give one example to illustrate these points, taken from a process that we have met before and will shortly meet again—animal learning. Often, when animals, inclusive of ourselves, learn they are predisposed to notice and register events that occur closely together in space and time. These events can become associated in the individual's brain. For instance, in what psychologists call "associative learning," animals learn most rapidly about events in their environments that regularly reoccur in close spatial and temporal contiguity. Animals also learn best when the intervals in space and time between contiguous events remain constant, and less rapidly when those intervals are irregular. Animals can also learn that a pattern of events that they have previously learned has changed. They can unlearn what they have previously learned, through a process known as "extinction" (though here, it is knowledge rather than the species that is lost). The parameters of these learning processes have been studied by psychologists over many years, mainly by the artificial manipulation of stimuli in the environments of laboratory animals, such as rats and pigeons. The point that I want to make is that these same parameters of learning in the lab are relevant to how animals learn about cause-and-effect relationships that they encounter in the wild in their particular environments.

What about the other two fundamental requirements for knowledge gain at level 2? They comprise the registration of previously gained knowledge in memory systems and the subsequent communication of the retrieved knowledge from memories to individual organisms in populations. Here, I use the term "memory" broadly, to encompass any information accrued by knowledge-gaining systems at any level.[8] To register information in memory, individual organisms need retain only some trace of their past developmental experiences. For instance, when they learn, animals typically do this by registering knowledge that they have previously gained in their central nervous systems or brains. They may subsequently communicate some of the information that is registered in their brains to later or older versions of themselves by recalling the information from their brains and applying it to their current interactions with their environments. In animal learning at level 2, there is no need for any additional communication channel, apart from the knowledge-gaining system itself (i.e., the central nervous system in the case of learning, or the immune system in the case of adaptive immunity).

Now let's turn to the mechanisms that operate at level 3. How do socio-cultural groups satisfy the same three fundamental requirements that are common to all knowledge-gaining systems? I will try to answer this question first, relative to other animals, starting with some well-known examples. In the simplest case, a reoccurring cause-and-effect relationship may be discovered, possibly by chance, by a single individual in a group. Initially, the discovery will be registered only in the brain of that single animal, but knowledge of the same cause-and-effect relationship may spread, possibly rapidly, to other animals in the same group (e.g., if they copy the original innovator). The information will then be registered in their brains too. Copying is an elementary form of communication between animals in social groups.

There are many examples of this kind of phenomenon occurring among diverse animal species. A celebrated example is the invention of potato washing by an individual called Imo among a group of Japanese macaques, a habit that spread throughout the troop.[9] Another is varying techniques that different populations of chimpanzees use to fish for termites and ants in different regions of Africa, which are known to be cultural traditions.[10] Yet another example is provided by the habit of drinking milk from milk bottles exhibited by various species of birds.[11] The individual animals that may have initiated each of these behaviors are not often known—Imo was an exception. In other, more complicated cases, an innovation, based on a newly discovered cause-and-effect relationship, may be made by a group of animals rather than by a single individual in a social group. For example, during the acquisition of cooperative hunting skills by killer whales, the initial recognition of a cause-and-effect relationship during the hunt would have required more than a single innovating animal: it would have required a cooperative group, or team.[12]

Now let's turn to humans. The ways in which human beings in socio-cultural groups satisfy the same three fundamental requirements of knowledge gain are so diverse that it is not feasible to summarize them briefly. I will only try to point out some of the salient similarities and differences between knowledge-gaining human sociocultural groups and the other kinds of knowledge-gaining systems that we have been discussing. Up to a point, humans behave in similar ways to other social animals. We imitate or copy each other and tend to copy perceived successful or high-status members of our social groups. It's also possible that many human discoveries may be made by accident. I sometimes wonder how many of the world's favorite recipes were originally discovered by mistake. But at the other end of a spectrum of alternatives, humans have developed during the last few centuries the most advanced way of telling the difference between the reoccurrence of true cause-and-effect relationships in nature and everything else: science.

The human cultural practices of science are the most powerful knowledge-gaining processes that have ever evolved on Earth. Potentially, they are the most potent sources of adaptive know-how available to human sociocultural groups. For this reason, I will treat science as the paradigm example of knowledge-gaining processes in human sociocultural groups, while not forgetting that there are many other kinds of human knowledge-gaining processes too. We have already considered how scientists discriminate between true cause-and-effect relationships in nature, especially in our natural sciences (see chapter 4). We have also considered how scientific knowledge is registered and communicated today, not only among scientists, but potentially to everyone. In the first instance, scientific knowledge is usually registered in the brains of scientists. However, it is registered in innumerable kinds of human-constructed artifacts as well.

Both the registration and the communication of human knowledge, however, whether scientific or otherwise, are critically dependent on another human adaptation, language. Many other animals in social groups can communicate with each other by smells, sounds, and even gestures, and even via some of their own artifacts (e.g., via scent marks or pheromone trails).[13] But no other animal can communicate information with the same efficiency and precision that is afforded to us by language. Language is a unique human adaptation.

We don't know when language first evolved. But it is safe to assume that, at least until the last few thousand years, the primary way in which humans communicated with each other was by talking and maybe by gesturing, and perhaps even singing.[14] This would have restricted our ancestors to communicating to only a few people at a time, who were physically close to them. That changed with the invention of writing, approximately 6,000 years ago.[15] Writing then enabled the development of additional languages, such as notations in mathematics and musical scores, apparently quite quickly. I will not attempt to consider how writing affects communication between people in other domains, such as the arts and humanities. It clearly does.

Of greater immediate interest in the present context is that writing makes it possible for a single human being to communicate to many other people, whom they do not know and who may be far away in space and time. Writing even makes it possible for some individuals to go on communicating to other members of their sociocultural groups long after they are dead. We can still learn from Aristotle, and we can still read and go to plays by Shakespeare. Language and the technologies enabled by language and mathematics are largely responsible for how we register and communicate culturally acquired information today.

We can register all kinds of cultural knowledge, including scientific knowledge, in multiple artifacts. They range from private notepads, books, journals, and newspapers to magnetic tapes, musical scores, paintings, and photographs, and vast data sets stored in the digital memories of computers. They also include physical institutions, such as libraries, museums, art galleries, research laboratories, and university departments, all of which store and make available to large numbers of people culturally acquired scientific data and much else, including works of art. We have also invented other kinds of artifacts for broadcasting information to each other. They include radio, film, television, and recorded music. More recently, we have added the World Wide Web and social media. Equipped with all these artifacts, how do human sociocultural groups inform individual human beings at level 3, compared to the ways in which individual organisms are informed by different biological systems at levels 1 and 2?

First, I'll compare human sociocultural groups at level 3 with the primary knowledge-gaining process of population-genetic evolution at level 1. At both of these levels, individual organisms can only gain samples of the totality of the information accrued by the particular collectives to which they belong. In this respect, levels 1 and 3 are similar. But there is also at least one significant difference between them. As we saw before, the primary knowledge-gaining process of population-genetic evolution can inform individual organisms only once, via genes that they inherit from their parents at the start of their lives. Similarly, at level 1, individual organisms can contribute to the evolution of their populations only at the moments when they reproduce. In contrast, individual animals or humans in sociocultural groups can learn continuously from the sociocultural groups that they belong to throughout their lives. Potentially, individual animals or humans can also contribute information that they learn themselves back to their sociocultural groups at any time during their lives. This is the major difference between the knowledge-gaining processes at level 1 and those at level 3 in bootstrapping populations.

Now I'll compare levels 1 and 3 with level 2. In one respect, level 2 is more similar to level 3 than to level 1. Individual developing organisms can potentially gain additional knowledge at any moment during their lives by interacting with their local environments, unlike at level 1. But the totality of the information that can be gained by individual organisms at level 2 is far less than the totality of the information that is gained by their collectives at levels 1 and 3.

There is also one other interesting difference between evolving populations at level 1 and sociocultural groups, particularly those of humans, at level 3. Sociocultural processes introduce an additional concept of fitness, known as "cultural fitness."[16] Cultural fitness is sometimes contingent on biological fitness at both levels 1 and 2, but it is not identical to biological fitness and indeed may conflict with it. Fitness at levels 1 and 2 refers to the achievement by individual organisms of their fitness goals of survival and reproduction. It primarily refers to the rate of transmission of naturally selected genes to offspring in the next generation.

At the cultural level, however, cultural fitness means survival and transmission of culturally acquired know-how, or R_i. It refers to the differential propagation of diverse cultural traits and cultural legacies to cultural offspring in the next generation of a sociocultural group.[17] Cultural fitness also implies the prior selection of transmitted cultural assets by cultural selection processes. The cultural offspring of an individual member of a sociocultural group need not be identical to their biological offspring. They need not even be close genetic relatives, as implied by Hamilton's inclusive fitness theory.[18] They could be the biological offspring of other members of the same sociocultural group. They could even be members of different species.[19] There is also a difference between the natural selection pressures that operate at level 1 and the cultural selection pressures that operate at level 3, which can conflict, particularly in human sociocultural groups.[20] For the moment, I will ignore the effects of cultural selection on developing individual humans at level 2. Clearly, cultural influences must be present too (e.g., during the education of children in schools). I will only consider cultural evolution relative to level 3.

It is also of interest to compare natural selection at level 1 with cultural selection at level 3 because they differ in several respects. For instance, natural selection can favor deception, at both levels 1 and 2, such as when cuckoos produce eggs that mimic their hosts and dump them in the hosts' nests for the hosts to rear them. Natural selection is likely to do so if the deceiving organisms outcompete their more honest rivals by their dishonesty. But the natural selection pressures themselves necessarily stem from true reoccurring cause-and-effect relationships in nature. The natural selection pressures themselves cannot lie. The same is not always true of cultural selection pressures at level 3. Cultural selection pressures stem from a variety of sources, which may be honest or deceptive. For example, they may stem from political or religious ideologies or social or commercial vested interests in subgroups of people in a sociocultural group. It is certainly possible that cultural selection pressures could stem from deliberate lies, as well as deliberate misinformation, promulgated by dishonest members of a sociocultural group. When this happens, there is a danger that cultural selection processes might sometimes select for maladaptive traits in sociocultural groups relative to the natural selection pressures that operate on populations at level 1. Since all members of a sociocultural group are also members of their evolving populations at level 1, this sets up the possibility of additional antagonisms between natural selection at level 1 and cultural selection at level 3.

The different operating characteristics of these biological systems, combined with their different histories, means that it is almost inevitable that each biological system will "acquire" qualitatively different adaptive know-how, or R_i, for individual organisms in bootstrapping populations. The qualitatively different forms of R_i will then translate into different adaptations at each level.

Level 1 adaptations are phenotypic traits in individual organisms, informed by R_i previously gained by population-genetic processes. Level 2 adaptations are phenotypic traits in individual organisms that are informed by adaptive know-how gained by the developing organisms themselves as a result of their prior interactions with their local environments. Level 3 adaptations are phenotypic traits in individual organisms previously informed by adaptive know-how gained as a function of their membership of a socio-cultural group.

BETWEEN-LEVEL INTERACTIONS

These different levels of adaptations, informed by different forms of adaptive know-how, can also interact with each other in individual organisms. They may interact positively or negatively. Usually, they can be expected to interact positively, or synergistically. If there were not a net benefit to biological fitness, the supplementary knowledge-gaining process at levels 2 and 3 presumably would never have evolved in those populations. But negative or antagonistic interactions are also possible, as we've just noted.

Organisms can experience positive or negative interactions in at least two ways. They may exhibit different-level adaptations in different phenotypic traits in the same individual organism. For example, a case with negative interactions occurred when the Fore people of Papua New Guinea had a cultural tradition for funerary cannibalism, which led to the spread of the deadly neurodegenerative disease kuru.[21] Or they may express more than one level of adaptation and more than one kind of adaptive know-how, or R_i, in the same phenotypic trait in the same single organism. For example, rhesus monkeys' fear of snakes is socially learned, but individuals are prepared by prior genetic evolution to be fearful of snake-shaped objects.[22] It is seldom easy to distinguish these different levels of adaptations in the phenotypes of individual organisms. Many phenotypic traits expressed by individual organisms are a mixture of plural levels of adaptation, informed by more than one kind of knowledge-gaining process.

At present, the problem of dissecting the processes at the different adaptive levels, responsible for the etiology of specified phenotypic traits in individual organisms, is often called the "nature-nurture" problem. Nature is commonly loosely thought of as traits determined by genes as a consequence of population-genetic evolution, and nurture as traits determined by environmental factors. Applied to individual traits, this simple dichotomy is highly misleading, though, since *all* traits are shaped by both nature and nurture.[23] However, variations among individuals can be attributed to different sources: for instance, the differences between individuals can be due to differences in their genes, their individual experiences, their culture, or a combination. To understand adaptations and phenotypic traits in organisms, I suggest that what we should be trying to do is to dissect the influences of each of the kinds of adaptive-know-how-gaining processes that are informing variations in the adaptations of organisms. That means dissecting the influence of genes and population-genetic evolution at level 1, the influence of supplementary developmental knowledge-gaining processes at level 2, and in some animals, especially ourselves, of the supplementary cultural knowledge–gaining processes at level 3.

Where the interactions between these levels of adaptation in organisms are positive or cooperative, there can be interest among biologists in dissecting these levels of adaptation. For example, there is extensive current interest in the topic of plasticity-led evolution. Conversely, where they become antagonistic or pathologically negative, these interactions may also become intensely interesting, such as to medical veterinary scientists (e.g., where intensive farming led to bovine spongiform encephalopathy) and to plant scientists and ecologists (e.g., where pesticides are overused). Then the relevant question will not be "Is the pathology due to nature or nurture?" but rather "Which knowledge-gaining process at which level is maladaptive?"

In this connection, phenotypic traits in organisms due to the level 1 process of population-genetic evolution may be well or badly adapted, relative to the organism's fitness goals of survival and reproduction. The same could be true of cultural phenotypic traits, relative to cultural selection pressures and the cultural fitness goals of organisms at level 3. However, even where level 1 and level 3 adaptations are well suited to their respective fitness goals, it is still possible that in conjunction with each other, they may not be well adapted, as they may not be compatible. In combination, they could be maladaptive for individual organisms and detrimental for their populations.

For example, a level 1 adaptation may precede a cultural adaptation at level 3, possibly by millions of years. Many generations later, a level 3 cultural phenotypic trait may take the form of cultural niche construction. The cultural niche-constructing activities of organisms in a population may then change the natural selection pressures that previously favored their level 1 adaptations. The changed natural selection pressures may then render the level 1 adaptation maladaptive, in an environment that has been changed by cultural niche construction. The changes may override natural selection pressures at level 1. For instance, by constructing environments replete in sugar and fat, humans have created a context in which adaptations for sugar consumption and fat storage, which evolved in environments where sugar and fat are comparatively rare, are commonly no longer adaptive, and lead to disease. However, there is one sense in which, in all organisms (including humans), cultural selection pressures at level 3 are subordinate to and contingent on biological fitness goals at level 1. Organisms have to survive and reproduce at level 1, before they can invest in any kinds of adaptations at levels 2 or 3. It is possible that our species is more prone than most other species to causing negative interactions between these levels of adaptations, including for other species. For instance, human cultural niche construction is largely responsible for the Anthropocene, as well as the actual or potential extinction of countless species who struggle to adapt to a human-modified world. Without wishing to be alarmist, in the longer term, humans may be just as vulnerable to the changes in selection pressures that our cultural activities are generating, and may themselves go extinct. The main reasons why these observations may be true are the potency of our human cultural niche-constructing activities in pursuit of our cultural fitness goals, in addition to our biological fitness goals. I will return to this hypothetical scenario later.

THE LIMITATIONS OF STANDARD EVOLUTIONARY
THEORY REVISITED

We need to update the old nature-nurture arguments along the lines that I've just outlined. There is a key question: Is it possible for any phenotypic traits, acquired by individual organisms during their lifetimes, to influence the subsequent evolution of their populations at level 1? SET clearly recognizes that many developmental knowledge–gaining processes, such as animal learning at level 2, are products of the primary evolutionary process of population-genetic evolution at level 1. SET also recognizes that human sociocultural processes are products of both the primary processes of population genetics, acting on our human ancestors at level 1 and developmental processes in individual organisms at level 2. Nevertheless, SET still falls short when it comes to fully acknowledging the implications of these supplementary knowledge-gaining processes at levels 2 and 3 for the primary adaptive-know-how- or R_i-gaining process at level 1. SET bypasses these implications in two ways. First, it assumes that any characteristics acquired by individual organisms during their lives cannot influence the genetic evolution of their populations at level 1, as, human culture aside, those characteristics are assumed to be lost when they die, and in any case, they are viewed as being unable to influence inherited genes. Such traits are assumed not to leave any traces that can be genetically inherited by their descendants.

These assumptions have ruled out any role for Lamarckism in evolution. Recall the textbook comparison of the theories of Jean-Baptiste Lamarck and Darwin. For Lamarck, giraffes acquired their long necks by stretching out their necks when trying to reach additional sources of food at the top of trees, for generation after generation. No, said the Darwinians. Giraffes got their long necks because in each successive generation, those individual giraffes that happened to have the longest necks were the fittest relative to natural selection pressures in their environments. They therefore contributed more of their genes to the next generation than their less-fit, shorter-necked competitors. Their genes would have included some that translated into longer necks. Initially, Darwinians won their argument. Darwin provided a clear and testable mechanism to explain how evolution works; Lamarck did not.

A second, closely associated reason for the success of Darwinism over Lamarckism was an experiment carried out by August Weismann. It led to widespread acceptance of what is often referred to as the "Weismann Barrier." Weismann's experiment appeared to demonstrate that phenotypic traits acquired by individual animals during their lives could not be passed to their offspring. Weismann cut off the tails of individual mice in successive generations and found that after many generations of doing so, the offspring of his "tailless" mice still had tails. Subsequently, Ernst Mayr (1961) claimed that Weismann's experiment was one of the most important ever done in biology because it separated out what Mayr called, "How questions" from "Why questions" in biology. "How questions" refer to mechanistic or developmental issues—they ask how things work. "Why questions" refer to evolutionary questions—why traits evolved. Mayr's paper was influential for many decades among evolutionary biologists, but it is increasingly seen as problematic.[24] Weismann's barrier, combined with Mayr's dichotomy, appeared to suggest that developmental experiences, which include level 2 knowledge-gaining processes, were irrelevant to evolution. To be fair to Weismann, it is doubtful if he ever gave as much weight to his own experiment as Mayr did later. In the light of contemporary molecular biology, as well as our contemporary understanding of inheritance systems, Weismann's experiment now appears rather unconvincing. It is now well established that a wide variety of developmental resources, including hormones, symbionts, epigenetic marks, antibodies, and learned knowledge, are passed from one generation to the next (see chapter 9).[25]

There is another reason why SET downplays the influence of the supplementary knowledge-gaining processes at levels 2 and 3 on the primary process of population-genetic evolution at level 1. SET assumes that these supplementary processes are largely, if not wholly, determined by the naturally selected genes that individual organisms inherit from their ancestors. This is where NCT, extended evolutionary synthesis, and several other contemporary approaches to evolution part company with SET.[26] It has also led to two of the most controversial areas in evolutionary biology. One is the relationship between evolutionary biology at level 1 and developmental biology at level 2.[27] The other is the controversial relationship between evolutionary biology at level 1 and human sociocultural processes at level 3.[28]

TEGMARK REVISITED

The more comprehensive theory of evolution that I propose here may be of greater interest to Tegmark and his colleagues at the FLI. More to the point, it should make an understanding of evolution of greater interest to anyone interested in the future of life on Earth. This more comprehensive theory implies that Tegmark's future Life 3.0 would not just be a product of level 1, population-genetic evolutionary processes, but also of all knowledge-gaining processes operating at all three levels. In particular, evolution would depend on human sociocultural knowledge–gaining processes at level 3. There would be a continuum between Life 1.0 and Life 3.0 in Tegmark's scheme. This approach also implies that Life 3.0 would be another product of the capacity of evolution to bootstrap, relative to yet another kind of life (in this case, artificial, abiotic life based on robots and AGI). It would be a new level of evolution.

In what circumstances should we expect the primary evolutionary process to bootstrap, and when should we expect it not to do so? We need to go back to the fundamental problems of life again. All living systems are very improbable systems that are far from equilibrium relative to their environments. Minimally, all organisms have to resist the second law of thermodynamics by active, purposeful physical work. But organisms cannot do this work without being informed by adaptive know-how, or R_i, gained by one or more process in evolution.

These thermodynamic requirements take us to the R_p versus R_i relationship again. Organisms have to invest energy and matter (R_p) resources to pay for whatever adaptive know-how (R_i) they need from all the knowledge-gaining processes that are informing them. In the light of this mutual dependency of energy and matter (R_p) resources and information (R_i) resources, it is possible to come up with some tentative rules about when an evolving population should bootstrap, as opposed to when it should not. If individual organisms gain R_i resources from any supplementary knowledge-gaining process at either level 2 or level 3, then the organisms must be able to gain more R_p from their environments via these supplementary processes than the R_p costs of those processes. There must be an R_p profit to make it worthwhile for their populations to bootstrap by investing in the supplementary processes.

If organisms satisfy this criterion, then their populations should bootstrap—that is, level 2 or level 3 knowledge-gaining processes will be adaptive and should evolve through natural selection. Conversely, if the energy and matter (R_p) costs of the supplementary knowledge-gaining processes at level 2 or level 3 exceed the value of the extra R_p resources that are gained by individual organisms via these processes, then their populations should not bootstrap. When calculating these R_p costs and benefits, the costs should include those of any between-level conflicts that may arise.

HOW TO AVOID BOOTSTRAPPING AT LEVEL 1

One way in which the primary process of population-genetic evolution can avoid bootstrapping is by improving the thermodynamic efficiency of each of the three subprocesses that are responsible for evolution by natural selection at level 1. Recall that these subprocesses comprise the generation of variety, the generation of fitness differences in the interactive organism-environment (O-E) niche relationship, and the inheritance of fit variants in successive generations of populations.[29] It now appears possible that each of these subprocesses is amenable to becoming more thermodynamically efficient as a consequence of natural selection.

Until recently, this was not thought to be true of the first subprocess of adaptive genetic evolution, the generation of variety in populations. It was assumed by modern synthesis, and also by neo-Darwinism, to depend on de novo random mutations or sheer chance. It now appears that natural selection can (at least sometimes) bias the rates and directions of mutation in populations. For example, in the region of the genomes of organisms where we find genes that translate into particularly important proteins, the genes appear to be intolerant of mutation.[30] In other regions of the genomes of organisms, however, the rates of mutations appear to speed up rather than slow. Increased mutations have the effect of increasing the genetic and phenotypic variance in organisms. Even when most of the extra mutations are deleterious, the implication of both these points is that natural selection must have some capacity to bias the rates of mutations in different regions of an organism's genomes, even though we still don't completely understand how it does so.[31]

The second subprocess of adaptive evolution is the one responsible for the generation of fitness differences. How might natural selection bias itself? One way is by selecting for niche-constructing activities in organisms that modify their own natural selection pressures. We have also already met one of the principal ways by which natural selection may bias inheritance. It can do so by favoring sexual reproduction over asexual reproduction (see chapter 4). There are limits to how much genetic evolution at level 1 can increase the thermodynamic efficiency of itself by responding to natural selection. Once a population has reached its limits in this respect, it should be expected to either go extinct or to bootstrap.

THE RECURSIVE ALGORITHM

There remains one more issue that we have not yet discussed. We have seen how each of the biological systems at levels 1, 2, and 3 can inform individual organisms in populations with adaptive know-how, or R_p, that they have previously gained from their past histories. But we have not yet discussed how each of these knowledge-gaining processes can gain new information. We know how to answer this question at level 1 relative to natural selection. The knowledge-gaining process works via the three subprocesses of adaptive evolution that we've just been discussing: generation of variety, generation of fitness differences in the interactive O-E niche relationship, and inheritance.

These genetic inheritances, bequeathed by fit organisms to their offspring, always include some novel genetic mutations, as well as some novel genetic recombinations in sexually reproducing organisms. Many such genetic novelties will never have been subject to natural selection. But there is always at least a slim chance that these novel genetic mutations and recombinations in offspring may express phenotypes that are fortuitously adaptive relative to novel natural selection pressures in their environments. In this way, the evolutionary algorithm at level 1 is able to acquire novel adaptive know-how, or R_i.

I now want to explore the hypothesis that this evolutionary algorithm at level 1 is recursive at levels 2 and 3. The same three subprocesses of adaptive evolution that operate at level 1, reoccur at levels 2 and 3. However, the recursive evolutionary algorithm uses different biological mechanisms to inform organisms at these two supplementary levels. Analogous mechanisms, corresponding to these same three subprocesses, support the recursions of this algorithm at levels 2 and 3. These analogous mechanisms are likely to be products of the primary process of population-genetic evolution at level 1. This is not a new hypothesis. An early advocate was Donald Campbell,[32] while more recent advocates include David Hull, Daniel Dennett, John Gerhart and Mark Kirschner, and others. The psychologist Henry Plotkin and I also put forward a version of this idea, as far back as 1979.[33] Yet, currently, the recursive nature of this algorithm is not widely recognized and is still controversial.

In spite of this recursive algorithm depending on different mechanisms at each level, the logic of how these mechanisms conserve adaptive know-how from the past and gain new information about the present is the same at every level. We discussed this logic, relative to level 1, in chapter 4. Information is deduced by individual organisms in the current generation of a population, from whatever information it has been supplied with, about what was adaptive for organisms in the same population in the past. Individual organisms then use this adaptive know-how, deduced from the past, in combination with novel variance due to mutations and sexual recombinations, to make inductive gambles about what may be adaptive for them again in their "unknown but not wholly unforecastable futures."[34] This recursive algorithm might be described as trial and error, with selective retention and reuse of what works.

SOME EXAMPLES

I now want to support this claim that the evolutionary algorithm at level 1 is recursive at levels 2 and 3, with some examples. I'll start with the vertebrate immune system and animal learning at level 2. I'll postpone talking about epigenetic inheritance until chapter 9.

The vertebrate immune system is not often thought of in terms of a recursive supplementary evolutionary process, although sometimes components of this recursion have been recognized before (e.g., by Macfarland Burnet).[35] The logic of how the vertebrate adaptive immune system works suggests that it is a fully recursive, supplementary evolutionary process. It combines the generation of variants with selection and the retention of selected variants. If the preceding argument is correct, the vertebrate adaptive immune system qualifies as a supplementary evolutionary process at level 2, even though it works only within the lifetimes of individual organisms.

In recent decades, the immune system has attracted a great deal of research. This is for several reasons. One has been human immunodeficiency virus (HIV) infections, which have the capacity to overwhelm human immune systems. Another has been a consequence of organ transplant technology, which tackled the problem of how to counter the natural tendency of the human immune system to reject foreign transplanted organs. A third has been the application of some recent advances in molecular biology, some of which have been highly relevant to immunology. I have no intention of reviewing this research here, nor do I want to describe how the immune system works in any detail. All I want to do is to describe some of the principal mechanisms whereby the vertebrate adaptive immune system works. First, I will indicate how the adaptive immune system brings together the three subprocesses of adaptive evolution—namely, the generation of variety, the selection of variety, and the transmission of selection of that variety into the future, thereby qualifying it as a supplementary evolutionary process. Second, I'll also indicate why this supplementary evolutionary process apparently evolved in support of the primary process of population-genetic evolution.

The vertebrate immune system has two components. The first is an innate immune system, which is shared with invertebrates. It is a relatively crude immune system. The second is the adaptive immune system, which is a more sophisticated system that occurs only in vertebrates. It works almost exclusively on a within-individual and within-lifetime basis, although in mammals, it is possible for mothers to transmit some immunity, in the form of antibodies, to their newborn offspring via the uterus, the birth canal, or lactation. In contrast, the innate immune system works primarily on a population basis between generations. These two processes interact. For instance, in vertebrates, the innate immune system calls the adaptive immune system into play when it is needed. Here, I will concentrate only on the adaptive immune system. That is because it is more clearly differentiated from the primary process of population-genetic evolution than the innate immune system. It is also a better example of a completely recursive supplementary evolutionary process.

The subprocesses that generate variance in the adaptive immune system are triggered by different antigens. Antigens come in a number of classes, but for my purposes, everything can be regarded as a pathogen. Pathogens can infect normal cells and include nonsymbiotic viruses, bacteria, fungi, and parasites. Antigens vary based on the proteins with which they bind. Antigen receptors that bind with infected cells or pathogens trigger the production of antibodies by B-cells and T-cells in the immune system. B-cells and T-cells do a similar job, but via slightly different mechanisms relative to different targets. I'll focus on the B-cells here.

Collectively, the B-cells in any individual organism have the potential to generate vast numbers of antibodies, which are a further source of variance in the immune system. However, a crucial restriction is that any single B-cell can trigger the production of only a single antibody in the immune system— namely, the antibody that can destroy the particular pathogen presented by a particular antigen-presenting cell. Once triggered in this way, the B-cell then causes the production and proliferation of the specific antibody that can counteract the specific pathogen that triggered it. The proliferation involves clonal selection, leading to the production of huge numbers of the appropriate antibody. That particular antibody is then carried around the organism's body by the bloodstream. When the antibody encounters any cell that is infected by that particular pathogen, it either kills the cell or prevents it from replicating, thereby halting the spread of the infection in the organism.

In addition to this projection of antibodies into the immediate future of an organism, the adaptive immune system has the capacity to remember episodes of prior infection for varying periods of time during the future lifetime of the organism. These memories are highly adaptive because they can protect an organism from most reoccurring infections. If an infection reoccurs during the life of an organism, it can be dealt with far more rapidly by a retained or "remembered" antibody. Sometimes this kind of memory of past infections can grant lifetime immunity. For example, in humans, recovery from measles ensures subsequent lifelong immunity from future measles infections. The same is not true of chicken pox, which in humans grants variable periods of immunity from subsequent chicken pox infections.

These examples of how subprocesses of evolution, comprising the generation of variety, the selection of variety, and the projection of selected variety into the future, are not exhaustive. However, they should suffice to indicate that a supplementary evolutionary process really is at work in the adaptive immune system of vertebrates.

The second question raised by the adaptive immune system is why it should have evolved in support of the primary population genetic process of evolution, as well as in support of the innate immune system in vertebrates. This question is not difficult to answer, at least at the hypothetical level. First, any multicellular organism that lives for any significant period of time is liable to encounter huge numbers of pathogens during its life. These numbers are so great that it would not be possible for an unaided primary population genetic process to cope with the innumerable selection pressures imposed by innumerable pathogens by conventional evolutionary responses at level 1. A back-of-the-envelope calculation suggests that to do so, the genomes of vertebrates would have to contain millions of genes. In fact, they typically contain less than 25,000, and the human genome contains only 20,000. There is a severe mismatch between these numbers. The implication is that the primary genetic process was forced to evolve the supplementary evolutionary processes of the adaptive immune system to cope with the sheer diversity of the pathogens that any long-lived organism is likely to encounter during its lifetime.

There is a second, closely related point too. Many pathogens encountered by long-lived organisms, such as viruses or bacteria, may not even have existed when the organisms were born. They may have evolved only much later during the life of any long-lived organism. Since the primary population-genetic processes of evolution cannot foresee any future selection pressure ahead of time, including any novel pathogen that does not even exist yet, it cannot supply long-lived organisms with defensive adaptations in advance. It therefore must evolve fast-acting supplementary evolutionary responses (in this case, the adaptive immune system) to enable long-lived organisms to survive relative to unpredictable pathogens.[36]

LEARNING IN INDIVIDUAL ANIMALS

My second example of a supplementary evolutionary process is learning in individual animals, which is another level 2 process. Once again, I won't attempt to describe how learning works in detail, but instead draw attention to the nature of the principal biological mechanisms that operate. I'll also briefly consider why learning evolved in the first place.

Learning qualifies as a supplementary knowledge-gaining evolutionary process because it involves a recursion of the same three subprocesses of population-genetic evolution: generating variety, selecting between variants, and projecting selected variants into the future. The biological mechanisms via which learning work are all brain based. However, because the brains of animals are themselves species-specific products of their prior evolution, as well as of intermediate developmental processes, all the mechanisms of learning are in a nested relationship with the underlying primary process of population-genetic evolution. They may also be in a nested relationship with underlying developmental processes.

The relevant basic subprocesses of learning are as follows. The first subprocess for generating variance involved repertoires of variant behaviors, both covert and overt. Animals inherit these repertoires as a function of both the prior evolutionary history of their populations and their own prior individual developmental histories, including their own prior learning.

The second subprocess of selection is derived from the capacity of animals to respond to what psychologists call "reinforcement." Reinforcing events are positive (i.e., rewarding) or negative '(i.e., punishing) events that occur in the environment of the animal. The capacity of learners to respond to such events as if they were rewarding or punishing ultimately relates to the biological fitness goals of organisms. These fitness goals are derived from the primary processes of population-genetic evolution and from prior developmental processes.

The third subprocess of learning, whereby selected variants are projected into the futures of individual animals, are memories. The relevant brain-based memory mechanisms record which events were associated with rewards and which events were associated with punishments in the past. Animals may also remember which events in their environments were neither rewarded nor punished in the past, and therefore were irrelevant to their fitness goals; and also which events occurred together, regardless of whether or not they were subject to reinforcement.

In general, a new episode of learning in individual animals is triggered by what psychologists who study learning frequently call "surprise." Surprise is generated when an animal encounters either an unfamiliar environment or, more often, a novel stimulus in a familiar environment. The novelty apparently generates uncertainty in the animal's brain. The animal seems to be threatened by the unpredictability of the novel environment or novel stimulus, as well as by not knowing the likely outcomes of its own variant behaviors in that environment. Surprise motivates it to acquire more adaptive know-how, or R_p, to help it to achieve one of its fitness goals. It does this by learning.

The three most basic kinds of learning in animals, most frequently studied by psychologists and to a lesser extent by behavioral ecologists, are habituation learning, Pavlovian conditioning, and instrumental conditioning. All these kinds of learning also involve sensory and perceptual mechanisms, attentional mechanisms, and nascent cognitive mechanisms. But I won't discuss any of these extra mechanisms here, nor will I discuss any further the learning about what is irrelevant in the animal's environment or habituation. Instead, I'll concentrate on one of the most familiar learning paradigms: instrumental conditioning. That should suffice to illustrate the recursive evolutionary logic and the mechanisms of learning in individual animals.

Instrumental learning concerns the learning by an animal about the consequences of its own overt behaviors in specific environments. An animal learns about what happens if it does behavior X in a particular environment Y. This kind of learning was originally investigated in the lab of an American psychologist called Edward Thorndike. It was later further investigated by another more famous American psychologist, B. F. Skinner. Skinner developed an apparatus known as a "Skinner box," comprising a restricted environment in which he introduced an animal, typically a hungry rat. The rat's hunger motivated it to explore its environment in a search for food. During its searches, the rat typically emitted a variety of alternative overt behaviors until, partly by chance, it would stumble against a protruding bar on one of the walls of its box. When pressed, the bar would cause a food reward, usually a food pellet, to appear in a tray for the rat to eat. The still-hungry rat would then be motivated to learn which of its behaviors had caused the arrival of the food pellet. It would repeat many of its previous diverse behaviors until it stumbled on the successful behavior again. After that, most rats would zero in on the successful behavior pretty rapidly. Eventually, they could feed themselves by repeatedly pressing the bar in their Skinner boxes.

In later tests, rats demonstrated what they had learned earlier by immediately repeating the same successful behavior when they were reintroduced into the same environment. Also, if the experimenter ceased to reward bar-pressing with food, then the animal's learned response would eventually cease. It would be extinguished. By stopping its behavior in this way, a rat would demonstrate the sensitivity of learning in individual animals to rapid environmental change. Psychologists subsequently explored both the instrumental and other learning paradigms in depth, and by doing so, they learned much more about animal learning. But what I have described should be enough to indicate the recursive evolutionary logic of the supplementary evolutionary process of learning, as well as its associated brain mechanisms.

The terminology used by psychologists to describe their conditioning experiments is forbidding for most people. I suggest it might have been better to refer to all kinds of learning as "causal learning" rather than as "associationist learning" or "conditioning."[37] More recently, psychologists have also concentrated far more on human cognition and learning at the cognitive level than on animal learning. But I've used animal learning in this discussion because it shows more clearly how the recursive evolutionary algorithm operates at level 2. Behavioral ecologists have been more interested in making connections between learning and evolutionary biology. They may eventually correct this inbalance, but probably not until evolutionary theory has been extended. At present, behavioral ecologists are usually held back by their adherence to the assumptions of SET. This causes many behavioral ecologists to follow Mayr's distinction between ultimate and proximate processes and how and why questions in biology.[38]

The reason why a capacity for learning evolved in animals is probably very similar to why immune systems evolved. It is not possible, particularly in any long-lived organism, for the primary population-genetic processes to anticipate in advance all the novel circumstances and stimuli that the organism may encounter during its life. Instead, the primary process of population-genetic evolution appears to have bootstrapped by evolving the supplementary evolutionary process of learning at level 2 in response to rapidly changing environments. I suggest that the fundamental evolutionary role of learning is to enable animals to achieve their biological fitness goals of survival and reproduction in rapidly changing, unpredictable environments. It does so by allowing animals to gain additional adaptive know-how, or R_p, for themselves rapidly during their lifetimes. Learning allows individual animals to do this as a function of their interactions with their own particular local environments.

It is also worth pointing out that animals, unlike plants, move around in their environments. They can relocate. Relocation is itself a form of active niche construction. Other kinds of niche construction by individual animals, such as innovative, perturbational niche construction, may also increase the amount of novelty that animals are liable to encounter in their environments during their lives. In turn, that should increase the demand for animals to learn more about the novelties that they themselves, as well as other organisms, introduce into their environments. In effect, niche-constructing animals may, by their own activities, modify natural selection pressures in their environments in favor of the selection of the supplementary process of individual learning.

SOCIAL LEARNING IN ANIMALS

My next example of a supplementary process stems directly from the previous one. Given that individual animals can learn, it is scarcely surprising that animals can learn from each other. There is considerable overlap between the three subprocesses of individual learning and the three subprocesses of social learning.[39] However, there are significant differences too. In social learning, variance is group-generated rather than being generated by individual learners.

The selection of that variance by reinforcement—that is, by rewarding or punishing events—may be determined by social fitness goals rather than by (or as well as by) biological fitness goals. For instance, it may be more rewarding for an individual animal in a social group to invest in behaviors that improve its status in its social group, or to conform to the norms of group behavior relative to its social fitness goals, than to invest in behaviors that advance its biological fitness, at least temporarily. Retained, reinforced behaviors will still be remembered in the brains of individual animals in a social group, but they may be imitated or copied by many other members of the social group, including juveniles in the next generation, as well as individuals outside the social group, or even in different species.

In humans, the unique adaptation of language, and later writing, affects all three of the subprocesses operating at level 3. I therefore need to say a bit more about each of these subprocesses in the special case of the gaining of human cultural information. The generation of variety by human cultural groups is greatly enhanced by the much more efficient communication provided by languages. I use the term "language" broadly in this discussion to include mathematics and music. Selecting these variants by rewarding or punishing events is often determined by reinforcing events that are derived from cultural fitness goals, either as well as or instead of biological fitness goals, at least in the short run.

Language, whether oral or written, also makes human cultural communication and inheritance far more potent, both within and between human generations. This potency is recognized by the theory of human cultural evolution and gene-culture coevolution.[40] Gene-culture coevolution recognizes that human cultural inheritances at level 3 ultimately rely on human genetic evolution at level 1. However, human cultural inheritances often appear to be determined at least as much by cultural fitness goals as by biological fitness goals. Because of the interactions between human genetic evolution at level 1 and human cultural knowledge–gaining processes at level 3, the adaptations will be informed by qualitatively different adaptive know-how, or R_I, at these two levels. Their interactions are likely to be a salient feature of human evolution, whether they are positive or negative.[41]

Positive interactions between level 1 and level 3 should enhance both their biological and cultural fitness, but negative interactions are also possible. If the adaptive know-how that informs human genetic evolution at level 1 and human cultural evolution at level 3 are incompatible, or just badly integrated with each other, then human cultural inheritances could degrade, rather than enhance, human biological fitness. They could degrade human cultural fitness too. I will illustrate these two alternative kinds of interactions with the same example. It is the human cultural niche-constructing activity of dairy farming among pastoralist communities.

Dairy farming depends on adaptive know-how gained by human socio-cultural processes at level 3. It appears to confer the advantage of providing pastoralists with additional food resources in the form of milk or processed dairy products such as cheese. However, originally there must have been a downside too. Like all mammals, human infants can digest their mother's milk because they are lactose tolerant, but when the cultural practice of dairy farming was first introduced in the Neolithic era, human adults were not lactose tolerant. They probably could not digest milk, at least not efficiently. Probably it made them sick. Subsequently, and apparently in response to natural selection pressures modified by the cultural niche-constructing activity of dairy farming at level 3, although possibly only at times of famine or disease,[42] pastoralists evolved lactose tolerance at level 1 in human adults, as well as in human infants. A genetic change occurred in chromosome 2 in the human genome, which enabled lactose tolerance to persist in adult humans.[43]

Today, those of us who had pastoralist ancestors, such as most northern Europeans, are usually lactose tolerant in adulthood, as well as in infancy. Others, such as many people in East Asia or Africa who apparently never had pastoralist ancestors, continue to be lactose intolerant in adulthood. It follows that originally, dairy farming probably introduced a maladaptation, at least during periods of stress, as a consequence of both negative and positive interactions between level 1 and level 3 knowledge-gaining processes. Only after the subsequent evolution of lactose tolerance in adults would dairy farming have become an unalloyed, consistent adaptation.

There must have been a time lag before this happened, and this is sometimes called an "adaptive lag." Even in pastoralists, it must have taken some time before level 1 population-genetic evolution in humans in the relevant population could have caught up with the new adaptive problem set for them by their own level 3 dairy farming activities. During this adaptive lag, the positive benefits of cultural adaptations of human pastoralists (i.e., consuming milk) would have been at least partly offset by the negative ramifications of the genetic maladaptation, comprising the inability of human adults to digest milk properly.

Thus there are risks, as well as advantages, in evolving supplementary adaptive know-how- or R_i-gaining processes by bootstrapping. These risks are not confined to dairy farming; rather, they are more general. They are always liable to occur, particularly when human cultural niche-constructing activities at level 3 modify natural selection pressures relative to human population-genetic evolution at level 1. Some time ago, my colleagues and I reviewed many other putative examples of human gene-culture coevolution in a paper in *Nature Review Genetics* entitled, "How Culture Shaped the Human Genome: Bringing Genetics and the Human Sciences Together."[44]

For now, this is all I want to say about the recursive algorithm at levels 1, 2, and 3. Because of the complexity of human cultural knowledge–gaining processes at level 3, it may be quite hard to detect the same three subprocesses of generating variety, selecting that variety and the retention of fit variance that occur at level 1 and reoccur in the supplementary processes at level 3, particularly in the case of humans. But I hope that I have done enough to show that the same recursive algorithm and its associated knowledge-gaining logic really does reoccur in human cultural evolution at level 3, as well at levels 1 and 2. I'll return to these negative and positive possible interactions between level 2 and level 3 adaptations in chapter 10.

Paradoxically, it may be easiest to detect the logic of this knowledge-gaining algorithm in science. We have already seen that in science, carefully collected data generate new ideas, concepts, conjectures, and finally testable hypotheses about cause-and-effect relationships in nature. Hypotheses are then tested empirically. Hypotheses that survive this selective testing may then give rise to major theories about nature and about our place in nature.

Many philosophers of science have described one or more of these subprocesses of science. For instance, Karl Popper, David Hull, and Daniel Dennett have described how science operates through a Darwinian algorithm of generating hypotheses, subjecting them to empirical tests and retaining the ones that have not been falsified.[45] The same logic of recursion observed in adaptive population-genetic evolution at level 1, is also clearly manifest at levels 2 and 3, including in the advance of science and technology.

THE RELEVANCE OF BOOTSTRAPPING
TO TEGMARK'S LIFE 3.0

Rather than asking questions about a possibly predetermined destiny for evolution, a different and more tractable question is to ask how, in the light of this new understanding of evolution proposed here, might it be possible to say anything about what evolution might do next, at least for our own species? One obvious possibility is that a species that has already evolved supplementary processes by bootstrapping might bootstrap again. This, after all, is what Tegmark's Life 3.0 amounts to. But his speculations led to the idea of a possible abiotic future kind of life.

Need that be the only possibility, or could humanity bootstrap by inventing one or more new kinds of biotic life? Here, a recent book by David Goldstein is suggestive, as well as alarming. Called *The End of Genetics*,[46] it warns us about the implications of one imminent and biotic future based on current genome editing. Goldstein describes a future in which it will be possible for human parents to start designing their own babies. These parents will need a lot of help from geneticists, medical specialists, and molecular biologists, especially experts in genome editing and CRISPR9-Cas technology.

A benign aim of this technology is to reduce the burden of genetic diseases on humanity. Few would quarrel with that goal. But Goldstein warns that there will be unpredictable consequences when it comes to designing babies. Benign eugenics, designed to eliminate genetic diseases, are possible. But unfortunately, malign eugenics, such as those advocated by the Nazis, are also possible. Also, we won't just want to control human reproduction and genetic inheritances. We will also want to control reproduction in many other species, with unknown consequences for ecosystems and the biosphere. Nevertheless, we are clearly on the threshold of a future in which design and control of reproduction in our own and in other species is fast approaching. Might that represent another bootstrapping event in the evolution of life on Earth?

LIFE 3.0 REVISITED

I would like to close this chapter by returning to Tegmark's Life 3.0 once again. Were Life 3.0 to happen, it would constitute another major transition in evolution, comparable only to the transition from abiota to biota at the origin of life. It would now be a transition from biotic life to some future abiotic life. However, it would still have to take the form of another supplementary evolutionary process, once again in a nested relationship with all the past underlying processes of evolution, including population-genetic evolution, even if those living forms of evolution no longer exist in the future. As Tegmark himself indicated, it would be based on another recursion of the fundamental evolutionary algorithm. But this time, all three of the subprocesses of the basic evolutionary algorithm would depend on abiotic mechanisms operating at the level of a novel-independent abiotic platform.

In the light of the more comprehensive theory of evolution proposed here, as opposed to SET, does a forthcoming Life 3.0 seem plausible? Does it seem more or less likely than it did before? Also, is there anything that a more comprehensive theory of evolution could teach us about how a putative life 3.0 might work? In this respect, is there anything that the advocates or prophets of Life 3.0 could learn from a more comprehensive theory of evolution that they have not already learned from SET?

The first point to acknowledge is that the very idea of abiotic life sounds like an oxymoron to cap all oxymorons. For anyone who still believes that humans at least owe their existence to supernatural processes, governed by one or more supernatural, creative deities, the concept of abiotic life must seem totally implausible. From that starting point, it may be difficult or impossible to consider or even think straight about Life 3.0. But given the bootstrapping capacity of the fundamental evolutionary algorithm discussed here, a further bootstrapping step, comprising the origin of Life 3.0 can at least be considered. It might happen along the lines sketched by Tegmark. It is certainly not totally implausible from the point of view of a more extended theory of evolution. In this context, how might we expect Life 3.0 to work? Alternatively, might we discover a fundamental obstacle that could prevent 3.0 from ever happening? Let's consider each of these questions in turn.

First, if there is an abiotic recursion of the evolutionary algorithm in Life 3.0, then this novel abiotic evolutionary process would still have to work in much the same way as all the other supplementary processes of evolution. It would still rely on each of the same three subprocesses of evolution again—namely, generation of variety, selection of variety, and the projection of selected variants into the future. The great change, however, would be that this time, each of these subprocesses would be based on abiotic rather than biotic mechanisms. Also, the way in which these subprocesses interact might be fundamentally different than in SET's characterization of biological evolution. For instance, each subprocess might do more than simply flow into each other; it might interfere with the others' inner workings to render them interdependent, just as niche construction does through contributing to the generation of phenotypic variety, the modification of fitness differences, and, through ecological inheritance, the projection of those variants into the future.[47]

We have already seen that one of these subprocesses—the projection of selected variants into the future—is already present in the form of computer-based memory systems. Selected variants stored in computers are projected into the future whenever humans retrieve information from them. But the other two subprocesses have not yet been fully replaced by abiotic mechanisms. Therefore, as yet, an independent abiotic platform to enable Life 3.0 to exist is not in place. Could it ever be? Tegmark and his colleagues at the FLI provisionally suggest that the answer is "yes." AGI carried by abiotic robots might acquire and deploy additional know-how, or R_i, beyond anything that humans can currently achieve. It might do so by some kind of abiotic learning. In principle, inductive machine learning of this kind is within the range of some of the artificial systems now being built.

But what other evolutionary problems would Life 3.0 have to solve? For example, what should be the fitness goals of a superintelligent robot? I suggest, because we are still talking about evolution, that any robot's fitness goals would have to be logically equivalent to, but distinct from, the fitness goals of any biotic organism. In simple terms, they should relate to the capacity of variant technology to generate copies of itself. This should be true of both individual robots and populations of robots. We should therefore expect the fitness goals of abiotic artificial systems to be survival and reproduction, and of their populations to be their own survival and regeneration and the spread of abiotic life into as many diverse niches as possible. Remember, however, that this time, these interactive niche relationships would comprise robot-environment (Rob-E) interactions with their local selected environments. These Rob-E niche relationships would now become the fundamental units of abiotic life at the level of Life 3.0.

The dispersal of abiotic life into diverse niche relationships could eventually be on a much grander scale than has been achieved by biotic life as far as we know. For instance, as Tegmark points out, abiotic life might well be better equipped to colonize the solar system and other solar systems elsewhere in our galaxy. That could be a prize worth having. It could be something to which contemporary human beings might wish to contribute.

There is also another point to establish here. I suggest that the fitness goals of all artificial systems would be constrained not only by the same laws of thermodynamics that apply to living systems, as originally described by Schrödinger (see chapter 1), but also by the bioenergetics versus bioinformatics dilemma discussed in previous chapters. Artificial life would still have to cheat the second law of thermodynamics, but without ever violating it, and it would still have to continuously resolve the R_p versus R_i dilemma to do so. For example, artificial abiotic life would still have to pay for any increased adaptive know-how that it might acquire by generating detritus of some kind in its environment. It could not escape these basic laws of physics any more than can any form of biotic life.

This raises a final fundamental question: Could abiotic life be equipped with the appropriate motives or emotions, or even sufficient active purposeful behaviors, to achieve these fitness goals and the continuous resolution of the R_p versus R_i dilemma? Or, to repeat a former question, may a fundamental obstacle exist to stop them from doing so? Is there any barrier to stop the full realization of Life 3.0?

This is perhaps the major remaining unknown, but my intuition is to say "no." In human evolution at Life 2.0, human behaviors are motivated by both emotions and consciousness. Of course, not all human behaviors are consciously motivated, nor are they consciously controlled. But many of our most significant behaviors clearly are. So, must artificial life in the form of robots, equipped with AGI, be equipped with motives and consciousness as well before Life 3.0 can exist? This is a question that Tegmark himself raises in his book. If the function of consciousness in humans, and by implication in other animals too, could be better understood, then it might be possible to make an artifact that could carry out the equivalent conscious functions in robots at the level of Life 3.0. However, the function of consciousness in humans is still very far from being understood.

In both individual and social learning, one of its functions appears to have something to do with a kind of higher-order attention control, which facilitates the learning process. But this is not much more than an educated guess at the moment. Consciousness is often described as a hard problem when it comes to understanding human evolution, human behavior, and human social life. It may prove to be the barrier that prevents Life 3.0 from ever happening, whether here on Earth or anywhere else in the universe. But I suspect that it will not be, and something akin to consciousness might be engineered into robot minds one day.

8 THE ORIGIN AND EVOLUTION OF ECOSYSTEMS

If Lewontin (1983) is correct about the logic of niche construction, evolutionary theory should really be about the coevolution of organisms with their environments rather than just about the evolution of organisms. In which case, evolutionary theory should also concern the evolution of ecosystems. The coevolution of organisms with their external environments invariably occurs in the context of a hierarchy of ecosystems, shared with multiple other species of organisms. This is true on all possible scales, ranging from a rotting acorn to Earth's biosphere.[1] The coevolution of organisms with their local environments inevitably contributes to the evolution of the ecosystems that they share. In this chapter, I want to consider both the origin of ecosystems and their subsequent evolution.

For most of the twentieth century, there was a disconnect between evolutionary biology and ecosystem-level ecology. There were two main reasons for this. One was a relatively trivial, but nonetheless understandable mistake about the respective timescales of evolutionary and ecological processes. The second was a more profound problem arising from the conjunction of populations of living organisms and nonliving abiota in ecosystems. Obviously, evolutionary theory can be applied to living organisms, but how can it be applied to abiota? In this chapter, I will consider both of these problems, the first only briefly, and the second in greater depth.

INCOMPATIBLE TIMESCALES?

Following the publication of Darwin's *The Origin of Species* in 1859, for the rest of the nineteenth century and most of the twentieth, it was assumed that evolution is always a very slow process. This idea was encouraged by Darwin, who frequently stressed the gradualism of evolution by natural selection, based on the steady accumulation of multiple small changes in the characteristics of organisms over considerable periods of time. It was also encouraged by the fossil record which, as evidence accumulated, appeared to indicate that evolution was a very slow process, taking millions of years for interesting things to happen.

Sometimes this is true. Sometimes it really has taken millions of years for a significant evolutionary event to appear in the fossil record.[2] In the twentieth century, it was also assumed that natural selection is frequently stabilizing rather than directional. "Stabilizing selection" refers to selection against extreme values of a character, which was thought to oppose evolutionary change rather than favor it. In fact, the evidence for stabilizing selection is surprisingly weak, and it is detected in natural populations comparatively rarely.[3] Despite this, the assumption that stabilizing selection is frequent persists even today, which is another reason why the evolutionary process was thought to be slow.

For all these reasons, evolution was assumed for many decades by biologists and naturalists to be a slow process. In contrast, ecological events typically happen on much shorter timescales, ranging from days to years and sometimes a few centuries, but not much longer. There are some exceptions that relate to macroevolutionary events. There have been long-term ecological changes, as well as short-term ones. Examples include the accumulation of shell beds on the ocean floor and the bioturbation of burrowing organisms.[4] For the moment, I'll leave these to one side, in which case, given the apparently contrasting timescales of evolutionary and ecological events, how can evolutionary and ecological processes interact with each other at the ecosystem level?

In the twenty-first century, on the basis of new evidence, some of it from molecular biology, we now know that evolutionary events can sometimes be just as rapid as ecological events. Evolutionary changes can track rapid ecological changes at comparable rates, such as in microorganisms with rapid generational turnover. One example is the very rapid evolution of resistance to antibiotics in bacteria, as discussed earlier in this book. Evolutionary change can also be fast in metazoans. A good example was provided by Peter and Rosemary Grants's forty-year study of the continuously and unpredictably evolving finches on the Galapagos Islands.[5] Another example is provided by the rapid evolutionary response of diverse species to human urban environments, inclusive of the recent innovation of street lights.[6] Further examples include the rapid evolution of smaller tusks in elephants and of slower growth in fishes in response to hunting and fishing by humans.[7] One consequence of these data is that the problem of ecology and evolution working on different timescales may no longer be a major obstacle, standing in the way of applying evolutionary theory to ecosystem-level ecology.

NONEVOLVING ABIOTA IN ECOSYSTEMS

The more profound question is: How can evolutionary theory be applied to ecosystems, given that ecosystems contain abiota that apparently cannot evolve? Abiotic components in ecosystems are connected to organisms by energy and matter, or R_p flows, that pass through both abiota and biota in ecosystems. It is clearly possible for energy and matter to flow continuously through both the living and nonliving components of ecosystems in biogeochemical flows and cycles, such as the nitrogen cycle and the carbon cycle. This has long been recognized by ecologists. These flows are captured by what the Oakridge ecologist Robert O'Neill and colleagues previously called "process-functional ecology."[8] However, from a traditional perspective, there seemingly cannot be an equivalent flow of evolutionarily significant information, or R_i, through all the components of ecosystems due to the presence of abiota in ecosystems, which don't evolve.

There can be continuous flows of evolutionarily significant information between coevolving populations in communities, of course. This is possible because, for example, population A can act as a source of natural selection pressures for population B, which may evolve in response to these selection pressures. Subsequently, a changed population B may act as a source of natural selection back on population A, which in its turn can exhibit another evolutionary response. In this manner, there can be a continuous flow of evolutionarily significant information between populations

A and B, causing them to coevolve with each other in the context of their community, such as what is seen in predator-prey or host-parasite coevolutionary interactions.[9]

The same is not true of interactions between populations of organisms and abiotic components of their ecosystems. Abiota can act as sources of natural selection for populations of organisms, in the same way that populations of organisms can act as sources of natural selection for each other. Abiota can thereby cause populations to exhibit evolutionary responses to the natural selection pressures that originate from them. But according to standard evolutionary theory (SET), it is not possible for abiota to respond in a reciprocal fashion to the activities of the organisms with which they interact. That is because abiota do not carry genes, so whatever changes do occur in abiota do not inherently lead to changes in a genetic inheritance system. Given that SET only recognizes genetic inheritance in evolution, that would appear to rule out any kind of biotic-abiotic coevolution in ecosystems. The presence of abiota in ecosystems apparently prevents abiota from participating in any kind of continuous flow of evolutionarily significant information through both the biotic and abiotic components of ecosystems. Every time there is an interaction between any biotic and any abiotic component in an ecosystem, the abiotic component appears to act as an evolutionary dead end.[10]

Once again, these limitations do not rule out all links between evolution and ecology. They permit evolutionary theory to be applied to population community ecology and to assemblies of organisms in communities.[11] Populations in wider communities, such as populations competing for light or water in food webs, can and do coevolve with each other.[12] Therefore, there can be a flow of evolutionarily significant information through all the living components in communities, provided that all the abiotic components that also exist in their environments are ignored or treated as inputs to the system. For modeling purposes, the abiota usually have to be edited out. But that is not true of ecosystem-level ecology, where it is not possible to ignore the role of abiotic components in energy and matter flows through the biotic and abiotic components of ecosystems.

The net result is that, in addition to the problem of different timescales, there were two other obstacles rather than just one, keeping evolutionary biology and ecology apart for most of the twentieth century. First, there was a disconnect between evolution and ecosystem-level ecology, which was caused by the presence of nonevolving abiota in ecosystems. Second, there was also a disconnect between population-community ecology and ecosystem-level ecology within ecology itself, with the former studying evolution in food webs, and largely ignoring abiota, and the latter studying entire biogeochemical cycles, but largely ignoring evolution.[13] Both of these divisions have frequently been discussed by ecologists.[14] They have been discussed by evolutionary biologists less often.[15]

During the final decade of the last century, things began to change. Two ideas, one from ecology and one from evolutionary biology, appeared in the literature almost simultaneously. They both suggested that the activities of organisms not only affect abiota in ecosystems, but these effects also sometimes feed back to populations of evolving organisms in ways that can subsequently affect both the ecology and the evolution of populations in ecosystems. In ecology, Clive Jones, John Lawton, and their colleagues introduced the concept of "ecosystem engineering."[16] They documented how the activities of engineering species could affect energy and matter flows in ecosystems in ecologically significant ways. One of their salient examples was beavers building dams in rivers, thereby causing multiple subsequent changes in riparian ecosystems.[17]

In evolutionary biology, Richard Lewontin introduced the logic of niche construction in 1982, although he didn't use that term.[18] He just wrote about the "construction" of environments. I first coined the term "niche construction" in 1988.[19] The niche-constructing activities of organisms are those that modify biotic or abiotic components of their local environments. These modifications may then become the source of modified natural selection that feeds back to affect the subsequent evolution of either the niche-constructing population itself or other populations in its ecosystem, or both. For example, when organisms interact with abiota, the state of the abiota can be transformed to register the informed niche-constructing activities of the organisms that act on them. These changes in abiota then may become the sources of specific modified natural selection pressures, either by feeding back to the same niche-constructing population or by feeding forward to one or more other populations in an ecosystem. This accounts for the two categories of natural selection that I introduced in chapter 4. They are unmodified, purposeless natural selection pressures in the environments of organisms, as opposed to natural selection pressures that were previously modified by the active purposeful agency of niche-constructing organisms in evolving populations.

The recipient population, or populations, exposed to modified environmental conditions, may subsequently exhibit evolutionarily significant responses in the usual way.[20] In this manner, the abiotic components can participate, not only in energy and matter (or R_p) flows, but also in continuous flows of evolutionarily meaningful information (or R_i) through both populations of organisms and abiota in ecosystems. This is in spite of the fact that there are no genes in abiota. Abiota can do this by acting as intermediate sources of modified natural selection, connecting the evolutionarily informed niche-constructing activities of one population to usually adaptive evolutionary responses in another population.

The capacity of intermediate abiota to act as evolutionarily significant bridges between either different generations of a single population or diverse populations in ecosystems is captured by the concept of environmentally mediated genotypic associations (EMGAs) in niche construction theory (NCT).[21] More recently, and perhaps not entirely independently, similar ideas have been picked up by the emerging field of ecoevolutionary dynamics.[22] Despite the parallels between NCT and ecoevolutionary dynamics, there is one aspect of NCT that the latter has not yet greatly emphasized. That is that when changes caused in abiota by niche-constructing populations modify the natural selection pressures encountered by descendant organisms, they introduce a second general inheritance system to evolution, which I've called "ecological inheritance." This means that successive descendant generations of evolving populations are now recipients of both genetic inheritance and ecological inheritance. These two inheritances relate to each other in evolving populations.

Earlier in this book, I described "ecological inheritance" as comprising natural selection pressures arising from either biota or abiota that have previously been modified by the niche-constructing activities of earlier generations of organisms in either the same or different populations (see chapter 1). Ecological inheritance seldom carries evolutionarily significant information directly between ancestral and descendant generations in populations in the same way that genetic inheritance does. Instead, it carries modified environmental states, and hence modified natural selection pressures between generations, relative to whatever historical information is encoded by previously selected genes. It thereby influences the fitness of the inherited genes. In this sense, a population's genetic inheritance system and its ecological inheritance system are complementary. It is possible for evolving populations to inherit information about something in their environment via genetic inheritance (e.g., a predisposition to grow well in a nest). It is also possible to inherit the "something" in the environment that the information is about via ecological inheritance (i.e., the nest itself).

I have now indicated how both the principal barriers that previously prevented the integration of ecosystem-level ecology and evolutionary theory can be surmounted. Recent insights in ecology and evolutionary biology are now allowing biologists to start exploring new ways of bridging the gaps between ecosystem-level ecology and evolutionary biology.[23] For example, biologists have begun to explore the ways in which ecoevolutionary feedback occur in ecosystems,[24] including ways in which this feedback can be detected and ways in which it contributes to the structure, function, robustness and resilience of ecosystems.[25] Other authors have concentrated on investigating specific cases, in a bottom-up attempt to understand the details of how ecoevolutionary feedback works in ecosystems.[26]

For the rest of this chapter however, I want to take a different approach to the problem of how to integrate evolutionary biology with ecosystem-level ecology—namely, a top-down approach. First, I want to consider the origin of ecosystems on Earth as an almost immediate consequence of the origin of life on Earth (see chapter 6). Then I want to consider the subsequent evolution of ecosystems as a function of the evolution of life on Earth. I am not going to consider the evolution of ecosystems in the light of SET, or neo-Darwinism, but rather in the light of NCT and of the more comprehensive theory of evolution offered in the preceding chapters. My aim is to show how the additional phenomena incorporated by NCT, as well as some other contemporary approaches to evolution, may be able to throw more light than SET can on both the origin of ecosystems and their subsequent evolution.

THE FUNDAMENTAL UNITS OF LIFE

The fundamental units of life are not just cells, or organisms, *in vacuo*. They are organism-environment (or O-E) interactive niche relationships, where O is any organism and E is that organism's environment (see chapter 6). This point traces to Lewontin's (1983) observation that organisms cannot exist, except relative to their selective environments, and reciprocally that selective environments, as opposed to just surroundings, exist only relative to their organisms. The reason why interactive niche relationships are the fundamental units of life is due to the physical thermodynamic requirements of life, as described by Maxwell, von Neumann and Erwin Schrödinger (see chapter 1).

To recap, organisms are very-far-from-equilibrium systems relative to their environments. The energetics of life is doubly dependent on the interactions between organisms and their environments. As Schrödinger (1944) originally put it, organisms must continuously import "negative entropy,"[27] in the form of physical resources that are relatively high in free energy, from their environments, and they must continuously export entropy, in the form of physical resources that are lower in free energy, back to their environments. To allow organisms to live, their local environments must both be open, capable of absorbing the detritus that organisms inevitably generate by living without posing a threat to them, and be able to supply the organisms with sufficient free energy, or R_p, to meet their needs.

Organisms also have to control a flow of energy and matter between themselves and their local environments in ways that allow them to resist the second law of thermodynamics, without violating it (see chapter 1). That introduces a third way in which organisms depend on their environments. Their environments must be sufficiently lawful, as specified by the laws of physics and chemistry and the putative "laws" or regularities of biology, to make it possible for organisms to acquire adaptive know-how (R_i) from the causal textures of their environments. Organisms in evolving populations must acquire sufficient R_i to enable them to control a flow of energy and matter between themselves and their environments, in ways that allow them to survive and reproduce (see chapter 4).

Here, NCT emphasizes how the relationships between organisms and their environments are necessarily two-way-street interactions. Populations of organisms are changed by natural selection in their environments, as both SET and NCT acknowledge. But populations of organisms also niche-construct to modify environmental states. SET recognizes that niche construction occurs, but it does not recognize it as a cocausal process in evolution, alongside natural selection. Conversely, NCT stresses that organisms have to niche-construct to control the energy and matter flows between themselves and their local environments on which their lives depend. Moreover, when they niche-construct, organisms inevitably cause some changes in their local environments, some of which feed back to modify natural selection again.

For NCT, organisms are goal-seeking systems. Minimally, they have to act purposefully in pursuit of their fitness goals of survival, growth, and reproduction. Individual organisms must be purposeful and active systems to achieve these fitness goals. Hence, when organisms change components of their external environments through their niche construction, the changes that they cause should reflect their goal-seeking purposes. This implies there must be an orderliness to the environmental change that arises from the niche-constructing activities of organisms, as well as an orderliness to the modified selection that follows—a prediction that I made several years ago[28] and that has subsequently been confirmed.[29] These two-way-street interactive niche relationships, according to which evolving populations of organisms and their environments change each other, means that populations of organisms and their respective environments must coevolve. Moreover, both the biotic and abiotic components of the environments of organisms must participate in this coevolution. This point applies to all populations, in all ecosystems.

BIOENERGETIC AND BIOINFORMATIC FLOWS IN ECOSYSTEMS

All populations are bound to participate in energy and matter, or R_p, flows in ecosystems, and to contribute to the generation of flows of evolutionarily significant information, or R_i, in ecosystems. Each population utilizes and potentially contributes to the biogeochemical flows and cycles that they share with many other populations in their ecosystem. All populations niche-construct, and therefore each must have a minimal capacity to modify natural selection, not only for themselves but also for other populations in their ecosystems. This happens as a consequence of the overlapping but not identical niches of diverse populations in ecosystems, and it should eventually affect all the populations in an ecosystem.

This point is consistent with Lewontin's (1983) comment that evolutionary theory should really be a general theory of organism-environment coevolution, rather than just a theory about the evolution of organisms. Since all populations in an ecosystem are coevolving with their own, and with some other organisms', selective environments, it follows that all the components of ecosystems, both biotic and abiotic, must be connected to each other by energy and matter, or R_p flows, as has long been recognized by ecologists. However, they must also be connected by flows of evolutionarily significant information, or R_i, that are generated by the coevolving populations of organisms with their environments. Like R_p, R_i flows through both biota and abiota in ecosystems.

This second point has not been previously recognized by either ecologists or evolutionary biologists. It implies that the integration of evolutionary biology with ecosystem-level ecology is going to depend on our ability to understand how these two flows, the bioenergetics (R_p) flow and the bioinformatics (R_i) flow, interact with each other and change each other in ecosystems.

THE ORIGIN OF ECOSYSTEMS

I'll start by considering a lifeless planet. In principle, it could be anywhere in the universe. The ones that we know most about are the other planets in our own solar system. On a lifeless planet, energy and matter flows can exist in the form of various kinds of geophysical and chemical interactions among the nonliving abiotic components of such planets. But there are no flows of evolutionarily significant information on planets that have hitherto been lifeless.[30] It is therefore possible to consider some of the basic properties of energy and matter flows without them being affected at all by even a rudimentary flow of evolutionary significant information.

Many geophysical activities, plus their associated dynamics, can occur on lifeless planets. Such changes are ultimately driven by the same four forces of nature that are apparently responsible for all the changes that happen throughout the universe. They are the force of gravity that works on the largest scale and the three other forces that stem from the properties of atoms and subatomic particles (namely, electromagnetism and weak and strong nuclear forces). On the scale of a lifeless planet, these forces can generate changes that can be observed at a distance (e.g., by human observers on Earth). Today, we can observe other planets, including other planets orbiting different stars elsewhere in our galaxy, by both Earth-bound telescopes and telescopes in space. Nowadays, we can send well-equipped spacecraft to all the other planets in our own solar system. This has already yielded a wealth of information about the contemporary state of these other planets and the dynamics of some of the geophysical changes that are currently occurring or did occur on these lifeless planets in the past. For instance, astronomers and astrophysicists have observed seasonal changes and changes of temperature on other planets. More dramatically, they have also observed volcanic activity, jets and flows of water and other liquids, storms, and interactions of these planets with their moons and satellites, including the rings of Saturn. Some of these planetary satellites, such as Europa, one of Jupiter's moons, are also very active. The net result is that today, we probably know more about all the other planets in our solar system than we know about Earth before the origin of life. That refers to approximately the first 500 million years of Earth's existence, starting with the Hadean geological era.[31] We obviously cannot send "inquisitive" spacecraft back in time to visit the prebiotic Earth, but we do have one advantage. As we live on Earth, we can investigate whatever residual hints remain (e.g., in early rocks) about the probable or at least the possible state of Earth before life began.

After the initial formation of Earth's first prebiotic atmosphere and the formation of its original oceans, there continued to be considerable geophysical activity during the Hadean era and the subsequent Archaean era. The principal drivers of this activity probably included both the direct and indirect consequences of the radiant energy from a younger Sun (e.g., storms, winds, waves, and a primitive hydrological cycle). After the birth of the Moon, they presumably also included interactions between the Earth, Moon and Sun, which should have resulted in the first tides in the oceans. Because the Moon will have been very close to Earth initially, the first tides would have been colossal. They would have included, not only tides in the oceans but "rock tides" on land as well, that heaved up and then released Earth's surface, with each rotation of the Moon. There was also a great deal of volcanism, although maybe less than was thought earlier, and there were collisions between the early Earth and asteroids and meteorites.[32]

What scientists have learned from observing other planets, as well as from inferences about our own prebiotic Earth, tell us something about how purposeless abiota interact with purposeless abiota on lifeless planets. When abiota react to other abiota, they merely react to each other, as well as to their surroundings. These changes may appear to be surprisingly creative. They can sometimes create Stuart Kauffman's "order for free."[33] In chemistry, they can also promote energy-producing endergonic reactions, as well as energy dissipating exergonic reactions (see chapter 4). Nevertheless, they are still just reactions.

All abiotic-by-abiotic interactions are purposeless in the same sense that physicists have learned to assume that the whole universe is purposeless. The universe may or may not be purposeless, but insofar as our incomplete understanding of it takes us, it does appear to be so. The cosmologist Stephen Weinberg once summed this up in a celebrated remark about the cosmos: "The more the universe seems comprehensible, the more it also seems pointless."[34] Yet we don't understand the cosmos enough to know if this is true. The only interactions that can occur on a dead planet are abiotic-by-abiotic interactions. Here, I'll assume that these abiotic interactions are not only reactions, but they are also completely purposeless.

All abiotic-by-abiotic interactions on lifeless planets have a common attractor. They all obey the second law of thermodynamics. They do so by contributing to a net increase in entropy. Their attractor is greater thermodynamic stability, and ultimately greater disorder. Over vast spans of time, this implies the destruction of all lifeless planets. Any lifeless planet, or for that matter any planet with life, orbiting any star, is likely to be destroyed by the eventual breakup of its star. For example, as stars approach the end of their lifetimes, many of them morph into red giants. Our own Sun is expected to do just that. When it does, it will engulf and consume most of its planets, including Earth—but it's not supposed to do that for a few billion years yet.

That's a short but probably sufficient list of the principal features of the geophysical events arising from abiotic-by-abiotic interactions on lifeless planets. The only reason for drawing attention to them is to contrast them with what can happen on a living planet.

THE EARTH WHEN LIFE FIRST ORIGINATED

What happens to these geochemical energy and matter, or R_p flows, when life first originates on a previously lifeless planet? How and why are these energy and matter flows changed by the presence of life? As we've already seen, the properties of all living organisms are very different from the properties of nonliving systems. To stay alive, organisms have to actively resist the second law of thermodynamics for the duration of their lives. No abiotic system has to do the same.

One consequence of the origin of life on any planet is that it introduces two novel kinds of interactions into these geochemical flows, neither of which can occur on a lifeless planet. They are biotic-by-abiotic interactions and biotic-by-biotic interactions. The first novel interactions introduced by the origin of life on Earth would have been biotic-by-abiotic interactions between the first purposeful living organisms and their purposeless abiotic environments. According to NCT, but not SET, these biotic-by-abiotic interactions would have initiated a flow of evolutionarily meaningful information between the first organisms and their environments.

This flow of meaningful information, in the form of adaptive know-how (R_i), between the first organisms and their environments would have been generated by the first variant organisms responding to abiotic sources of natural selection in their environments. Crucially, it would also have depended on the feedback generated by the subsequent modification of natural selection pressures by the elementary niche-constructing activities of the first organisms. The initial niche-constructing activities of the first organisms, or cells, would have affected only their own immediate micro-environments. They probably amounted to no more than the extraction of specific molecular energy and matter resources from their environments, combined with the dumping of specific molecular detritus back into their environments. However, when organisms niche-construct, they don't just modify sources of natural selection in their own environments, they are also likely to modify some sources of natural selection in the environments of other organisms in their vicinity.

The addition of this new flow of evolutionarily meaningful information, generated by the first biotic-by-abiotic interactions between the first organisms and their abiotic environments, would have had a further consequence. It would have triggered the first coevolutionary relationships between those organisms and their initial abiotic environments. These first coevolutionary relationships should have led these organisms and their environments to coevolve with each other by causing changes in each other by their two-way-street interactions. Organisms will have modified their local environments, most likely pushing them into states that they could not have occupied on a lifeless planet. These interactions would have also have caused the new flow of evolutionarily meaningful information on a now-living planet to interact with the original geochemical energy and matter, or R_p flows, on the previously lifeless Earth. In combination, these two flows would then have merged to become the origin of the first biogeochemical flows on Earth.

The initial two-way street biotic-by-abiotic interactions between the first organisms and their initial abiotic environments should have been enough by themselves to introduce the first ecosystems on Earth. It should also have been enough to have converted the geochemical energy and matter flows on the previously lifeless Earth to biogeochemical flows of both energy and matter (R_p) and meaningful information (R_i) on the now-living planet. In their turn, these biogeochemical flows would have affected the flows of both R_p and R_p in all ecosystems, on all scales, up to and including the biosphere itself. This implies that there was no need to wait for the second kind of novel interactions, biotic-by-biotic interactions, between different populations of organisms for the origin of the first ecosystems. If this hypothesis is correct, ecosystem-level ecology, based on no more than the coevolutionary interactions of the first organisms with their abiotic environments, should have preceded population-community ecology.

Now we can turn to the second kind of novel interaction introduced by the origin of life on Earth. These biotic-by-biotic interactions would probably have been introduced quite rapidly after life appeared. They would have comprised the interactions between different organisms in different evolving populations and different species, or perhaps cell lines. The fundamental origin of these biotic-by-biotic interactions would have been the reproductive capacity of the first organisms. By reproducing, organisms immediately change their own interactive niche relationships with their environments by introducing new organisms (namely, their daughter cells, or "offspring") into their own and each other's environments. Reproduction is hence another kind of elementary niche construction (see chapter 6). The reproductive capacity of organisms represents another way in which living organisms are unlike the abiotic components of their environments. The biogeochemical flows of energy and matter that organisms encounter in their environments will now include other purposeful, active, fuel-consuming, detritus-generating agents like themselves, as well as abiota.

The presence of other organisms in the environments of all organisms is then responsible for another distinction between biotic-by-abiotic and biotic-by-biotic interactions. Abiotic sources of natural selection, even if they have been previously modified by the purposeful niche-constructing activities of organisms, are completely indifferent to the subsequent fates of the organisms on which they act. For example, rainfall is a significant source of natural selection in the environments of multiple species of organisms, but the rainfall itself is indifferent to the subsequent fates of the organisms that it affects.

The same is not true of biotic sources of natural selection. For example, if two species of organisms are competing for the same environmental resource, then the organisms in both competing species will themselves be affected by the natural selection that each is generating for the other. When organisms interact with each other, they are likely to affect each other's fitness goals.

These differences between biotic-by-abiotic and biotic-by-biotic interactions are likely to be both evolutionarily and ecologically significant. For example, on average, organisms are likely to need relatively less adaptive know-how, or R_i, to respond adaptively to abiotic sources of natural selection than they do to respond to biotic sources of natural selection. To respond adaptively to abiota, organisms need to "know" only two things. They need to know some relevant lawful properties of the abiota that they are interacting with, as described by the laws of nature, and they need to know about the consequences of their own actions on the abiota with which they are interacting. Organisms need to be able to anticipate or predict, or at least prepare for, the consequences of their own actions on other abiota. But they don't need to know much more than that.

However, when organisms interact with other organisms, they do need to know more. Ideally, they need to know something about the fitness goals and purposes of the other organisms they are interacting with, as well as some relevant lawful properties of the biota they are interacting with as described by the laws of nature. For instance, they may need to know something about how these other organisms can niche-construct. They also may need to know the likely consequences of their own purposeful actions on both the purposeful fitness goals of the organisms that they are acting on, as well as on their own fitness goals. This means that the adaptive know-how, or R_i, that they acquire from natural selection arising from other organisms should reflect something about the purposeful fitness goals of those organisms.

It is usually more difficult to predict how the purposes and niche-constructing activities of other organisms are going to play out than to predict how purposeless abiotic systems may change in the future. It is likely to be more difficult to anticipate or predict the activities of active purposeful systems than the reactions of passive purposeless objects. Biology is harder than physics, as implied by Schrödinger's analysis of what more is needed, beyond the known laws of physics and chemistry, to understand life?[35]

ADDITIONAL CONSEQUENCES OF BIOTA

Another consequence of these novel interactions between biota and abiota or biota and biota is that informed purposeful organisms may be able to recruit and harness some of the laws and forces of nature for their own purposes. For instance, organisms can sometimes drive abiotic components of their environments into new physical states through biotic-by-abiotic interactions that could never occur on a lifeless planet. For example, this is true of animal artifacts such as beavers' dams or human houses, partly or entirely constructed out of abiotic materials but which could never arise without the purposeful activities of organisms. Biotic-by-biotic interactions can also drive novel changes in evolving populations of organisms that are different from any changes induced by abiotic-by-abiotic interactions. For instance, reed warblers would probably never have started to hide their nests, nor would cuckoos have started to disguise their eggs by making them look more like reed warblers' eggs, in the absence of each other.[36]

I also need to say a bit more about the biogeochemical cycles, introduced by the origin of life on Earth. Rather than considering origin-of-life biogeochemical cycles, about which little is known, I'll focus instead on the contributions of contemporary organisms to contemporary large-scale biogeochemical cycles, such as the nitrogen, carbon, and hydrological cycles. For example, the contemporary nitrogen cycle on Earth begins with the active agency of bacteria that converts atmospheric nitrogen (N_2) into ammonia (NH_3), which can subsequently be used by plants in contemporary ecosystems. Nitrification is the process that converts ammonia into nitrite ions, which the plants can take in as nutrients. Similarly, the last step in the nitrogen cycle also depends on bacteria actively converting the residual nitrogen compounds left over by living organisms, typically in the form of their waste products, back into nitrogen gas, which is then returned to the atmosphere, possibly to be recycled.

THE CONSTRUCTAL LAW

Before leaving this hypothesis about the origin of the first ecosystems on Earth, I would like to draw attention to the apparent consistency of my hypothesis with the constructal law, described by Adrian Bejan and Peder Zane.[37] Bejan and Zane propose that, in conjunction with the laws of thermodynamics as well as the other laws of physics and chemistry, the constructal law is responsible for the designs of all inanimate and animate systems in nature.

Bejan and Zane propose that all systems in nature that demonstrate design facilitate the many kinds of dynamic flows that occur in nature. The flows that they talk about include abiotic flows of lava and flowing rivers and biotic flows of blood through the vascular systems of animals. I assume that all these flows are ultimately derived from the fundamental flow of negative entropy to entropy, as Schrödinger (1944) put it, or from lower-entropy states to higher-entropy states, as described by the second law of thermodynamics. I also assume that these flows include both the energy and matter (R_p) flows and the flows of evolutionarily meaningful information (R_i) between the first organisms and their first local environments, that were responsible for the origin of both the first biogeochemical cycles and the first ecosystems on Earth.

The authors' approach implies that, in spite of its gloomy prognosis about the ultimate dissipation of everything at the end of time, in the meantime, the processes described by the second law of thermodynamics must actually be very creative. Is the function of all design in nature, inanimate as well as animate, to accelerate the flow of lower entropy to higher entropy by its creativity of everything everywhere in the universe? That is an astonishing question, based on an intriguingly counterintuitive idea. It goes well beyond the hypothesis that I am advocating here. Nonetheless, I suspect that it might be true, and if so, that it might one day lead physicists to a significantly better understanding of the laws of thermodynamics.

THE SUBSEQUENT EVOLUTION OF ECOSYSTEMS

After considering the origin of ecosystems, I want to turn to their subsequent evolution. What drives the evolution of ecosystems? I suggest that the evolution of ecosystems must have begun with individual organisms having to resist the second law of thermodynamics, by energy and matter, or R_p, consuming work. Organisms have to work to oppose the flow of energy and matter between themselves and their environments that is favored by the second law to stay alive. But, as we also saw before, organisms can achieve their fitness goals of survival and reproduction only by actively protecting their essential variables. In this instance, to do so, organisms have to keep a third variable—namely, the varying relationships between their own varying phenotypic traits and the varying natural selection pressures that they encounter in their environments, which are continuously adaptive (see the discussion of Ashby in chapter 3). However, organisms cannot be adaptive, relative to either abiotic or biotic sources of selection, unless they are sufficiently well informed by evolution with appropriate adaptive know-how, or R_i.

This R_i must be about how specific organisms can adapt to specific sources of natural selection in their specific environments. But this takes us back again to the R_p–R_i relationship and its associated dilemma. At the origin of life, since both R_p and R_i depend on the prior acquisition by organisms of the other, the dilemma is: "Which came first, R_p or R_i, and how did life on Earth get started?" The prevailing hypothesis is that both R_p and R_i resources must have been acquired by the first organisms almost simultaneously, possibly in a ribonucleic acid (RNA) world (see chapter 6).[38]

However, this dilemma is not just an origin-of-life dilemma. It never goes away. The R_p–R_i dilemma applies to all organisms in all evolving populations, at all times and in all places. It therefore has to be resolved by specific individual organisms interacting with specific sources of natural selection in their idiosyncratic environments. The dilemma reoccurs every time that there is a change in the relationship between organisms and their environments. Organisms can resolve this reoccurring "Which came first, R_p or R_i?" dilemma only insofar as their interactions with both abiotic and biotic components of their environment are continuously adaptive. They must be adaptive, not only relative to specific natural selection pressures that individual organisms encounter, but also relative to a set of rules that govern the relationship between R_p and R_i. We saw some of these rules in chapter 7, but here is a more comprehensive list:

Rule 1: R_p is relative to R_i. Organisms need R_i, a priori, to gain R_p. Unless organisms already carry the appropriate adaptive know-how, or R_i, to harvest specific sources of free energy, or R_p, from their environments, they will not be able to acquire the R_p that they need to live, nor will they be able to dump their detritus back into their environments adaptively. They will therefore not be able to achieve their fitness goals.

Rule 2: The acquisition of R_i costs R_p, a priori. This is because all evolutionary and developmental processes are energy-consuming and detritus-generating processes. According to this rule, the acquisition of sufficient R_p has to occur before further R_i can be acquired. If acquiring both R_p and R_i depends on the acquisition of each other a priori, then that accounts for the origin-of-life dilemma.

Rule 3: The R_i carried by organisms here and now is relative to prior natural selection pressures or other prior selective processes (e.g., cultural selection processes in humans). It is not necessarily adaptive relative to contemporary natural selection pressures in the contemporary environments of organisms.

Rule 4: Sources of natural selection for organisms may change either because of autonomous environmental events or because they have been modified by the prior niche-constructing activities of organisms. The changes then reintroduce the R_p–R_i dilemma. The reoccurrences of this dilemma then requires new resolutions from organisms.

Before describing any further rules, I'll consider some of the possible ways in which organisms may resolve the R_p–R_i dilemma within the scope of these first four. The principal possibilities are as follows. After any change in the interactive relationship between the organisms and their environments, given their current R_i, organisms may no longer be able to harvest sufficient R_p to resist the second law. This means they may no longer be able to survive. Some organisms may be able to survive for a time with less R_p, but over time, provided that the population does not go extinct, the changes in their environments are likely to favor the selection of novel phenotypes, corresponding to additional R_i. This will require further evolution. However, any given population may not be able to pay for the extra R_p demanded by further evolution. If a population lacks sufficient physical resources to evolve, it may become extinct. To survive, it has to resolve the R_p versus R_i dilemma again to adapt to its changed environment.

Another possible problem for evolving organisms is that they may fall into an historical trap. Because organisms can evolve only from where they are at the moment, they may get stuck with the same body-plan, or the same entrenched behaviors, irreversibly, lacking the evolvability to respond fully to the changed natural selection pressures in their environments.[39]

Alternatively, they may be able to evolve in some directions but not others. One example is bilaterian organisms, which having once evolved, can never shed their bilaterian body plan during the course of their subsequent evolution, even though in other respects, they may have high evolvability. A second example is the respiratory system in insects, which breathe through trachea on the surfaces of their bodies. But this means they can never grow larger in volume than their surface area and its associated respiratory system allows. Insects are enormously evolvable in most directions, but not in the direction of larger body size.

There are also potential solutions to these problems compatible with these rules. For instance, in response to changing sources of natural selection in its environment, organisms may use the adaptive know-how, or R_i, that they already possess to harvest novel sources of free energy, or R_p, in their environments. This amounts to evolutionary opportunism, or perhaps to Gould and Vrba's concept of exaptation.[40] Another possibility is that individual organisms in a population may modify a source of natural selection in their external environment in ways that benefit them through their niche-constructing activities. For instance, organisms can oppose some changed sources of selection in their environments, by counteractive niche construction, such as when an organism digs a burrow or builds a nest to counter the extremes of temperature.[41] But it will still cost them some extra R_p to do so.

A third possibility is that two or more evolving populations may be able to pool their adaptive know-how in mutual or collaborative relationships. They will then gain the extra R_i they need to harvest more R_p by sharing and combining some of their R_i with that of a mutualistic partner. A well-known example is the relationship between fungal mycorrhizae and the roots of plants. In this case, each of the mutualistic partners supplies a nutrient to the other that the other needs but cannot harvest for itself on the basis of its own R_i alone.

A fourth possibility is for a population, confronted by new selection pressures, to retreat to a more restricted niche in time and space to avoid the changed natural selection, or for a population to evolve to become more of a specialist. In the latter case, a specialist population might be able to outcompete more generalist competitors relative to a reduced number of energy and matter, or R_p, resources in its environment. In this way, it might be able to retain its viability. In general, if organisms are to resolve the R_p-R_i dilemma following changes in their local external environments, they will have to do so by changing their interactions with either abiotic or biotic sources of natural selection in ways that are consistent with the rules that govern the dilemma.

We have already considered biotic-by-abiotic interactions at some length. These interactions are comparatively straightforward. Abiotic sources of natural selection in the environments of organisms stem from purposeless reactive systems, which have no capacity for niche construction. In contrast, biotic sources of natural selection in the environments of organisms are potentially much more complicated. Biotic sources of selection stem from other purposeful organisms that are capable of purposeful niche-constructing activities and of pursuing their own fitness goals. These differences between abiotic and biotic sources of selection then affect how organisms either do or do not manage to resolve the R_p-R_i dilemma.

If, in imagination, we now return to an early near-origin-of-life scenario, it seems likely that the first biotic-by-biotic interactions between organisms, in different but overlapping interactive niche relationships, would have been competitive. In ecological shorthand, competitive relationships are referred to as "minus-minus" relationships since each population has a negative impact on the other's fitness. These minus-minus relationships can refer to competing organisms in either the same or different evolving populations. Ultimately, competition between organisms would have been a consequence of the capacity of organisms to reproduce. This reproductive capacity of the first organisms would have led to the multiplication of organisms, equipped with either exactly or nearly the same adaptive know-how, and occupying adjacent and probably strongly overlapping niches in environmental space and time.

Presumably different organisms, with nearly the same R_i but in slightly different but overlapping niches, would have attempted to harvest the same or similar energy and matter resources, or R_p, from their shared local environments. They must also have dumped nearly the same detritus back into their environments. In both these respects, closely related organisms in almost identical niches should have competed with each other. Competition between organisms sometimes may have been made worse by a growth in the size of the population of organisms to which they belong. Typically, in contemporary populations, it is made worse when the size of evolving populations start to exceed the carrying capacity of their environments relative to specific sources of R_p. It may also be made worse when a growing population of organisms exhausts the capacity for its environment to act as a sink for its detritus.

A possible solution to excessive competition between coevolving populations of organisms that is consistent with the abovementioned rules could have occurred if some individual organisms had discovered that they could exploit a different source of energy, or R_p, in their shared environments, from any exploited by their competitors. They could have done this by utilizing the same adaptive know-how, or R_i, in a novel way. The general point here is that competitive relationships among some of the first organisms on Earth should have acted as a spur toward their subsequent diversification.

The second classical ecological relationship that could have evolved soon after the origin of life may have been some kind of cooperative or mutualistic relationship (in ecological shorthand, a "plus-plus" relationship), among coevolving populations. Initially, the first cooperative relationships among the earliest organisms would have been elementary. For example, individual microorganisms may have assisted each other by combining their structural features to withstand or dissipate material energy and matter forces in their environments that may have threatened them. An early example of this kind of elementary cooperation is provided by the stromatolite bacterial mats that are still found in Australia and are known to date to more than three billion years ago.[42] More complex mutualistic relationships involving an exchange of different resources among organisms could not have appeared until some diversity among early organisms had evolved. Many mutualisms depend on the ability of organisms in different niches to trade with each other.[43] They may trade either in informational (R_i) resources, in energy and matter (or R_p) resources, or both.[44] But if they are to cooperate, they must trade.

The third classical ecological relationship between different organisms in different interactive niche relationships is the asymmetrical (or "plus-minus") relationship between predators and prey or parasites and their hosts. Predator-prey relationships could not have appeared in ecosystems until at least some organisms were able to gain energy and matter by consuming other organisms. That would probably have required the evolution of some kind of phagocytosis whereby microorganisms engulfed and then consumed other microorganisms.

The predator-prey relationship between metazoan organisms that occur in contemporary ecosystems could only have evolved much later. An earlier asymmetrical relationship might have been that between parasites and their hosts, such as between viruses and bacteria or archaea. Unfortunately, we still know far too little about either the origin or the subsequent evolution of viruses, or even whether viruses can be described as alive, independent of their host. Other kinds of parasites probably evolved only after viruses had already appeared.

At this point, all the classical ecological relationships that occur in contemporary ecosystems, including commensal relationships (+/0 and −/0), would probably have been in place. In commensal relationships, one population of organisms typically acts as a catalyst for the evolution of another population, but the second population is irrelevant to the first population. Collectively, all these ecological relationships between evolving populations would have formed networks of populations or species, interacting with each other in ecosystems.[45]

Another difference between SET and the present more comprehensive theory of evolution is also relevant. The difference also has its roots in the relationship between the bioenergetics and the bioinformatics of evolution, or the R_p and R_i relationship. Evolutionary and developmental processes should always be searching for the most thermodynamically efficient solutions to the problem of supplying organisms with adaptive know-how (R_i). This is because more thermodynamically efficient solutions are always likely to outcompete less efficient solutions in any particular evolving population.

This point is connected to Richard Watson's ideas about more-parsimonious evolutionary algorithms prevailing over less-parsimonious algorithms.[46] It is as if Occam's razor applies to both evolutionary and developmental processes, demanding that they are constantly searching for greater thermodynamic efficiency when supplying organisms with their adaptive know-how. The putative Occam's razor refers to the universal demand for all organisms to continuously solve the R_p–R_i dilemma in the context of their changing environments. It also applies to the structurally and functionally adaptive know-how, or R_i, needed by particular organisms to enable them to adapt to the idiosyncrasies of their particular local external environments.

What is meant by greater thermodynamic efficiency here? One way to answer this question is to introduce some extra rules that describe how organisms might achieve greater thermodynamic efficiency during their evolution:

Rule 5: Organisms should use whatever adaptive know-how, or R_i, they already possess with maximum efficiency. They should therefore either maximize the amount of free energy, or R_p, that they can harvest from their environments, or possibly maximize the rate of energy gain, by utilizing whatever adaptive know-how, or R_i, they currently possess. They should also minimize the destruction (or perhaps rate of destruction) that they cause to their environments by emitting detritus.

Rule 6: If organisms increase their adaptive know-how, or R_i, by further evolution, they should minimize the energy and matter or R_p cost that they have to pay for their further evolution. This implies that natural selection should favor levels of robustness, plasticity and bet-hedging that optimize the population's evolvability.

From rule 6, it is possible to derive two further rules:

Rule 7: Organisms should always be prepared to pay an extra R_p cost for extra R_i, provided that by doing so, they are enabled to gain more R_p from their external local environments than the R_p cost of their extra R_i.

Rule 8: If the free energy, or R_p, cost of expressing the R_i that organisms already carry is greater than the payoff in R_p that they gain by expressing that R_i, then they should either switch off the R_i or maybe edit it out. They should no longer express it. They should only carry and express profitable R_i, relative to both the harvesting of R_p from their external environments and the dumping of detritus back into their external environments.

These rules are not exhaustive. It's likely that further rules may be generated if and when the centrality of the reoccurring R_p–R_i dilemma is recognized more fully.

In the light of this discussion, I again suggest that the subsequent evolution of ecosystems could not have occurred in the context of organisms responding to autonomous sources of natural selection in their environments, as SET suggests. I propose that the subsequent evolution of ecosystems must have occurred in the context of organism-environment coevolution, driven by the two-way-street interactions between niche-constructing organisms and abiotic and biotic sources of natural selection in their environments, as NCT specifies. These two-way-street interactions between organisms and their environments must also have provided the links between the energy and matter (i.e., R_p) flows and the flows of evolutionarily meaningful information (i.e., R_i), that characterize ecosystems.

These links would have converted the geochemical flows of energy and matter that occur on lifeless planets, and were present on Earth before the origin of life, to the biogeochemical cycles combining flows of both energy and matter and meaningful information that now occur on Earth. These biogeochemical flows and cycles would then have become additional sources of change, driving the subsequent evolution of ecosystems on Earth. I also propose that the evolution of ecosystems must occur in the context of a reoccurring R_p–R_i dilemma, which has to be continuously resolved by evolving populations in ways that are compatible with all these rules.

Finally, I want to consider the trophic pyramids that occur in different ecosystems. These pyramids must have evolved long after the origin of life on Earth. They go some way toward answering the question "Who eats whom?"

in ecology. The following, crudely oversimple description of a trophic pyramid will suffice. I deal more fully with trophic pyramids in chapter 10. At its base, a trophic pyramid typically includes primary producers, which gain their energy directly from sunlight via photosynthesis. At the next level up (which I'll call "level 2"), the pyramid typically comprises herbivores, which feed on the primary producers at the base level. At level 3, above the herbivores, small carnivores typically feed on herbivores. At level 4 in the pyramid, large carnivores feed on small carnivores.

Trophic pyramids describe what is sometimes called the "grazing path" in ecosystems. However, there is also a "detritus path" comprising the detritivores, which feed on the detritus generated by organisms on the grazing path at all levels in these pyramids. The detritus path is subordinate to the grazing path and could evolve only after at least a rudimentary grazing path already existed. It is also intimately connected to the grazing path at all levels in a trophic pyramid.

This intimacy is captured by Hamlet in the play of that name by Shakespeare. In act IV, scene iii, Hamlet is talking to his mother, Queen Gertrude. Hamlet suddenly realizes that he is being overheard by someone behind a curtain. Hamlet draws his sword and kills the man. He discovers that the man is a courtier called Polonius, sent by King Claudius to spy on him. He then removes Polonius's body. Enter the king: "Now Hamlet, where's Polonius?" he asks. "At supper," replies Hamlet. "At supper! Where?" asks the king again. "Not where he eats, but where he is eaten," replies Hamlet. For Polonius, the detritus path was only a sword thrust away from his grazing path.

One easy way to illustrate the validity of these rules is to reconsider some of the major transitions in bioenergetics of evolution described by Judson and discussed in chapter 5, while simultaneously considering how they interact with the bioinformatics of evolution.[47] In Judson's scheme, the five major transitions in bioenergetics are geochemical, sunshine, oxygen, flesh, and fire. Let's consider sunshine first. The Sun is a huge external source of energy, or R_p, that must have been present at the origin of life on Earth. But no organisms could harvest this source of energy until they had evolved sufficient adaptive know-how, or R_i, to enable them to do so by photosynthesis (see rules 1 and 2).

What about the next transition, oxygen? Most organisms need extra adaptive know-how, or R_i, to defend themselves against the toxicity of oxygen, as well as to benefit from it (again consistent with rules 1 and 2). The origin of eukaryote cells,[48] a major transition in bioinformatics,[49] also merits consideration. Extra energy, or R_p, was acquired by single-cell organisms when they first engulfed mitochondria organelles by the process of endomorphism.[50] Instead of consuming the mitochondrial organelles, the cells harnessed the extra energy and matter, or R_p, provided by the mitochondria. The cells were then able to utilize the extra energy provided by the mitochondria that they had engulfed to acquire more genes, and therefore potentially more R_i, during their subsequent evolution.[51] More to the point, single-cell organisms now had enough energy to acquire and express extra genes and novel adaptations relative to their environment. The extra energy provided by mitochondria subsequently led to the evolution of multicellular or metazoan organisms, which for more than two billion years had not been possible because of a lack of energy.[52]

SUMMARY

We cannot do much more with this top-down approach to the evolution of ecosystems that I've just been sketching. We need to return to bottom-up approaches as well. Possibly the main contribution of the present top-down approach has been to emphasize the role of the two-way-street relationship between niche-constructing organisms and abiotic and biotic sources of natural selection in their environments. These two-way street interactions are responsible for the coevolution of environments with their organisms. They were also responsible for converting the original geochemical flows of energy and matter on Earth before the origin of life to the biogeochemical cycles that combine the flows of both energy and matter, or R_p, and meaningful information, or R_i, relative to each other that occur on the living Earth. They were, and still are, responsible for the ongoing evolution of ecosystems on Earth today, driving change not just in biota, but in abiota too. As Lewontin (1983) intuited, evolution is a process of organism-environment coevolution.

9 EXTENDING THE SYNTHESIS

In this penultimate chapter, I want to consider the case for extending the modern synthetic theory of evolution. I propose to do that by reconsidering some of its core assumptions, which have now been adopted by contemporary evolutionary theory, here labeled "standard evolutionary theory (SET)." What else is needed beyond the assumptions of SET to enable us to understand the evolution of life better than we do at the moment? How can evolutionary theory be extended to the point at which it can incorporate the new data accumulating in all the subdisciplines of biology, ranging from molecular biology to ecosystem ecology and paleobiology, as well as other disciplines?

I'll start by considering two of the principal assumptions of SET, which we first encountered in chapter 1. One is that genetic inheritance is the only inheritance system that matters in the evolution of populations. The other is the closely associated assumption that there is no such thing as the inheritance of acquired characteristics, such as that described by Jean-Baptiste Lamarck (see chapter 5). According to SET, it is not possible for evolving populations to be influenced by any characteristics that individual organisms acquire during their lifetimes. According to SET, all the characteristics an individual organism acquires during its life are erased when the organism dies. Consequently, genetic inheritance becomes the only inheritance system recognized by SET that is relevant to evolution.

The principal effect of these two assumptions of SET is to downgrade the role of phenotypes in evolution. From the SET standpoint, phenotypes are relevant in evolution only insofar as they survive and reproduce and pass on their naturally selected genes, along with chance-based mutations, to viable offspring in the next generation.

These two assumptions of SET do not mean that phenotypes in evolving populations are necessarily determined by naturally selected genes and chance-based mutations. But given that, according to SET, individual organisms cannot transmit any of their acquired characteristics or traits to their offspring, individual organisms might as well be treated as if they were genetically determined. This accounts for the gene's-eye view of evolution, as described by Richard Dawkins in his book *The Selfish Gene*.[1] In SET, the phenotypes of organisms become little more than transitory, throwaway "survival machines" for their potentially "immortal genes."[2]

In spite of this downgrading of the role of phenotypes in evolution, the modern synthesis, and later SET, were enormously successful in expanding the comprehension of evolution in the twentieth century. SET led to numerous advances in the understanding of evolutionary processes, which benefited animal and plant breeders. The breeders' equation, $R = h^2 S$, encapsulates SET. This equation specifies that the evolutionary response (R) of a population is given by the product of the narrow heritability (h^2), which is the fraction of phenotypic variance that can be attributed to variation in the additive effects of genes, and the selection differential (S), which specifies the relationship between trait values and fitness. The equation greatly advanced the scientific use of artificial selection by plant and animal breeders. As well as discrete characteristics such as eye color, it can handle continuous characteristics, or phenotypic traits such as height or birth weight, by assuming that there are many genetic loci, each of which has a small effect on a specific phenotypic trait.[3] However, this equation excludes any cryptic inheritance, such as recessive genes in heterozygotic organisms that natural selection cannot "see" because they are not expressed in the current generation by the phenotypes that the selection is acting on. Perhaps more important, it excludes extragenetic forms of inheritance.[4]

The problem with this gene's-eye view approach to evolution is that SET's assumption that genetic inheritance is the only inheritance system that matters in the evolution of populations can no longer be defended. We now know that there are many nongenetic or extragenetic inheritance systems that play a role in evolution, in addition to the genetic inheritance system described by SET.[5] The second assumption—that organisms can never transmit any phenotypic traits that they may acquire during their lifetimes to their offspring—can no longer be defended either.[6] We now know that many species of organisms can transmit some of the phenotypic traits that they acquire during their lives to their offspring—ranging from prions, to hormones, to antibodies, to symbionts, to learned knowledge—via one or more of these nongenetic inheritance systems in evolution.[7]

There is one caveat. Although SET rejects Lamarckian inheritance, to some extent it can still explain some acquired characteristics and phenotypic plasticity in organisms quite well without abandoning its assumptions. This is sometimes referred to as "phenotypic plasticity," which is determined by open gene programs in evolution.[8] For example, the reaction norms of phenotypes in response to different environments in both plants and animals can often be explained satisfactorily by SET. Reaction norms that depend on contingent, genetic instructions (such as if in environment A, a genotype expresses phenotype X, but if in environment B, the same genotype expresses phenotype Y) are compatible with the assumptions of SET. These kinds of contingent genetic instructions can explain many of the reaction norms expressed by both plants and animals, including traits that change continuously with an environmental variable without contradicting the assumptions of SET.

For example, plastic developmental responses allow some plants to experience an enhanced oxygen environment in the event of flooding. Plants in flood-prone habitats suffer a severe drop in oxygen if their shoots become submerged. However, in some species, the submerged shoots produce leaves with enlarged surface areas and thinner epidermal cell walls, which provide the plant with a supply of carbon dioxide. Together with a reorientation of chloroplasts toward the epidermis, the net effect of these changes is a higher underwater photosynthetic rate that results in elevated tissue oxygen concentration.[9]

Another example of phenotypic plasticity that appears to be compatible with SET's assumptions, this time in an animal, is the coat-color markings of the Himalayan rabbit.[10] The extremities of this animal, such as its feet and nose, are colored dark brown or black, but the rest of its body is white. This is due to a contingent genetic instruction relating to ambient temperature. Any part of the rabbit's body that is regularly below a temperature of 32°C during the animal's development turns black. This accounts for its extremities being black, as they are usually the coldest part of the animal. Conversely, any part of its body that is above this temperature turns white because the mechanism for turning black breaks down. This coloration is not only attractive, it is probably adaptive as well. In a snow-and-ice landscape, the black extremities of the animal absorb heat better, and the main part of the body being white is better for camouflage.

Not all kinds of phenotypic plasticity can be satisfactorily explained by SET. Consider, for instance, epigenetic processes. Biologists use the word "epigenetics" in many ways. One common way is to describe the mechanisms of the molecular modification of gene expression. Bonduriansky and Day[11] describe two kinds of epigenetic mechanism: "obligate" and "facultative." Obligatory epigenetic processes are controlled by genes and specify molecular attachments to DNA, histones, or RNA, which alter patterns of gene expression and allow cells to acquire alternative molecular cell profiles. For example, obligatory epigenetics is responsible for the differentiation of various tissues in developing embryos. Leaving symbionts to one side, all the cells in multicellular organisms contain the same genomes, and therefore the same genes, but obligatory epigenetic processes determine which genes are expressed and which are not in particular cells and particular tissues. For instance, epigenetic processes determine which tissues will eventually turn into muscle, skin, neural, or heart tissue as the embryo develops. According to the current understanding, obligatory epigenetic processes make no contribution to transgenerational inheritances in evolving populations. For these reasons, obligatory epigenetic processes are ostensibly fully compatible with the assumptions of SET.

Conversely, Bonduriansky and Day define facultative epigenetic processes as either partly or wholly independent of DNA sequences in the genomes of organisms. They either can involve spontaneous changes or be induced by environmental factors. Sometimes these changes can also be passed on to offspring through transgenerational inheritances in the form of an epigenetic inheritance. In the latter case, they are incompatible with the core assumptions of SET.

There are three main ways in which epigenetic processes can interact with genetic processes, which correspond to three principal kinds of molecular mechanisms that can either turn off or regulate the expression of particular genes.[12] These are DNA methylation, the modification of histone proteins, and noncoding RNAs, including interference RNA. Each of these mechanisms modifies the expression of the base-pair sequences of DNA in the genomes of different organisms.

Typically, in organisms such as mammals, the more methylation takes place, the less particular DNA sequences are expressed. Histone proteins affect chromatid structure via tightly bundled regions that shield DNA from expression. Small, noncoding RNAs do not make proteins. They appear to have regulatory roles, influencing which genes are expressed when, where, and by how much. They work primarily by affecting the regulation of specific genes in gene regulatory networks (GRNs). All these mechanisms are highly dynamic and sensitive to environmental conditions.

One example occurs in toadflax (*Linaria vulgaris*), a small, flowering plant. Two variants of this plant exist, with strikingly different petal structures. The more common variety has asymmetric flower structures; the other is symmetric. Throughout most of the twentieth century, it was assumed that these differences were due to a genetic mutation in the less common variety, in line with the assumptions of SET. In 1999, it became possible to sequence the genomes of both varieties of this plant. It was then discovered that the two variant toadflax plants had identical genes with respect to the determination of their flower structures, but they differed in their epigenetics. It was later discovered that their variability was due to a difference between their epigenomes, with the symmetrical form exhibiting extensive methylation of a gene that affects flower shape.[13]

This "epi-mutation" must have been transmitted across multiple generations of toadflax, independent of their genetic inheritances, since it was first noticed by the Swedish taxonomist Carl Linnaeus in 1744. It therefore demonstrated transgenerational epigenetic inheritance across multiple generations, in addition to genetic inheritance.

Many other examples of transgenerational epigenetic inheritance have now been documented in animals as well as in plants. Sometimes heritable epigenetic changes arise as a consequence of the interactions of developing organisms with specific factors in their external environments. Other investigators have demonstrated that epigenetically determined phenotypic variations in populations can respond to natural selection and can influence the subsequent evolution of populations by doing so.[14]

Many researchers have started to list other kinds of nongenetic inheritance systems in evolution, as well as in epigenetics.[15] I won't describe these here; instead, I'll substitute a list of my own by citing all the different kinds of inheritance systems, genetic as well as nongenetic, that can occur in human evolution. I've chosen humans because our species is probably on the receiving end of more different kinds of inheritances than any other species.

But before that, let's go back to the origin of life. It is possible that the first significant inheritance system in evolution predates the origin of life itself. One hypothesis is that prebiotic protocells generated daughter protocells by cell division. In doing so, they transmitted much of their protocell architecture, including their cytoplasm and self-assembled membranes, to their daughter protocells.[16] When the first true living cells appeared on Earth, it is likely that they did the same. They very probably transmitted components of their cell architecture, including cytoplasm and their membranes, by acting as physical templates for their daughter cells. This kind of elementary inheritance system may predate both natural selection and genetic inheritance in evolution, and it remains a feature of many single-cell organisms today.[17] However, it may not predate all kinds of elementary prebiotic selection processes, such as the selective properties of protocell membranes (see chapter 6). Although this prebiotic inheritance system may seem remote from humans, its template-copying characteristics remain relevant to humans today (e.g., arising in inherited prion-related diseases).

The first noncontroversial inheritance system in evolution is genetic inheritance, based on the inheritance of specific sequences of DNA. Humans inherit two kinds of DNA: nuclear and mitochondrial. Nuclear DNA refers to the vast majority of human genes and their regulatory machinery located in our genomes within the nuclei of all the eukaryotic cells that make up our bodies. We also inherit a few mitochondrial genes, carried by the mitochondrial organelles in each of our cells. There is a significant difference in the ways in which these two kinds of DNA are inherited. We inherit our nuclear genes via the sexual reproduction of our male and female parents, as originally described by Gregor Mendel. In contrast, our mitochondrial genes are transmitted asexually only from our female parent. The relationship between these two kinds of DNA inheritance, sexual and asexual, is still far from being fully understood.[18]

Recently, we have also had to come to terms with the fact that two other kinds of DNA are inherited in the microbiomes of infants, mainly from their mother's birth canal when they are born. Their microbiomes typically include a small number of single-celled archaea and a very large number of single-celled bacteria, as well as protists, viruses, and fungi. These symbionts have their own complements of genes. This means that humans also inherit their DNA, in addition to our own nuclear and mitochondrial DNA. It is now known that bacterial DNA, at least, does affect the development of individual humans, sometimes considerably. For instance, bacterial genes affect our digestive systems, our nervous systems, and even some of our behaviors.[19] In addition to inheriting their microbiomes from their mothers, human infants inherit aspects of immunity (such as antibodies) from their mothers, which may prepare their own immune systems for coping with some of the infections that their mothers previously experienced during their lives.

Further, like so many other plants and animals, humans also acquire epigenetic inheritances that are governed by the same molecular mechanisms that occur in all the other organisms that we know about. There is growing evidence that human epigenetic inheritances can be transmitted across multiple generations and are likely to contribute to the evolution of our species.[20] One of the best-known examples of epigenetic inheritance in humans occurred in response to a famine known as the "Dutch Hunger Winter" at the end of World War II. A large proportion of the descendants of starving pregnant women developed health disorders, as their malnutrition modified epigenetic factors. For instance, famine led to higher body mass index by triggering the methylation of a gene called *PIM3*, leading to health problems such as obesity and heart disease in both offspring and grand offspring.[21]

There is one more kind of unusual inheritance system that was discovered only in the late twentieth century, which operates in humans and in some other organisms. It refers to the capacity of certain proteins called "prions" to impose their three-dimensional shape on other proteins with similar amino-acid sequences. They do so by acting as physical templates for other proteins. It originally came to light when scientists were puzzled about how a disease called kuru was transmitted across generations, apparently only between females, which could not be explained by conventional genetic inheritance. It was eventually discovered that it was transmitted primarily by women and children eating the brains of deceased loved ones in a local ritual in Papua New Guinea. It turned out that it was responsible for a neural degenerative disease that devastated some communities in that country. A number of associated diseases have now been found in humans and other animals. They include mad cow disease, caused by feeding cows with food that contained animal tissue, and Alzheimer's and Creutzfeldt-Jacob disease.

Niche construction theory (NCT) then assigns a variety of ecological inheritances to humans. These ecological inheritances comprise environments that have been modified by the prior niche-constructing activities of their parents and other ancestors at several evolutionary levels. They may include modified cultural environments, modified developmental environments for individual organisms, and modified natural selection pressures in the external environments of human populations (see chapter 7).

In general, ecological inheritance does not transmit information directly to successive generations of organisms. Rather, it modifies environmental states and thereby changes what existing information is about. Specifically, it modifies what the adaptive know-how (R_i) inherited by children from their parents is about, relative to natural selection or developmental or cultural pressures in their environments (see chapter 7). It therefore modifies the fitness value of the information that children inherit via plural inheritance systems (see chapters 3 and 4). For instance, it is well documented that slash-and-burn agriculture among some West African communities created puddles of standing water when it rains, which became breeding sites for mosquitoes, facilitating the spread of malaria and triggering the selection of the sickle cell (*Hbs*) allele, which confers resistance to malaria in heterozygous carriers.[22] Here, the genetic evolution of generations of humans is being affected by the activities of their forest-clearing ancestors generations before.[23]

Human beings are also social animals, which means that we are likely to inherit socially transmitted behaviors and traditions from whatever social groups we belong to.[24] For instance, for millennia humans have inherited foraging and tool-making skills and technical ecological knowledge from other members of their local populations.[25] Finally, there is one inheritance system that is probably unique to humans. It refers to the unusually potent cultural inheritances transmitted by languages.[26] We inherit our languages and the information that they carry via many routes, ranging from our parents, siblings, and friends to our more formal education when we go to schools, universities, and other institutions in later life.

That is a formidable list of both genetic and nongenetic inheritance systems in human evolution. It supersedes the traditional nature-nurture dichotomy to some degree. Most of these inheritance systems occur in the evolution of many other species too, and some occur in all species. It is not even an exhaustive list. For example, in bacteria and other microorganisms, genes are shared and exchanged among organisms during their lives by horizontal or lateral gene transfer (see chapter 6). This form of horizontal genetic transmission (HGT) preceded vertical inheritance between generations by perhaps two billion years.[27] In humans, HGT is restricted to the microorganisms in human microbiomes, but it probably should be listed as yet another type of inheritance system that affects human evolution. For instance, the ability of the Japanese, but not Westerners, to digest agar arose because they have a bacterium in their gut that contains a gene that codes for agarose. Apparently, eating seaweed led to the transfer of enzymes from marine bacteria to the gut microbiome of Japanese populations.[28]

None of these nongenetic inheritances are rare; all arise in traits that affect fitness and all can be inherited independent of genetic variation. As a result, these nongenetic inheritances challenge the assumptions of SET. Most obviously, they challenge the assumption that genetic inheritance is the only inheritance system that matters in evolution. They also challenge the assumption that random mutations are ultimately the only source of novel variations in evolution. They challenge the assumption that phenotypes cannot transmit any of the characteristics that they may have acquired during their lives to their offspring.

Now let's turn to another of SET's core assumptions. The theory assumes that natural selection is the only process capable of directing the adaptive evolution of populations. Everything else, including genetic mutations and drift, is due to chance. Only natural selection can catch chance-based mutations and use them to direct the evolution of populations of organisms in nonrandom ways.[29] SET also assumes that the sources of natural selection are autonomous events in the local environments of organisms. It recognizes that organisms can be sources of natural selection for other organisms, such as in coevolution or sexual selection, but in such cases, adaptive evolution is explained through natural selection alone. In contrast, NCT argues that niche construction is not just a product of earlier natural selection, but rather a cocausal evolutionary process in its own right. Niche construction both creates and modifies natural selection.

Now I want to raise two queries about SET's assumptions. The first concerns the origin of natural selection; and the second concerns how and why organisms contribute to their own and each other's natural selection. Before I do that, however, I want to remind the reader of the distinction between systems that are, versus systems that are not, subject to natural selection in the Darwinian sense. Recall from earlier chapters that abiotic systems are not subject to natural selection. Nonliving, purposeless abiotic systems do not have fitness goals that are mediated and evaluated by natural selection. The same is true of organisms after they are dead. Dead organic matter (DOM) has the same status as inorganic matter relative to natural selection. Whether abiotic systems or DOM persist or not, in the context of their surrounding environments, is governed by the second law of thermodynamics.

Living systems are different, however. Living organisms are very improbable, far-from-equilibrium systems relative to their environments. Unlike abiota, living organisms are also purposeful systems in pursuit of their fitness goals of survival and reproduction. Instead of only reacting passively to the second law of thermodynamics as abiota do, organisms have to actively oppose the second law to achieve their fitness goals. They have to oppose the flow of energy and matter, or R_p, between themselves and their environments at each moment during their lives. Organisms can only oppose the second law by energy and matter, or R_p, consuming work. Organisms have to harvest the energy and matter resources that they need to fuel their work from their local environments. They also have to dump the detritus that they generate by living back into their environments. Organisms can do this only if they are informed with appropriate adaptive know-how, or R_i, by the processes of evolution.

Relative to these thermodynamic requirements of life, individual organisms in evolving populations are bound to encounter obstacles and threats, as well as opportunities, in their environments. The threats and opportunities, afforded to organisms by components of their environments are the sources of the natural selection pressures that they confront during their lives. Organisms encounter these natural selection pressures as a consequence of their niche interactions with their local environments, moment by moment, throughout their lives. It follows that natural selection originates from the demands that the second law of thermodynamics impose on organisms.

In 1859, when he released *The Origin of Species*, Darwin could not have anticipated that natural selection stems from the requirements of organisms to oppose the second law. The laws of thermodynamics were not yet known in his day, but they are known now. I suspect that in spite of Darwin's discovery of natural selection, we will never fully understand either natural selection or evolution, except in the light of the demands made on life by the second law. Natural selection is subordinate to thermodynamics. Specifically, it is subordinate to the demands made on living organisms by the second law. When we realize this, what difference could it make to our understanding of natural selection and evolution by taking us beyond what Darwin knew in 1859 and beyond SET today?

I suggest that there is a relationship between the degree of improbability or complexity of various species of organisms and the demands made on them by the second law. The more complex or improbable evolving populations of organisms become, the greater the demands made on them by the second law. If this is true, then it has several implications. First, if organisms cannot acquire the R_p and R_i that they need to oppose the second law, they will not be able to maintain their own improbability, relative to their environments, and will die. Second, if individual organisms in populations do manage to maintain their current level of improbability, they will not be able to evolve further improbability or complexity unless they can acquire additional R_p or R_i resources to pay for it. If organisms cannot acquire the additional R_p or R_i resources that they need to become more improbable, then they will get stuck at their present level of improbability, behind improbability barriers. Third, if evolving populations of organisms do manage to acquire additional R_p or R_i resources, then this may open up opportunities for them to break through whatever improbability barriers were previously constraining them. Is there any evidence to support these conjectures?

It may be possible to get some clues about the answer to this question by looking again at some of the least-understood phenomena that we know occurred in evolution in the past. These phenomena include "punctuate" equilibrium and "saltatory" evolutionary events, comprising relatively sudden jumps, often toward greater complexity, in evolving populations of organisms. Plausibly, such periods of evolutionary stasis in the fossil record may correspond to instances in which populations failed to acquire the necessary R_p or R_i, and consequently are hindered by improbability barriers, while the rapid evolutionary change manifest in the fossil record may follow the acquisition of the R_p or R_i necessary to break improbability barriers. They may also include the phenomenon known as "exaptation,"[30] which arise from the capacity of organisms to evolve adaptive redundancy. Because of this redundancy, exaptations sometimes appear to anticipate future sources of natural selection, but this is an illusion. "Exaptation" may also refer to the capacity of organisms to use their current adaptations in new ways.

I have previously described how, through their niche construction, evolving populations can behave opportunistically to exploit new resources, thereby potentially breaking the improbability barrier. An example may be provided by the transition of animals from sea to land, led by tetrapods, about 375 million years ago.[31] Moving into a novel environment is an example of inceptive, relocational niche construction. Tetrapods' ability to invade the land appears to have depended on the aggregation and cooperation of diverse adaptations, inherited from ancestral species of fish and earlier aquatic tetrapods, adaptations that had served different functions for the aquatic ancestors of the tetrapods. However, on land, the diverse adaptations were gradually combined and made to serve new functions during the subsequent evolution of tetrapods, allowing them to break improbability barriers, such as by exploiting land-based food resources. Subsequently, fins evolved into primitive feet, which allowed tetrapods some capacity to crawl, and later walk and run, on land.

THE MAJOR EVOLUTIONARY TRANSITIONS

Perhaps the best evidence for evolving populations of organisms breaking through improbability barriers, however, comes from the major transitions in evolution in both the bioenergetic transitions[32] and the bioinformatics transitions.[33] The major transitions typically involve the aggregation and cooperation of lower-level biological entities into a higher-level entity. Examples include the aggregation of genes into chromosomes, the aggregation of single cells into multicellular metazoan organisms and the aggregation of individual animals into social groups. In the light of these and other examples of the major transitions, I want to consider what contributions natural selection, whether modified by niche construction or not, could have made to the engineering of these transitions. I propose to consider three major transitions in evolution again: the origin of life, the transition from prokaryotic to eukaryotic cells, and the transition from social animals to language-based, human sociocultural groups.

The Origin of Life

The origin of life transition must have required breaking through the most formidable improbability barrier of all time. At present, we know of nowhere else in the universe where the fundamental improbability barrier of life itself was breached, although it is certainly possible that that occurred somewhere else in the universe. However, from what we now know about life on Earth, three requirements appear to be paramount for the origin of life, whether on Earth or elsewhere in the universe. First, living systems need boundaries separating them from their external environments. Second, because all living systems have to oppose the second law of thermodynamics to stay alive by active, purposeful work, they need sufficient energy, or R_p, to do the work to maintain their improbable existence. Third, all living systems also need sufficient, meaningful information, or R_i, when interacting with their external environments to confer the know-how necessary to survive and reproduce. Only when these three paramount requirements are met could it have been possible for life to have originated by breaking through the enormous primary improbability barrier between nonlife and life.

We discussed each of these three requirements, relative to the origin of life on Earth, in chapter 6. I don't propose to discuss them again here, but a brief reminder may be in order. In the light of both the deep-sea-vent and the cycling "warm little pond" hypotheses, how far have we gotten toward understanding each of these three paramount requirements? The easiest of these to understand is the origin of the boundaries between protocells, and later the first living cells, and their external environments. The boundaries around protocells, and later living cells, self-assembled into membrane boundaries. They did so primarily as a function of the hydrophilic and hydrophobic properties of lipids at the two opposite ends of themselves. These boundaries appear to have self-assembled in the form of lipid and sugar membranes. The membranes separated the protocells, or cells, from their external environments. These membranes must also have encapsulated tiny packets of improbability, in the form of the internal environments of the protocells, and later living cells. At the origin of life, they would have been micropackets of improbability.

The second paramount requirement of life—namely, sufficient energy resources, or R_p, to allow living systems to oppose the second law—was probably initiated by the proton-motive force (see chapter 6). The proton-motive force, flowing from the external environments to the internal environments of protocells or cells, synthesized adenosine triphosphate (ATP) within protocells and cells. Recall that ATP is the universal energy currency for biological systems. The presence of ATP could have led to chemical reactions in the protocells subsequently kick-starting metabolic reactions in the first living cells.[34]

The hardest of the three paramount requirements of life to understand is the third one. It is the origin of sufficient adaptive know-how, or R_i, to enable the first living cells to interact adaptively with their external environments in ways that allowed them to oppose the second law. This paramount requirement of life seems to have depended on a combination of factors. The first was probably the chemical kinetic traps that enabled the assembly of simple monomers into more complex polymers. According to Deamer and Damer, the assembly of more complex polymers would have been made possible by the succession of wet-moist-dry cycles in the freshwater of Darwin's "warm little ponds" in the volcanic regions of the Earth. In these environmental circumstances, chemical kinetic traps could have enabled the assembly of more complex polymers during the dry phases of the cycles at a faster rate than they could be broken down again by hydrolysis in the next wet phase of these cycles. Polymer strings, generated in this manner, could have been the elementary precursors of DNA-based genes, which register and encode the meaningful information, underpinning the adaptive know-how (or R_i) carried by organisms in evolving populations. That still does not explain why this genetically encoded information can be adaptive for the organisms that carry it relative to their environments. To explain that, we need some understanding of the origin of the three subprocesses of adaptive evolution: the generation of variety, the generation of fitness differences between variants, and the differential inheritance of fit variants by later generations.[35]

The generation of variety was probably initially a by-product of the capacity of protocells, and later of living cells, to reproduce. The reproduction of daughter cells by elementary kinds of cell division would have had two effects. It would have multiplied the number of protocells, and then cells, in populations. In chapter 6, I suggested that reproduction is an elementary kind of niche construction that modified the environment of each individual protocell, and then cell, by adding other protocells, and later cells, to its environment. At the same time, reproduction introduces variety into all evolving populations, if only because of inevitable miscopies. The presence of variety in these populations would then have been sufficient to have triggered the sorting between fit and less fit variants, thereby initiating the processes of adaptive evolution. This account of the origin of the third paramount requirement of life (namely, the adaptive know-how, or R_i, that organisms need to oppose the second law) is still sketchy and speculative. However, it is likely that it depended on both natural selection and the modification of selection by the niche-constructing activities of protocells, and later living cells as well.

Prokaryotes to Eukaryotes

The second major transition that I want to reconsider is from simple prokaryotic cells to more complicated eukaryotic cells. It represents another breakthrough, penetrating another improbability barrier. The principal contemporary hypothesis is that this transition was caused by prokaryotic cells engulfing other prokaryotes that became mitochondrial organelles in their cellular environments. The mitochondrial organelles provided the engulfing prokaryotic cells with an increase in energy resources, or R_p (see chapter 8). The extra energy that the engulfing prokaryotic cells gained from the mitochondrial organelles then gave them the opportunity to acquire and express a greater number of genes during their subsequent evolution. That in turn allowed the cells to acquire more adaptive know-how, or R_i, relative to natural selection pressures in their environments, enabling them to harvest still more R_p from their environments. This eventually allowed the prokaryotic cells to evolve into more complicated eukaryotic cells.[36]

How does niche construction contribute to this transition? The most obvious point is that because choice is a hallmark of niche construction,[37] the choice made by the first innovating prokaryotic cells to retain rather than eat the prokaryotes that they engulfed makes that choice a candidate example of niche construction contributing to this transition. That choice may have relied on the plasticity of the individual prokaryotic cells, since as Watson and Thies suggest, the collective fitness of the package of interacting prokaryotes may require plasticity on the part of each cell.[38] However, Watson and Thies also stress that for the collective to exhibit heritable fitness differences, the individual elements—here, both the engulfing and engulfed cells—must exhibit niche construction that modifies their interactions.[39] Without that niche construction, the higher-level functionality of the eukaryote would not be possible. The significance of this transition from prokaryotic to eukaryotic cells can be estimated by the time that it took for it to occur. Eukaryotic cells did not appear for more than two billion years after the origin of life on Earth. For all this time, prokaryotic cells were stuck behind an improbability barrier that they could not cross. The preceding billion years before the first eukaryotic cells appeared is sometimes rather dismissively called the "Boring Billion" by paleobiologists who do not work on this time period.[40] The paleobiologists and geologists who do work on the Boring Billion complain that this is unjust. They argue that a great deal was happening during this period that affected the subsequent evolution of life on Earth. Following the Boring Billion, the transition from prokaryotic to eukaryotic cells eventually led to another major transition in evolution, the transition from single-cell organisms to the first multicellular or metazoan organisms. But I won't say anything more about that transition here.

Culture to Language

The third major transition that I want to consider is from the relatively limited cultural traditions of nonhuman social animals to the far more advanced, language-based cultural traditions and institutions of human sociocultural groups. Here, various contemporary accounts of the origins of language already emphasize the construction of cultural, symbolic, collaborative foraging or pedagogical environments.[41] In addition, this transition, relative to the bioinformatics of evolution,[42] is roughly contemporaneous with the fifth bioenergetic transition, described by Judson as the "controlled use of fire by humans in human evolution"[43] (see chapter 5). The controlled use of fire is a good example of human cultural niche construction. It preceded the origin of our own species, *Homo sapiens*, perhaps by more than 100,000 years. The controlled use of fire could have been made possible only by an increase in adaptive know-how, or R_p, occurring among small groups of our ancestors, whether in our own species or in our earlier hominin ancestors. We don't know to what extent this increase in adaptive know-how depended on the ability of humans to exchange ideas and experiences by talking to each other in a shared language. We also don't know whether our earlier hominin ancestors had any linguistic abilities that could have contributed to their learning how to control fire. Fire is dangerous and can easily get out of control, if mishandled. On balance, the evidence of the apparent ability of our pre–*Homo sapiens* ancestors to control fire suggests they may have been aided by primitive language abilities that allowed them to communicate with each other.[44]

In the hands of our own language-speaking *H. sapiens* ancestors, the control of fire subsequently led to many other kinds of cultural niche construction, enabling them to harvest extra energy and matter (or R_p) resources from their environment. One early example is the use of fire by our ancestors for cooking.[45] Cooking allowed our ancestors to gain more food resources in two ways. It detoxified some foods that were toxic if eaten raw, and it increased the nutritional value of many cooked foods compared to raw foods, partly by aiding digestion.[46]

Later, the controlled use of fire enabled further innovations. For instance, it led to the forging of metals and metallurgy and to better tools and weapons.[47] Subsequently, in addition to the controlled use of fire, the diverse cultural niche-constructing activities of humans, in increasingly large socio-cultural groups, modified many more natural selection pressures in their environments (e.g., through aggregating into towns and cities, or through the domestication of plants and animals). This had the net effect of allowing our ancestors to harvest greater amounts of energy and matter from their environments. More recently, the controlled use of fire, in conjunction with many other kinds of cultural niche construction, contributed to both the Agricultural and the Industrial revolutions. They also contributed to the construction of the first urban settlements and urban-based civilizations (see chapter 10).

Shared Fate and Separate Fate

Thus, all the major transitions that I've mentioned here probably depended on both natural selection in the environments of organisms and on the niche-constructing activities of organisms modifying those selection pressures. But how does natural selection work after a major transition has occurred, compared to how it worked before the transition? This takes us into multilevel selection theory[48] and to a comparison between what I'll call "shared fate" natural selection and "separate fate" natural selection. I will illustrate this point by comparing the selective fate of a single-cell organism with the selective fates of various eukaryotic cells embedded in the body of a multicellular metazoan organism.

Natural selection works in the same way with single-cell organisms as it does with all organisms. But it doesn't work in the same way for different individual cells that are embedded in the bodies of multicellular organisms. None of the cells of metazoa have separate selective fates from others. In spite of the many differences between different kinds of cells in metazoan organisms, all of them share their naturally selective fates with other cells in the organism, for two reasons. First, because they are all component cells in one or another organ or system of a metazoan organism, the immediate source of natural selection for any individual cell is the microenvironment provided by the metazoan's organ or system to which the cell belongs. Second, provided that the cells of a metazoan organism survive long enough, they will ultimately share the naturally selected fate of the metazoan organism itself, relative to sources of natural selection in the metazoan's external environment.

There are advantages and risks of this relationship, both for individual cells and their host metazoan organism. A major advantage for both the individual cells and the metazoan organism to which they belong is greater thermodynamic efficiency with respect to opposing the second law of thermodynamics. This may take the form of better access to energy and matter (R_p) resources or a reduced cost of gaining those resources, both for the individual cells and for the organism as a whole. But the metazoan organism also suffers from a potentially significant risk, which we have seen before. If an individual cell or a few cells in one of the metazoan's organs or systems become severely maladapted relative to its local natural selection pressures, this may be detrimental to the organism itself. The entire system could crash and the organism could die.

There are some other risks too. For instance, there is a risk that the cooperative relationships between the aggregate of cells in the metazoan's organ may break down. The cooperative aggregate may fragment into cellular components that go their separate ways, as manifest in cancer or in autoimmune diseases. The possible links to cancer or to autoimmune diseases are beyond the scope of this book. However, at a completely different level, human sociocultural groups such as a nation could fragment in adverse environmental circumstances, and I will return to the formation and breakup of human sociocultural groups in chapter 10.

The transitions also demonstrate a variety of ways in which the acquisition of energy and matter resources (R_p) and the acquisition of meaningful information (R_i) can ratchet each other up during a major transition. By doing so, they can sometimes enable evolving populations of organisms to cross another improbability barrier that was previously constraining them, thereby causing a further transition in evolution.

HOW AND WHY ORGANISMS CONTRIBUTE TO NATURAL SELECTION

All of the information given thus far converges on my second query about natural selection: How and why do evolving populations of organisms contribute to their own and each other's natural selection in the context of their ecosystems? Recall from chapter 8 that the niche interactions between organisms and the sources of natural selection in their environments are two-way-street interactions. Organisms respond to both abiotic and biotic sources of selection in their environments, as both SET and NCT recognize. But they also modify both abiotic and biotic sources of natural selection in their environments by their niche-constructing activities, as NCT emphasizes, but SET does not. SET recognizes that organisms are capable of niche-constructing activities that modify components of their environments. But SET assumes that adaptive niche construction can always be fully accounted for by prior natural selection. SET does not recognize the reciprocal causal arrangement that natural selection can sometimes be a product of prior niche construction, and hence that adaptive evolution is determined by both natural selection and niche construction.[49]

The niche-constructing activities of organisms, as described by NCT, introduce feedback into evolution relative to both abiotic and biotic sources of natural selection, in their own environments and in the environments of other organisms in their ecosystems. This feedback may affect the future evolution of their own populations, as well as other populations. The feedback also has another more general implication. It implies that evolutionary theory should not be about the evolution of populations of organisms, but rather it should really be about the coevolution of organisms with their environments. Or, if looked at in the opposite way, it should be about the coevolution of environments with their organisms.

A general theory of organism-environment coevolution is alien to SET. That is because the theory does not recognize niche construction as a cocausal process in adaptive evolution, working in partnership with natural selection. Currently, SET recognizes that two or more evolving populations of organisms can coevolve with each other in the context of population community ecology. It also recognizes that different traits in the same population can coevolve, as observed in sexual selection. However, SET does not recognize the coevolution of organisms with abiotic components of their environments. In contrast, NCT recognizes both the coevolution of evolving populations with each other and the coevolution of evolving populations with specified abiotic components of their own and each other's environments.[50]

Purpose

The last assumption of SET that I want to consider concerns that word "purpose" again. SET acknowledges that organisms have fitness goals but denies that to achieve their fitness goals of survival, growth, and reproduction, organisms must be active, purposeful agents relative to their interactions with their external environments. Rather, SET assumes that the apparent purposefulness of organisms is misleading.[51] The proponents of SET argue that that is an illusion, which can be fully explained by prior natural selection. Biologists are often happy to talk about the purposes of organisms, when talking informally either to each other or to laypeople. For instance, when talking informally, biologists tell us that plants "seek" light and their roots "seek" moisture, birds and mammals "choose" mates, and some Japanese macaques wash potatoes "to" clean them. The easiest way to talk about all these actions is in terms of the purposes of the organisms that emit the actions. But when talking formally and in the context of SET, biologists switch to talking about the "structures" and "functions" of the phenotypic traits of organisms instead of talking about their purposes. SET assumes that the phenotypic traits of organisms are consequences of prior natural selection, often with reference to the naturally selected genes that organisms inherit from their ancestors. Perhaps biologists are closer to the truth when they talk informally about the purposes of organisms, rather than when they are talking formally.

I think biologists are unnecessarily coy when it comes to their unwillingness to describe organisms as active, purposeful agents in evolution. I suggest that this coyness ultimately has two sources: fear of God and physics envy. Fear of God reduces to fear of the Creationists. In her book *The Restless Clock*, Jessica Riskin draws attention to what appears to be an inadvertent collusion between the neo-Darwinists on the one hand and the creationists and the mystical vitalists on the other.[52] She makes this collusion sound both irrational and comic. Both the neo-Darwinists and the creationists appear to think that if organisms really are purposeful agents, then their purposes can be attributed only to supernatural processes, or deities, or God. Creationists like to argue just that, but the neo-Darwinists do not.

Ever since Darwin, there has been a tension between the narratives of various religions and evolutionary biologists. Maybe it is because of this tension that many evolutionary biologists feel threatened by the creationists. Some seem to want to put as much distance as possible between themselves and the creationists. In the light of the evidence provided by nature and in the light of reason, evolutionary biologists are clearly right to oppose the creationists. They do not need to "bother God" to explain either the evolution of life on Earth, or the active, purposeful agency of organisms. It therefore seems irrational for SET to claim that the active, purposeful agency of organisms is just an illusion.

So much for the fear of God; what about physics envy? Physics envy starts with the point that physicists don't have to deal with such messy things as purposeful systems. Were biologists to start referring to organisms as purposeful systems, they might run the risk of not being regarded as serious scientists, at least not by physicists. I think that is an inappropriate fear, too. I suggest that Schrödinger, in his book *What Is Life?* was right when he implicitly suggested that biology is harder than physics by his analysis of what else is needed, beyond the laws of physics, to understand life.

There is also one interesting exception to the general rule that physicists don't have to deal with purposeful systems. Physicists, engineers, mathematicians, and programmers sometimes do have to deal with the purposes and the goals of some of the artificial systems that they themselves construct. One example is provided by the software in some modern cars, which can decide for themselves whether their drivers are drunk or not, and if so, pull to the side of the road. Some autonomous decision-making capacity has to be delegated by the designers and programmers of these cars to the cars themselves.

A more dramatic example is provided by the Rovers that humans are now sending to the neighboring planet of Mars. To what extent must the human constructors of a Mars Rover equip it with some autonomous capacity to make some decisions for itself as it trundles across the surface of Mars? This question is made more relevant by the approximately twenty-minute delay that is required for a single communication between the Mars Rover and its ground controllers, owing to the distance between Mars and Earth. As robots get more advanced, the degree of autonomy that is assigned to them by their constructors is likely to increase greatly. For example, even at present, if a Mars Rover runs into an obstacle, it needs the capability to decide for itself whether to surmount or circumvent the obstacle, on the basis of its own built-in sensory and decision-making capacities. It may make mistakes of varying seriousness. One serious mistake, sometimes referred to by the National Aeronautics and Space Administration (NASA) as "high-centering," occurs when the Mars Rover drives into a mound of sand that collects under its belly, lifting it up to the point that its wheels no longer touch the ground. This mistake can be fatal to the Rover. Even if not fatal, it may still take the Rover's ground controllers on Earth weeks to get it moving again, thanks to the necessarily slow two-way communication between the Rover on Mars and the controllers on Earth.

Mars Rovers are built by their human constructors for a human purpose. The constructors of the Mars Rover have to anticipate as many of the problems that the Rovers will encounter as possible in advance in order to enable the Rover to fulfill the purposes assigned to it by its constructors. But the Mars Rover's constructors must also anticipate the kind of problems that are best left to the Mars Rover itself to solve, independent of its ground controllers. Its constructors must also be able to equip the Mars Rover in advance with sufficient autonomous sensory and decision-making capacity to solve those problems for itself.

Given that it is not possible for any purposeless system to anticipate any future events, this takes us back to Conrad Waddington's (1969) question of how any system, living or artificial, can anticipate its own "unknown but usually not wholly unforecastable future'."[53] It also takes us back to the bootstrapping problem again, which occurs in evolving populations, where purposeful individual organisms are equipped by supplementary evolutionary processes, including both developmental and cultural processes, to make some decisions for themselves (see chapter 7). Unlike purposeless abiotic systems, purposeful living organisms can anticipate some future events, relative to their purposes, while pursuing their fitness goals.

There is now more than one level of control in the Mars Rover that affects its performance. It comprises the ground controllers on Earth and the Rover's autonomous capacity to control some of its own actions in its Martian environment independent of its ground controllers. However, this analogy between a Mars Rover and an individual organism in an evolving population breaks down in one crucial respect. The designers and controllers of the Mars Rover are cognitive, competent, intelligent human beings. According to both SET and NCT, there is no equivalent intelligent designer in biological evolution. There is only the laborious process of the natural design of organisms by the apparently purposeless and partly chance-based processes of evolution. Intelligent human designers can design artificial systems, including Mars Rovers, that are capable of taking some decisions for themselves, independent of their ground controllers. But how is it possible for an apparently purposeless evolutionary process to assemble organisms in populations with any degree of active, purposeful agency to enable them to oppose the second law of thermodynamics? This is equivalent to Paley's question (see chapter 3). If a complicated system such as a watch, whose purpose it is to tell the time, cannot exist in the absence of an intelligent watchmaker, then how can a much more complicated living organism, whose purpose it is to oppose the second law of thermodynamics, exist as a consequence of the apparently purposeless processes of evolution?

The philosopher of biology Denis Walsh[54] describes this dilemma in terms of one cause and two effects. The single cause of evolution is the continuous turnover of individual organisms in evolving populations. This turnover is due to the struggle for existence that each individual organism has to endure to survive and reproduce. The two effects occur at two levels.

The lower-level effect is the immediate struggle for existence of every individual organism in every evolving population to survive and reproduce. This effect is a manifestation of the active, purposeful agency that improbable individual organisms have to possess to resist the second law. By definition, living organisms have to be purposeful systems to exist in their environments in opposition to the second law. Purpose is a defining property of life (see chapter 6).

The higher-order effect (namely, the evolution of populations) is a consequence of enough individual organisms in evolving populations, achieving their fitness goals of survival and reproduction. But to survive and reproduce, individual organisms must be equipped with sufficient adaptive know-how about their local environments. This is possible only if organisms inherit sufficient meaningful information, in the form of adaptive know-how (R_i), as a result of their ancestors' struggles for existence in past environments. Evolving populations can appear purposeful only as a consequence of the active, purposeful agency of all their individual organisms.

This is also where the analogy breaks down between the natural design of organisms by the processes of evolution and the design of purposeful human artifacts, such as the Mars Rover by teams of humans. Another way of illustrating the difference between a living organism and a Mars Rover is with reference to time. It took approximately four billion years for the evolutionary process to come up with the cognitive, conscious, intelligent human designers who built the Mars Rover. In contrast, it only took a few years for teams of physicists, cosmologists, astrobiologists, engineers, mathematicians, and programmers to design and build the Mars Rover and send it to Mars. That is because the cultural evolutionary process is able to exploit adaptive know-how accrued through the process of biological evolution, which helps it to provide rapid solutions to adaptive challenges.

Another relevant question here is whether it is appropriate to use the same words to describe the cognitive, conscious purposes of human beings as it is to describe the purposeful phenotypic traits, motives, and activities of nonhuman organisms.[55] The incipient problem is that words such as "anticipate," "intention," "expectation," "cognition," and "perception," as well as the word "purpose" itself, have powerful connotations with human cognition and consciousness. This may be another reason why SET is so reluctant to accept that the phenotypes of living organisms are purposeful systems. The words that we use to describe purpose appear to anthropomorphize biology in ways that are completely inappropriate. But by these words, I am only seeking clarity by using plain English. Logically, none of these words need imply anything about the cognitive, conscious, or unconscious states of living organisms. They merely imply different capacities and system states, in different species, and at different places and moments.

Teleonomy versus Teleology

There is also one other possible reason why biologists are so reluctant to acknowledge that organisms are active, purposeful agents in evolution. It is an historical and philosophical reason. It goes back to Aristotle's concept of a final cause. This has frequently been interpreted as if Aristotle were implying that effects precede their final cause, but that would be nonsense, as philosophers and scientists have always pointed out. However, it is possible for an anticipated future goal, derived from the purposes of any kind of purposeful system, to motivate the achievement of future goals or subgoals. This is not equivalent to a teleological explanation—that would imply that the current activities of organisms are effects of some final cause. Here, they are only the effects of the goals anticipated by active, purposeful systems, including living organisms. A simple human example is athletes. An athlete may anticipate a future goal of winning a medal during some future Olympic Games. That anticipated future goal will then cause the athlete to train hard in the present in the hope of achieving the goal. More generally, the anticipated future fitness goals of contemporary organisms are likely to cause aspects of their behavior and other outputs, from searching for food in animals, to growing roots toward soil moisture in plants, to moving along chemical gradients in microorganisms.

This idea is closer to Pittendrigh's concept of "teleonomy" rather than Aristotle's "teleology." Pittendrigh introduced the word "teleonomy" in 1958[56] to refer to the goal-directedness of the adaptations of fitness-seeking organisms, which arises from the past and present functions of the adaptations of organisms. It is not to be confused with Aristotle's final cause, and therefore with the finalism of teleology. Instead, it seeks to explain the past and present activities of organisms in terms of their assumed, anticipated fitness goals. Aristotle's teleology implies that the final effects are the cause of their prior causes, which would require a reversal of the arrow of time. Teleonomy avoids this error by pointing out that an anticipated future goal may motivate (and therefore cause) some of the current behaviors and other activities of organisms. That does not demand a reversal of the arrow of time. All it does is explain the current activities of phenotypes by referring to what are assumed to be the "anticipated" future fitness goals of the organisms. In practice, this is a ploy frequently adopted by biologists when they are trying to understand the adaptations of organisms.

It is interesting to speculate whether Aristotle's final cause may originally have included this teleonomic component. It is also of interest that even though in physics and cosmology there is no such thing as a final cause, the third law of thermodynamics does describe a final destination. It is sometimes called the "heat death" of the universe and is defined as when the universe is due to achieve a final temperature of absolute zero. It represents the end of change in the universe and the end of time. It corresponds to the total dissipation of all order, the total dissipation of free energy, and the achievement of maximum entropy by the universe. Confusingly, this does not affect the universal conservation of energy, as described by the first law of thermodynamics. There is no change in the energy of the universe—only the total loss of free energy capable of doing work and of causing change and driving the arrow of time.

This final destination of the universe, which may or may not ever be reached in reality, does not cause anything teleologically. Rather, it is the consequence of all the changes that preceded it earlier in the history of the universe, in accord with the second law of thermodynamics. The ultimate attractor of all these changes is maximum thermodynamic stability, and therefore maximum entropy. This irreversible increase in entropy throughout the universe also accounts for the arrow of time itself, which proceeds from past to present to future. The arrow of time is compatible with conventional cause-and-effect relationships and with Pittendrigh's teleonomy. It is incompatible with Aristotle's teleology and his final cause.

CONCLUSIONS

The implication of this chapter is that all these core assumptions of SET need revising. We saw at the beginning of this book how these core assumptions were initially justified by the need to understand the fundamental evolutionary processes of Darwinian natural selection combined with Mendelian genetic inheritance. But now they are holding back an advance toward a more comprehensive theory of evolution that can account for new data, as well as for additional processes, such as evolutionarily relevant developmental processes, that are not yet fully recognized by SET.

I will conclude this chapter by summarizing the principal points raised in this and the previous chapters that will have to be addressed by a more comprehensive theory of evolution. I have emphasized the relevance of the second law of thermodynamics to the processes of evolution. All living organisms have to struggle for their existence throughout their lives, as Darwin pointed out. They do so to protect their own improbable status and to achieve their fitness goals of survival and reproduction, and in the process oppose the second law of thermodynamics.

At the start of their lives, organisms receive a niche inheritance as a gift from their parent cell or organism. This niche inheritance includes their initial location in environmental space and time, and minimally it must also include the energy and matter (R_p) resources and adaptive know-how (R_i) to support them at the start of their lives. After that, organisms have to utilize their initial niche inheritances to oppose the second law of thermodynamics in order to gain further energy and matter resources from wherever they are in their environments.

The capacity of the primary process of population-genetic evolution to bootstrap, then, makes it possible for many organisms in diverse species to gain additional adaptive know-how through developmental processes, or in the case of some social animals (notably humans) through sociocultural processes as well (see chapter 7). Many organisms may also be able to transmit some of their acquired knowledge or characteristics to their offspring through one or more of the nongenetic inheritance systems reviewed in chapters 8 and 9. In addition, offspring inherit acquired environments, in the form of modified sources of natural selection, due to the prior niche-constructing activities of their parents and earlier ancestors. The capacity of active, purposeful organisms to construct their niches changes the role of phenotypes in evolution, in ways that still are not fully recognized by SET.

This capacity of organisms to construct niches also changes what a more comprehensive theory of evolution should be about. Instead of being exclusively about the responses of organisms to natural selection pressures in their environments, as per SET, evolutionary theory should be about the coevolution of organisms with their environments.[57] If a more comprehensive theory of evolution recognized that the subject matter of evolutionary theory should be organism-environment coevolution, it would do justice to the active, purposeful agency of phenotypes in evolution, in contrast to the more limited role assigned to phenotypes by SET. It would also acknowledge the subordination of natural selection to the second law of thermodynamics. There needs to be a new synthetic theory of evolution, which includes ecosystem-level ecology, inclusive of the impact of niche-constructing organisms on their environments. Let's call it an "eco-evo-devo synthesis."[58]

10 HOW ARE HUMANS CONTRIBUTING TO THE EVOLUTION OF LIFE?

In this final chapter, I want to consider how the extended theory of evolution, sketched in the previous chapters, applies to human evolution and to our contemporary lives. I also want to consider how humans are currently contributing to the evolution of life on Earth. To what extent does the extended synthesis shed more light on who we are and where we have come from, and cause us to think again about how we are living our lives now? We seem to be standing on the threshold of two very different futures, one negative, one positive. I'll postpone talking about the positive future until we've considered the negative one.

We are all becoming more aware of a possible negative future. It takes the form of a human-induced mass extinction of life on Earth. Such a mass extinction could be achieved in at least two ways—by nuclear war or global warming—both of which are due to the potency of our cultural niche-constructing activities. Standard evolutionary theory (SET) does not yet acknowledge that niche construction is a cocausal process in evolution. It is therefore not well placed to understand why humans are currently facing the prospect of human-induced mass extinction by either of these two alternative routes, nor how it could be avoided. As a result, these issues are often left to the social sciences to worry about. Conversely, the proposed extended synthesis includes the recognition of cultural niche construction as a significant adaptive process in human evolution. This extended synthesis may thus be better placed to enlighten us about our possible negative future. It might also help us understand how we could avert it.

NUCLEAR WAR

Let's begin by considering nuclear war. One day of all-out nuclear war would be more than enough to trigger a human-induced mass extinction of life on Earth. Yet we are still developing progressively more dangerous nuclear weapons in high numbers. This is in spite of the fact that it might take only about 2 percent of the world's total arsenal of nuclear weapons to trigger a mass extinction of life on Earth.[1] Nuclear weapons are also proliferating among more nations. We seem unable to sufficiently inhibit our between-nation rivalries to avoid all possibility of a nuclear war. In the Cold War between the Soviet Union and the West in the 1960s, the policy of mutual deterrence by "mutually assured destruction" was nicknamed MAD!

Why have we not yet quelled this madness? It may be an extreme example of an adaptive lag in human evolution. Recall that nuclear weapons were first assembled by a relatively small number of talented physicists, mathematicians, and engineers working on the Manhattan Project in Los Alamos in the US, during World War II. It was an emergency project, designed to respond to an immediate threat. At the time, it seemed imperative to the Allies to acquire nuclear weapons before Hitler and the Nazis did. They succeeded in that, but not before the war with Germany was over, even though the war with Japan was still continuing. The first atom bombs were dropped on Hiroshima and Nagasaki in Japan, in August 1945, which ultimately ended World War II, but it introduced a new problem for humanity. The production of those atom bombs, by the scientifically enhanced niche-constructing activities of a few clever people, changed our world irreversibly. It introduced the possibility that humans could now cause a mass extinction of life on Earth by engaging in a nuclear war. Shortly after the destruction of Hiroshima, Albert Einstein made a memorable comment: "The release of atom power has changed everything, except our way of thinking."[2] The release of atom power is now demanding that we change the way we think, particularly about our relationships with each other. It requires most of us, not just those responsible for controlling nuclear weapons, to think differently.

How could we change the way we think? Since we are both social and cultural animals, we need to address that question with our "sociocultural" brains.[3] It is possible to detect three kinds of adaptation in our human sociocultural brains and behaviors. They are derived from the three levels of evolutionary processes that we met in chapter 7. They are the primary processes of population-genetic evolution at level 1, the relevant supplementary developmental processes in individual organisms at level 2, and the additional supplementary sociocultural processes at level 3. As we saw in chapter 7, the processes that operate at each of these three evolutionary levels acquire qualitatively different adaptive know-how, or R_j, on behalf of organisms, human and nonhuman alike, contributing to social adaptation in each case.

In the case of humans, that means that our sociocultural brains and behaviors are informed by three qualitatively different kinds of meaningful information, which, to be consistent with chapter 7, I'll call "level 1 adaptations," "level 2 adaptations," and "level 3 adaptations," respectively. If the various kinds of adaptation, acquired at each of the different evolutionary levels, enhance or are fully compatible with each other, then humans are likely to remain well adapted to their environments most of the time. But if for any reason, the kinds of adaptations become incompatible with each other, then, even if they were originally well adapted to specific components of their environments at each separate level, in combination, they may become maladapted relative to their environments. This is not only true of humans but of all animals that depend on more than one evolutionary level for their brains and behaviors.

Humans are more prone to these between-level incompatibilities than other animals. This is because of the potency of our cultural evolutionary processes at level 3 relative to our level 1 and level 2 adaptations. For example, our level 1 and level 3 adaptations can become incompatible with each other for two reasons. The first is that the cultural selection processes that operate at level 3 may select for adaptations (e.g., birth control) that are incompatible with our adaptations selected by natural selection at level 1 (e.g., genetic propensity to reproduce). The second reason is that our potent cultural niche-constructing activities, at level 3, may change our human environments (e.g., through industrial pollution) in ways that make them incompatible with our adaptations previously gained via population-genetic evolution at level 1 (i.e., leading to health problems). Either or both of these outcomes may then introduce an adaptive lag between out-of-date adaptations, encoded in our naturally selected genes at level 1, and our sociocultural adaptations, registered in our brains at level 3.

For the moment, I'll simplify my discussion about how these processes at different levels interact by ignoring the supplementary individual developmental processes at level 2, and focusing on incompatibility problems arising between level 1 and level 3 adaptations. I'll start with level 1. The principal legacy that humans have inherited from our social primate ancestors comprises within-group cooperative social alliances, which can also lead to between-group rivalries. Both the degree of cooperation within groups and the degree of hostility between groups can vary considerably in these relationships. I will call this legacy a "level 1 social adaptation." Organisms in the same species, including our own, are always potential competitors. They all need to acquire much the same energy and matter from their environments as each other in order to stay alive by resisting the second law of thermodynamics. However, individual humans, like many other social primates and social animals such as ants,[4] may do better in fitness terms by forming within-group alliances with some other members of their own species. Individuals in these within-group cooperative alliances may then contribute to competitive relationships between their own social groups and other rival groups of conspecifics. In humans, the net result is a pattern of within-group alliances and between-group rivalries, hostilities, and competition that we inherited from our social primate ancestors via our naturally selected genes at level 1.

Now let's turn to level 3 sociocultural adaptations, which are not strictly genetic adaptations but rather traits produced by a cultural evolutionary process. Although many other animals acquire social adaptations from cultural processes too, from here on I'm going to be talking about humans. I do not intend to do more than describe the principal alternative ways in which it is logically possible for individual humans to come together to form within-group alliances and to split up again subsequently. I will concentrate on how individual human decisions, choices, and interactions affect the formation of alliances, or their breakup, and ignore the many other factors that affect human groups and societies. These factors encompass multiple anthropological and sociological variables, including diverse ecological variables, such as the availability of energy and matter resources in human environments or the presence of parasites or infectious diseases. All these other variables are important, but they are not directly relevant to the particular issue that I want to focus on here.

Human sociocultural adaptations at level 3 include diverse rituals, laws, customs, and ceremonies, as well as political ideologies and religious beliefs. They may be supported, and are sometimes enforced, by social institutions, such as religions or legal systems. Sociocultural institutions may be required to enable humans to put their diverse sociocultural groups together in different ways in different parts of the world. These human sociocultural adaptations at level 3 do not replace the level 1 in-group versus outgroup adaptations inherited from our primate ancestors. Rather, they are superimposed on our level 1 adaptations (as discussed later in this chapter). But is this superimposition invariably adaptive?

The psychologist and primatologist Robin Dunbar addresses this question in his book *How Religion Evolved.*[5] Similar to primate social groups, human groups were probably very small for most of the time that our species has existed, as they remain among the few human hunting and gathering communities left today. Dunbar discusses what it takes for primate social groups, including humans, to grow larger than what is now called the "Dunbar number" of about 150 individuals. Dunbar claims that this number is approximately the maximum size of human social groups that depend exclusively on close kin, personal friendships, and useful acquaintances. The idea remains controversial,[6] but less contentious is the general idea that human social groups need mechanisms or structures that provide additional support as they grow larger, including some form of within-group policing to hold them together, as well as a capacity to defend themselves from competition or attacks by rival groups. The additional support typically involves a variety of social institutions, often arranged within more-or-less hierarchical social structures, which vary across sociocultural groups in different parts of the world. Leaders and other positions of authority may emerge within these social groups, with these individuals able to make some decisions on behalf of their groups. This risks introducing or exacerbating inequalities between different individual members of the social group; for instance, if those in positions of authority are given or demand extra privileges or resources at the expense of other members of their group. If the inequalities become excessive, that may cause larger social groups to split up.[7]

This is where our human sociocultural adaptations at level 3 come into play, as they allow human groups to get bigger and become potentially more valuable in fitness terms for individual members of their groups. Individual members of social groups may either support their groups in some way or they may parasitize them by becoming free-riders, to the detriment of others in their group. What any individual human may do is also likely to depend on his or her level 2 developmental adaptations, which encompass individual learning and personal experience, although I will not dwell on it here. On all scales, human groups are far more likely to endure if enough individuals in the group support their cultural institutions and their customs, laws, and governance.[8] If they do endure, the social group may still have to compete with rival outgroups for economic resources in their shared environments. Alternatively, a larger group may fragment into smaller subgroups, which may then start to compete with their original group or with each other. This is more likely to happen when individual members of the original group take much more than their fair share of the benefits conferred on individuals by membership of their group, at the expense of others.[9]

A third possibility is that two or more originally rivalrous social groups may settle their differences, form alliances with each other, and merge to form a single, larger sociocultural group. This could confer increased fitness benefits for individual members of the enlarged group, but it is likely to happen only through the application of level 3 adaptations, such as trade, negotiation, and social learning. If human sociocultural groups do get much larger, individual members of the group may form substructures, without threatening the unity of the wider group. Such substructures are boundless in modern societies. They can be based on shared interests and shared skills and often involve a division of labor between different members of a sociocultural group. Shared interests may stem from family groups, schools, businesses, professions, football clubs, dramatic societies, choirs, social media groups, and much else. But these kinds of structural organization within larger sociocultural groups are more likely to be an asset to the enlarged group, rather than a threat, so long as all the components remain affiliated to the larger group.

Historically, neither the splitting up of groups nor rivalries between groups, with or without armed conflict, would have mattered much during most of human existence. They would not have posed wider threats to the ongoing evolution of our species or of life on Earth. However, during the last few centuries, human warfare has become much more destructive. This has been mainly due to our increasingly powerful weapons. For any particular sociocultural group, developing more powerful weapons, via human cultural niche construction at level 3, has frequently been adaptive. For instance, insofar as it enabled a sociocultural group to outcompete a rival group, it may have enabled it to gain more resources from its environment at the expense of its rivals.[10] However, ever since the destruction of Hiroshima by a nuclear weapon, that is no longer the case. Nuclear weapons are unusable. If a sociocultural group or nation were to use nuclear weapons against a rival group or nation that also possessed nuclear weapons, the original sociocultural group would destroy itself, its rivals, and much of the rest of life on Earth too. Multiplying our stocks of nuclear weapons is risky in itself. Nor does it make any sense, given that it would take only a small percentage of the world's nuclear arsenal to induce a mass extinction of life on Earth, which would almost certainly include our own self-induced extinction. That is not adaptive; it's MAD.

From the point of view of evolutionary biology, this madness is due to an unusual kind of maladaptation. We usually think of maladaptations in terms of one or more phenotypic trait in organisms that are badly adapted to aspects of their environments. But the maladaptation this time takes the form of an adaptive lag, generated by the incompatibility of our level 1 social adaptations, derived from human population-genetic evolution, with our social adaptations, derived from the qualitatively different adaptive know-how that we acquire from our supplementary cultural evolutionary processes at level 3. Our genes do not confer adaptations that allow us to survive a nuclear war, but they may predispose us to outgroup hostility.

This unusual kind of maladaptation raises two points. The first is that, because the source of this maladaptation is the adaptive lag generated by the incompatibility of our ancient level 1 social adaptations with our much more recent level 3 sociocultural adaptations, it will not be possible for our species to correct for this particular maladaptation at a single evolutionary level alone. Given the immediacy of our present human predicament in a world that we have now changed irreversibly by inventing nuclear weapons, we cannot wait for population-genetic evolution at level 1 to catch up by diminishing our between-group rivalrous social behaviors.

Conversely, we will not be able to correct for this adaptive lag maladaptation by further cultural evolution at level 3 unless we also take our level 1 social adaptations fully into account. To do that, we will have to recognize the primacy and potency of our level 1 social adaptations and their persisting consequences for our species today. We are going to have to come to terms with the persistence of our level 1 social adaptations, which manifest as tendencies to generate both alliances within groups and hostilities or rivalries between groups. The prospect of a future catastrophic nuclear war is not just a political problem; it is also a major evolutionary problem for our species.

This brings me to my second point. Our current orthodox theory of evolution—neo-Darwinism or SET—cannot supply an adequate theoretical framework for understanding the adaptive lag between our level 1 and level 3 adaptations, and therefore the nature of our current predicament. This is because SET does not recognize niche construction as a cocausal process in the evolution of organisms. It only recognizes niche construction as a product of prior natural selection. Nor does SET recognize that the adaptive know-how- or R_i-gaining processes of development at level 2, or the sociocultural processes at level 3, can be supplementary sources of adaptation for organisms. Conversely, the extended theory of evolution explicitly recognizes both niche construction as a cocausal adaptive process and developmental and cultural processes as supplementary adaptive know-how gaining processes in evolution. Might this extended theory of evolution do a better job of enabling us to understand our present maladaptive predicament in an age when rival social groups or nations are equipped with nuclear weapons?

Correcting for this adaptive lag will not be easy. A major problem for all evolving populations is that evolution cannot go backward. Evolving populations can only go forward from where they already are, here and now. However, evolution can sometimes undo an adaptation that was previously acquired by an evolving population by further evolution. A well-known example is cavefish in Mexico, which have lost their eyes and have no residual vision after living in dark caves for generations.[11] Another example is naked mole rats, which now live underground in permanent darkness and have also lost their eyes.[12] However, level 1 evolution alone is unlikely to work here, both because there may be insufficient time and because it is unclear that any diminishment in the social adaptation is selectively advantageous.

It might be possible to get some hints about how we could correct for this particular adaptive lag from a surprising source: sex. Within our various cooperative social groups today, we attempt to civilize our sexual attractions, affections, desires, and appetites between individuals by imposing social constraints on them that are derived from human sociocultural processes at level 3. These constraints take the form of laws, customs, rituals, and ceremonies, supported by institutions and even social mores or gossip. For example, to the best of my knowledge, every human society recognizes marriage between men and women, although marriage can be defined differently in different societies. In some societies today, same-sex marriages are also recognized and legitimized by social institutions. Perpetrators of sexual violence are often subject to severe punishment. In these several ways, although the results are never perfect, to some extent we do manage to apply our sociocultural adaptations at level 3 to limit the extent to which our sexual adaptations at levels 1 and 3 are incompatible. Unlike for human sociobiology, when cultural niche construction and multiple-level adaptation are taken into account, there is no inevitable adaptive lag between our level 1 and level 3 adaptations with respect to human sexuality.

Might it be possible to use the theoretical framework offered by the extended theory of evolution to help develop analogous sociocultural mechanisms that are capable of constraining our between-group and between-nation rivalries in a way that permanently eliminates the possibility of a future nuclear war? Might we be able to regulate our relationships between rival social groups and nations, which have now become maladaptive because of the invention of nuclear weapons?

In the twentieth century, after both world wars, we did attempt to build international institutions designed to prevent future wars. They were the League of Nations after World War I and the United Nations after World War II. The League of Nations failed to prevent World War II, and so far the United Nations has not been effective or powerful enough to prevent innumerable wars from breaking out in the late twentieth and early twenty-first centuries, including between nuclear power nations such as India and Pakistan. If my argument in this chapter is correct, we will not be able to quell our nuclear madness or avert the negative future that it threatens by concentrating exclusively on our level 3 sociocultural adaptations.

The principal point, which we are not yet fully acknowledging, is that our level 3 sociocultural adaptations cannot replace our level 1 social adaptations—they can only be superimposed on them. Because of the immediacy of our present predicament, we can't wait for genetic evolution to solve the problem, nor can we guarantee that it will. This means that to avert the possible negative future of human-induced mass extinction by nuclear war, we will have to understand ourselves, as best we can, in the context of both our evolutionary and our cultural past. We will probably never get rid of our between-group rivalries, that we have inherited from our social primate ancestors fast enough. We are therefore going to have to accept our level 1 social adaptations as a long-term component of our contemporary lives, and try to tame them by further cultural evolution and social adaptations at level 3.

To understand ourselves better, however, we are going to need the aid of our biological sciences, our environmental sciences, and our human and social sciences. Unfortunately, the relationships between the biological and the human social sciences have not been good for decades. This has been partly due to the rejection of SET by the social sciences. Social scientists were and still are unhappy with human sociobiology[13] and the "selfish gene" approach to evolution.[14] It has also been due to a tendency for the social sciences to "throw out the baby with the bath water" by claiming that the biological evolution of our species is of little relevance to the phenomena that they now study. Ironically, the biological and human sciences are often rivals instead of cooperative partners when it comes to studying ourselves. This is where the extended evolutionary synthesis, including niche construction theory (NCT), might help to build better bridges between the biological and human sciences.

Could a better understanding of ourselves, in the context of our evolutionary past as well as in the context of our sociocultural past histories, eventually outlaw nuclear wars permanently? That sounds improbable—almost wishful thinking. It certainly sounds like another improbability barrier. But improbability barriers have been crossed before. If, in the future, human sociocultural processes did create institutions that effectively eradicated nuclear war, that would be nothing less than another major transition in evolution, this time in human evolution.

GLOBAL WARMING

The second way in which humans are currently threatening to induce a mass extinction of life on Earth is global warming. The principal drivers responsible for the ways in which humans are currently contributing to global warming are our economic activities and our reproductive activities, as well as their interactions.

Human Economic Activities

As with all other organisms, our economic activities occur in the context of the ecosystems in which we live and ultimately in the context of the entire biosphere. In their book *Into the Cool. Energy Flow, Thermodynamics and Life*,[15] Schneider and Sagan describe the relationship between ecosystems and the laws of thermodynamics. They claim that ecosystems, as a consequence of all the individual organisms in evolving populations in them, also have to resist the second law of thermodynamics in order to persist. Here, I'm going to adopt a more limited approach. I'll concentrate primarily on the ways in which our scientifically and technologically enhanced cultural niche-constructing activities are currently contributing to human economic activities at level 3.

There is something very peculiar about the way that humans relate to ecosystems compared to all other organisms. There seem to be too many of us. Why would that be? To answer that question, we need to reconsider some of the principal points about how ecosystems work.[16] The first of these is that typically, mature ecosystems comprise a series of trophic levels, which roughly describe who is eating whom at each level. Recall that the first trophic level comprises autotrophs; that is, organisms that can feed directly off chemical resources or on energy from the Sun. They are primary producers of the biomass upon which all other organisms ultimately depend.

At the next trophic level, we find heterotrophs; that is, organisms that gain their energy and matter resources by feeding on other organisms. Heterotrophs depend on autotrophs, the primary producers. Heterotrophs break down into the familiar pyramids of various heterotrophs, starting with herbivores, then small carnivores, and finally larger carnivores at the top of the food chain. In theory, we humans are large animals at the top of our food chain. In reality, however, this description is overly simple, as we have seen before in this book. Humans are omnivores, taking food from several trophic levels. Our ape ancestors might also be better described as frugivores because their diets depend so heavily on fruit. In addition to this "grazing path," ecosystems include a "detritus path," as discussed in chapter 8. Detritivores feed off the detritus generated by diverse organisms at each trophic level. Detritivores also recycle energy and matter resources throughout ecosystems.

The second point about ecosystems that we need to acknowledge is how they proceed from immaturity to maturity over time and space. They can do so over months, years, or centuries via the processes of ecological succession. Ecological successions are often described in terms of a succession of plants, where later plant species depend on the prior existence and activities of earlier plants. Often, this is because the earlier plant species eventually destroy their own habitats, while at the same time constructing a habitat for their successors. Typically, early successional plants are short-lived and rapidly reproducing organisms, while later successional plants are usually longer-lived and more slowly reproducing. Generally, ecological succession starts with primary producers, regenerating at a rapid rate and proceeding through sometimes multiple plant species, with slower generational turnover times, toward a climax forest. Different ecosystems are likely to proceed toward different climaxes. Animals go through closely related successions of their own.[17]

We can now go back to our question: Why are there more humans on Earth at present than we should expect to find from the ways in which ecosystems work? The relevant point is that at each higher trophic level in an ecosystem, there is a considerable loss of energy and matter resources compared with the resources available at the adjacent lower trophic level. The so-called 10 percent rule in ecology tells us that it is possible for only about 10 percent of the energy and matter available at a lower trophic level to be transferred to the higher trophic level immediately above it. Each higher trophic level should therefore only support a decreasing number of organisms, hence the classic trophic pyramid. The highest trophic level should support relatively few large animals at the top of their food chains, such as resource-hungry humans. But there are more than 8 billion people on Earth today. That requires an explanation.

Biologists refer to the number of organisms in a population that their environments can support as the "carrying capacity" of that environment, for that species. Yet the number of humans on Earth today far exceeds the expected carrying capacity of our human environments, supposing that humans were exploiting only the resources provided by the natural trophic levels that occur in ecosystems. How and why has this happened? I want to propose a hypothesis, drawn from NCT. I suggest that our potent cultural niche-constructing activities have enabled us to construct two artificial extensions to the classic trophic pyramid. These extensions were constructed by the human Agricultural Revolution and the human Industrial Revolution. Together, they greatly increased the carrying capacity of human environments. That permitted an exceptional growth in the size of human populations on Earth. For convenience, I'll call these two artificial extensions of the trophic pyramid two additional artificial trophic levels.

The Agricultural Revolution

We cannot claim to be the first agriculturalists on Earth. Ants beat us to that distinction by several million years. If we loosely define "agriculture" in terms of the way that one species looks after and feeds one or more other species to enable it to harvest their resources, then ants qualify as farmers. Two examples will suffice. One is the way that some ant species tend aphids, which allow ants to harvest the sucrose that the aphids produce.[18] Another much better-known example is the leaf cutter ants that tend fungus gardens, sometimes on an enormous scale. The fungus, constantly fed by the leaves that the ants cut and bring back to their nests, then becomes the ants' staple diet.[19] However, human agricultural activities are orders of magnitude more general and productive than those of the ants or of any other species.

The Agricultural Revolution was responsible for two major innovations, both of which increased the carrying capacity of human environments. The first was the introduction of artificial selection of the plants and animals that humans domesticated, in lieu of their natural selection. The second was the introduction of artificially constructed environments for these organisms, in lieu of their natural environments. Human agriculture began about 11,000 years ago in the Middle East, and then elsewhere.[20] Human agriculture replaced the hunting and gathering of wild animals and plants with their domestication.

From the start, our ancestors were probably motivated to domesticate plants and animals by a desire to increase their productivity. Maybe unconsciously, or inadvertently at first, our ancestors must have introduced an elementary version of artificial selection relative to the plants and animals that they were domesticating. They probably did so as a by-product of the early agriculturalists choosing to domesticate those plants that yielded the richest and most nutritious food crops and were the easiest to process. Similarly, they probably selected animals that promised the richest source of meat or dairy products and were the easiest to manage. But it would not have taken long for our ancestors to realize what they were doing.

Soon they would have started to breed from the most productive and easily processed plants and the most productive and docile animals, such as sheep and cattle. At some point, our ancestors would have realized that they could use other animals that they had previously domesticated and artificially selected to help them manage the animals that they were farming. Some used dogs to control flocks of sheep. Others used horses they could ride to help them control herds of cattle. They also used domesticated horses, buffaloes, and other animals to plow their land. Today, the artificial selection of plants and animals has become a science. It is based on our modern knowledge of genetics and a much better understanding of how artificial selection works in evolving populations.

The second major innovation introduced by the early agriculturalists was the construction of artificial environments for the plants and animals that they were domesticating. They modified the natural selection experienced by those organisms in ways that are fully consistent with NCT. By doing this, they increased the productivity of individual plants and animals and greatly increased the numbers that they could cultivate in their vicinities. For plants, the construction of artificial environments typically depended on the advance preparation of land for the cultivation of crops. It may also have involved protecting crops from being eaten by other animals (e.g., by erecting fences). On modern farms, it frequently involves the use of chemicals. Artificial fertilizers are spread on land to increase the productivity of crops. Herbicides are used to destroy weeds, and pesticides are used to protect crops from harmful insects.

For animals, the construction of artificial environments was probably minimal at first. It may have amounted to little more than the first agriculturalists choosing pastures for their animals and then moving them to new pastures at different seasons. It involves much more than that today. It often involves the artificial preparation, protection, and enrichment of pastures for grazing animals. Also, in northern Europe, cattle are typically taken off the land during the winter months, housed in large barns, and fed on artificial foodstuffs. In the case of the most intensively farmed animals, such as chickens or pigs, the artificial environments constructed for them are sometimes drastically different from their natural environments. For instance, intensively farmed chickens are often kept and fed in small cages, in large numbers, inside buildings. Modern farmers also use veterinary services to protect the health of their livestock, sometimes combined with a heavy use of antibiotics.

As a consequence of the replacement of natural selection by artificial selection and the replacement of natural environments by artificially constructed environments, the Agricultural Revolution introduced a third major innovation. It suspended the normal processes of evolution relative to the organisms that they were domesticating and replaced it with an "artificial evolutionary process" based on artificial selection and artificially constructed environments. The survival and reproduction of domesticated plants and animals now largely depend on how they are managed by human farmers. The purposes and fitness goals of domesticated organisms are overridden by the intentions, purposes, and fitness goals of humans. For instance, contemporary farmers usually change the natural sex ratios of the animals that they are farming in order to increase their net productivity. In herds of cattle, most males are prevented from reproducing by being castrated shortly after they are born. A few carefully selected bulls are retained for reproductive purposes. In contrast, almost all female cattle are retained for both reproductive and other purposes, such as dairying. By introducing an additional artificial trophic level, the Agricultural Revolution considerably increased the carrying capacities of human environments. That permitted an abnormal increase in the size of human populations. However, in return for overriding the fitness goals of domesticated organisms, farmers have to pay the thermodynamic costs incurred by the organisms that they domesticated (e.g. by feeding and preparing suitable environments for them to enable them to oppose the second law to stay alive).

The Industrial Revolution

The human Industrial Revolution is much more recent; it began only just over 300 years ago. The Industrial Revolution largely replaced human and animal muscle power with fuel-consuming machines and engines, eventually on a vast scale. Today, we get fuel-consuming machines of one kind or another to do most of our physical work. Fuel-consuming machines partially mimic the thermodynamic demands of living organisms. They mimic organisms by consuming energy and matter resources from their environments, which are relatively high in free energy. They also dump detritus, which is relatively lower in free energy, back into their environments. This was why James Maxwell was so interested in Victorian steam engines and why we were so interested in Maxwell in chapter 1. But our contemporary machines are not autonomous. Unlike organisms, they cannot supply themselves with the fuel that they need to do their work. They cannot control the detritus that they have to emit. Nor can they assign themselves purposes. They have to be given their purposes by human designers, constructors, and programmers that serve our human purposes. We can imagine some future robots that will be autonomous, perhaps assisted by quantum computers that will be massively more powerful than today's digital computers. These robots might write their own algorithms and acquire purposes of their own. They might discover how to feed themselves, probably by using sunlight. They might even manage their own detritus. But we are not there yet.

Meanwhile, the lack of autonomy of our present machines imposes additional thermodynamic costs on humanity. We now not only have to resist the second law of thermodynamics for ourselves by importing energy and matter resources from our environments and by exporting detritus back into our environments, but we have to do the same for our machines. We have to supply our machines with the fuel that they need to do their work, at present primarily in the form of fossil fuels derived from long-dead organisms. We also have to return the detritus that the machines emit to the external environments that we share with them.

These extra thermodynamic costs imposed on humanity by our machines account for the nature of the additional artificial trophic level constructed by the Industrial Revolution. It is trophic, primarily because of the energy and matter resources that we have to harvest on behalf of our machines to "feed" them. It is also artificial because it no longer resembles any of the naturally occurring trophic levels in ecosystems. It is a human-made and machine-made artificial trophic level, another product of our potent, human scientifically and technologically enhanced, and cultural niche-constructing activities.

The Industrial Revolution affects how we feed ourselves, as well as how we feed our machines. It does so by interacting with the artificial trophic level created by the Agricultural Revolution, primarily by mechanizing most of our modern agricultural practices. Apart from increasing the productivity of our agriculture, the Industrial Revolution has generated huge numbers of different kinds of artificial consumable goods. As a result, in the twentieth century, most human societies came to be called "consumer" societies. Humans acquired new appetites for these new kinds of goods, far exceeding the appetites of our ancestors. I won't attempt to describe all the ways that the Industrial Revolution has changed our human societies and our individual lives, but among many other things, it has equipped us with cars, chainsaws, televisions, computers, and innumerable powered household appliances.

These artificial trophic levels have not been without their costs and setbacks. For example, in the case of the Agricultural Revolution, it takes a lot of additional resources to set up a farm in the first place. A farm represents an inescapable capital cost. There were other costs too. Humans became overdependent on monocultured domestic crops. That initially led to poorer and less-balanced diets. Also, the monoculture crops that entire human communities depended on were sometimes devastated by catastrophic diseases. As a result, early agriculturalists suffered a slight decrease in human life expectancy, as well as various negative health conditions. Nevertheless, agriculture led to a net increase in the size of human populations from early on. This was because birth rates typically increased faster than death rates in agricultural communities. The costs and setbacks due to the Industrial Revolution were and still are primarily associated with the pollution caused by our human industries.[21] However, in combination, the two additional artificial trophic levels due to the Agricultural and Industrial Revolutions vastly increased the carrying capacity of human environments. That explains why there are so many people on Earth today.

NICHE CONSTRUCTION AND THE CARRYING CAPACITIES OF ENVIRONMENTS

How do the two additional artificial trophic levels relate to NCT? The relationship between the Agricultural Revolution and niche construction has been well documented, even though it is still controversial.[22] The connection between the much more recent Industrial Revolution and human cultural niche construction is not yet widely recognized or documented.[23] One author who has written about it is the paleobiologist Doug Erwin,[24] who criticizes the biological concept of the carrying capacity of environments.

He points out that the carrying capacities of environments are always changing, relative to particular species. At the same time, the adaptations of those organisms also change in ways that affect those carrying capacities.

Thus the carrying capacities of environments are not exclusively fixed by exogenous environmental variables that are independent of the organisms in question. They also depend on endogenous variables arising from the adaptations of the focal organisms. The latter includes the ecosystem-engineering and niche-constructing adaptations of organisms.[25] Erwin emphasizes that by their ecosystem-engineering and niche-constructing activities, organisms affect not only the carrying capacities of their own environments, but also those of many other organisms. That led Erwin to make a connection with economic theory. Until recently, biologists have usually assumed that the carrying capacities of environments are determined exclusively by exogenous variables. Economists usually do the opposite; they usually assume that the carrying capacities of human environments are determined exclusively by endogenous variables. The agricultural economist Ester Boserup provides one example.[26] She claims that human ingenuity will always be capable of offsetting any limitations imposed on human populations by their environments. She also claims that one way in which we do this is by substituting an alternative resource for one that is running out. Boserup implicitly assumes that human environmental resources are infinite.

That approach by Boserup and others gave rise to endogenous growth theory in economics.[27] This implies that the carrying capacity of human environments is exclusively determined by human cultural niche construction, although I know of few economists who are saying that yet. Erwin[28] draws attention to this implication by referring to the work of two well-known economists, Robert Solow and Paul Michael Romer. In 1956, Solow argued that economic growth always requires technological innovation.[29] Later, however, Romer showed that not all technological innovations cause economic growth.[30] He distinguished between "rivalrous" and "nonrivalrous" economic goods. Rivalrous goods are goods that can have only one user at a time, such as a conventional bicycle. In contrast, nonrivalrous goods can be used by many people at the same time. An example of such a good is a global positioning system (GPS), which can be used by huge numbers of people simultaneously. Romer also distinguished between "excludable" and "nonexcludable" goods. These terms refer to how easy or difficult it is for a user to exclude others from using the same good. Romer's insight was that human economic growth depends on technical innovations that generate nonrivalrous and nonexcludable goods, not rivalrous and excludable goods.

Erwin then applied Romer's insights into paleobiology and ecosystem ecology. He proposed that the greatest impact of ecosystem engineering or niche construction on ecosystems depends on organisms producing innovations that are analogous to nonrivalrous and nonexcludable goods. He cites the production of oxygen by photosynthesizing organisms as a good example of a nonrivalrous and nonexcludable resource in ecosystems. This may explain why photosynthesizing organisms generated so much biomass and diversity in ecosystems. Erwin contrasted oxygen production with another kind of resource, the living spaces provided by intertidal zones for organisms such as barnacles.[31] Barnacles survive by anchoring themselves to rocks and holding seawater in their enclosed shells to avoid drying out. The availability of these resources, therefore, may be partly determined by the niche-constructing activities of the barnacles, as well as of other organisms. They are highly rivalrous and highly excludable. Translated back into human environments again, relative to our human economic activities, the implication is that different human niche-constructing activities have different consequences in different environments. Although Erwin is not an economist, his arguments are sufficient to illustrate the links between economics and NCT.[32]

Given our present dependence on the two artificial trophic levels of agricultural and industrial production, are we still managing to resist the second law of thermodynamics without threatening ourselves by violating it? It is not possible for organisms to violate the second law and stay alive, but organisms can sometimes cheat the second law by sacrificing their own and their descendants' futures for the sake of the present. This takes us back to the here-and-now versus elsewhere-and-later adaptational dilemma that we first looked at in chapter 3. Recall that organisms always need to be adapted to their immediate here-and-now environments or else they won't have a future. But organisms also need to carry sufficient general-purpose adaptability to allow them to adapt to possible changes in their future elsewhere-and-later environments. It doesn't matter whether the changes are caused by autonomous environmental events or by the niche-constructing activities of the organisms themselves. Different species invest in different adaptational strategies.

Specialist species maximize their adaptations to their here-and-now environments at the cost of sacrificing their general-purpose adaptability. Generalist species invest more heavily in general-purpose adaptability at the cost of risking being outcompeted by specialists in their immediate here-and-now environments. If environments change, however, then generalists are likely to outcompete their specialist rivals in their elsewhere-and-later futures.

In the light of this here-and-now versus elsewhere-and-later dilemma, there are two ways in which organisms could resist the second law here and now, risking their future survival. Organisms could fail to harvest sufficient energy and matter resources from their environments, in which case they would starve to death. Or organisms could fail to return their detritus to their environments before it kills them. Generally, humans are not making either of these mistakes yet. We are continuing to import resources high in free energy from our environments, and we are continuing to dump detritus lower in free energy back into our environments before it kills us. But there are two other ways in which organisms can resist the second law by risking their future survival. Organisms can take excessive energy and matter resources from their environments here and now, beyond the amount of resources that their environments can sustain over time. Or organisms can dump excessive detritus into their environments here and now, beyond the amounts that their environments can absorb by acting as sinks for it in the future.

Humanity is already making both of these mistakes. This is because of the ways in which we are currently mismanaging the two additional artificial trophic levels that we inherited from our niche-constructing ancestors. There is growing evidence that we are currently overexploiting our environments by taking excessive resources from them and dumping excessive detritus back into them. We risk destroying the carrying capacities of our future environments for our later selves and our descendants by maximizing our immediate here-and-now gratification. We are also modifying the carrying capacities of the elsewhere-and-later environments of countless other organisms. We are thereby threatening life on Earth with a future human-induced mass extinction.

Let's go back to the evidence. We already know that we are currently destroying the Earth's biodiversity at an accelerating rate, primarily by destroying the habitats of so many other organisms.[33] We also know that at present, we are overexploiting our forests, overfishing our oceans, and probably overmining the Earth for resources, including so-called rare earth resources such as lithium and cobalt and more basic resources such as iron ore and oil. However, in spite of earlier predictions that we would be running out of oil by now, we have not done so yet. Instead, as predicted by many economists, we have used our technological expertise to extract oil in new ways and from new, previously inaccessible places. It is now harder and more expensive to extract oil than it was in the twentieth century, but the oil is still there. We have not yet reached the limit of the resources that we can take from our environments here and now.

In contrast, we may already be dumping too much detritus back into our environments as a consequence of our here-and-now economic activities. The amount of detritus that we have to return to our environments is closely correlated with the amount of resources that we are taking from our environments. At present, economists and politicians measure the amount of resources that we are taking from our environments in terms of the gross domestic product (GDP) of economies or nations. But GDP is closely correlated with another economic index, which is seldom or never mentioned. I'll call it "gross domestic detritus (GDD)," emitted by nations and economies. In general, the greater their GDP, or the more energy and matter (R_p) resources high in free energy that economies and nations take from their environments, then the greater their GDD; that is, the more energy and matter (R_p) resources lower in free energy that they will have to dump back into their environments. I suggest in future that whenever a GDP index is cited, we should also cite its correlated GDD. We have invested far more capital and scientific and technological know-how in increasing GDP than in reducing GDD. As a result, we are dumping detritus into our environments to an extent beyond our environments' capacity to act as sinks for it.

Thanks to our scientific know-how, we are manufacturing many new materials, artifacts, and products that have never existed on Earth before. They include plastic and electronic items. When we have finished with them, we discharge excessive amounts of them back into our environments. We are also doing so at a much faster rate than new detritivores could evolve that might be capable of degrading them and recycling them. We could delay the introduction of new materials into our environments until we have worked out and invested in ways in which it is possible to dump them back into our external environments without threatening our future. The most obvious detritus that we are dumping back into our environments to an excessive extent are greenhouse gases, notably carbon dioxide. We are generating excessive amounts of carbon dioxide by our industrial activities. This is the waste product that, more than any other, is contributing to global warming. We are beginning to recognize this and are starting to take corrective action, but we have not gone nearly far enough yet (see the latest reports from the Intergovernmental Panel on Climate Change).

I'll summarize the preceding argument in terms of a familiar equation in ecology, the logistic growth equation. This not only allows me to relate my hypothesis more formally to ecological theory, it also introduces the second driver of global warming, our reproductive activities. The logistic growth equation is $dN/dt = rN(1-N/K)$, where N is population size, r is the intrinsic growth rate of the population, and K is the carrying capacity of the populations' environments. Our economic activities refer to both K and N in this equation, while our reproductive activities refer to r.

According to the preceding argument, the value of K should be calculated both in terms of the energy and matter resources that an environment can supply to a population, and in terms of the environment's capacity to act as a sink for the detritus emitted by the population. This way of calculating K determines N, the size of a population, relative to the carrying capacities of its environment. The intrinsic growth rate, r, relates to the growth in the value of N over time. That is the usual way of calculating r in ecology, but this way of calculating the intrinsic growth rate r is misleading in the case of human populations. It fails to capture either the extra energy and matter resources that humans are taking or the extra detritus that we are returning to our environments as a consequence of our construction and management of those two artificial trophic levels.

We need to be able to measure our increasing impact on our environments. The parameter that we really need to calculate is the growth in the "effective size" of our growing human populations. This parameter refers not only to the extra people in a population, but also to the extra resources that the extra people are taking from their environments and the extra detritus that they are dumping back into them. Let's call this parameter N' and the intrinsic growth parameter associated with it, r'. The latter refers not only to the growth of a human population, but also to the growth in the resources that the population is taking from its environment and to the detritus that it is returning to its environment.

If we want to understand the thermodynamics of the impact of our growing populations on our environments, caused by our two additional artificial trophic levels, we need to substitute N' and r' for N and r, in a revised logistic growth equation that applies exclusively to humans. We also need to substitute K' for K in the same logistic equation. K' refers to the enhanced carrying capacity of human environments caused by the human Agricultural and Industrial revolutions. At present, the value of K' far exceeds the value of K for the reasons that we have just been discussing. It's even possible that the difference between K' and K is still increasing globally. K' should then affect the values of N' and r' in the revised logistic growth equation that applies to humanity.

But we are currently destroying biodiversity and dumping excessive amounts of detritus back into our environments. That means that the difference between K' and K may start to reduce rather than increase in the near future. Should it do so, the degraded human environments will then be able to support smaller human populations than today. In the case of a catastrophic human failure to prevent a human-induced mass extinction, the value of K' could sink far below the value of K. It could become far less than the original carrying capacity of human environments before the origins of either the human Agricultural or Industrial revolution. The difference between K' and K is the one to watch. It is bound to vary at different places and times, in different regions of the world, and across different countries. As predicted by NCT's emphasis on relocation, that will probably cause people in regions where K' is lowest to relocate by migrating to where K' is higher. The consequences of declining values of K' are likely to cause increasingly desperate migration. I'll illustrate the difference that a revised human logistic growth equation makes to our understanding of the relationship between our contemporary human populations and their environments.

In 1700, just before the start of the Industrial Revolution, the world's human population was approximately 600 million. It is 8 billion today—more than a 12.5-fold increase in the size of the total human population in just over 300 years. Let's now compare the estimated resources taken by humans in 1700 with the total resources that we are taking from the environment today, in light of this revised logistic growth equation. Let's also compare the correlated detritus that our ancestors were generating in 1700 with the detritus that we are generating today. This should enable us to capture the interactions of humans with their environments in terms of the true thermodynamic costs that we are imposing on our environments by the ways that we are living today. This measure is often called our "collective" or "individual" ecological footprint on our environments. It might be better to call it a "thermodynamic" footprint. But I will continue to call it an "ecological" footprint because that term is more familiar.

I'll arbitrarily assume that on average, every human being alive on Earth today consumes 25 times more energy and matter resources than the average individual human did in 1700, which various sources suggest is likely to be a conservative estimate.[34] We are not each eating 25 times more food than our ancestors; it is that the energy and matter costs of our individual lives have increased by more than 25 times. This is primarily due to our contemporary dependence on huge numbers of fuel-consuming and detritus-producing machines for so much of what we do today. For example, every time we travel by bus, car, rail, ship, or plane, our individual ecological footprint is increased by our individual share of the thermodynamic costs of those machines. The same is true every time we switch on the light, turn on a computer or television, or cook on a modern stove. It's also true of the roads, railroads, ports, and airports on which we depend and the buildings in which we live in, work in, and play.

The construction and maintenance of all these things required and continue to require fuel-consuming and detritus-producing machines to build them, heat them in winter, air-condition them in summer, and keep them in good repair. It is even true when we go to supermarkets to buy food. The food on the shelves may be local or may have come from distant places. But even food grown locally will now take up far more resources than the food consumed by our ancestors in 1700 because it depends so much on agricultural machinery. When you add all of that up, the increase of 25 times the energy and matter resources that we are using to live, compared with our ancestors in 1700, starts to make sense. It means that the effective human population size N', as opposed to N, must have increased more than 300 times (i.e., $12.5 \times 25 = 312.5$) in a little over 300 years. That is not sustainable.

How have we gotten away with that, and for how much longer can we do so? Let's consider some of the ways in which we have managed to continue to feed our rapidly growing populations since the start of the twentieth century. We've done so as a consequence of both the Agricultural and Industrial revolutions working together. But remember Malthus.

Malthus was one of those highly original clergymen who in the eighteenth and nineteenth centuries, owing to their light ecclesiastical duties, did a lot of thinking. He noted that human populations tend to increase geometrically (i.e., exponentially). While the food and energy resources that they need to live increase only arithmetically. Malthus inferred from this contrast that human numbers would always be limited by lack of sufficient environmental resources. He envisaged a nightmare scenario involving a lot of human poverty, famine, disease, and misery as a result.

Malthus's ideas contributed to Darwin's discovery of natural selection. Darwin realized that if Malthus were correct, then all organisms in all species would inevitably have to compete with each other, as well as with other species, for the energy and matter resources that they need to live. Darwin also realized that because organisms in the same species always vary, the fittest among them would be more likely to survive and reproduce in each generation, while the least fit would be eliminated by the competition. As the demand on the environment by any growing population increases, it introduces a category of natural selection called "density-dependent selection."[35] Density-dependent selection intensifies when an increasing number of organisms have to compete with each other for diminishing environmental resources. Today, we like to think that the nightmare scenario envisaged by Malthus no longer applies to us. This is because so far, we have always managed to offset the threat of too little food, and more generally of insufficient energy and matter resources, by increasing the productivity of our environments by our niche-constructing activities. But have we really left the Malthusian nightmare behind permanently? Or have we only postponed it?

Originally, neither Malthus nor Darwin realized that it was possible for organisms to change the carrying capacity (K) of their environments by their niche-constructing activities positively as well as negatively. However, it is possible for organisms to increase the carrying capacity of their environments from K to K' in the ways that we have just discussed. If we now go back to the beginning of the twentieth century, we come across several ways in which humans have continued to increase their food supply as a function of their management of the two additional artificial trophic levels. One major way was through a scientifically enhanced capacity to produce artificial fertilizers. That led to a considerable increase in the productivity of our agricultural activities. Then, in the 1970s, just as human population numbers were threatening to outstrip our food resources, we had the so-called Green Revolution.[36]

The Green Revolution introduced not only genetically superior and more productive plants and animals, but also a further intensification of our agricultural activities, leading to even greater productivity. But there was a cost. "Green Revolution" crops demanded more artificial fertilizers, more pesticides, and more fresh water than the crops that they replaced. Each of these additional requirements made further demands on our environments, costing additional energy and matter resources. That generated the next batch of adaptive problems for humanity. The increased use of fertilizers has become a major source of pollution. Fertilizers are currently polluting our rivers, lakes, and oceans.[37] The overuse of pesticides is killing so many insects that eventually, there may not be enough insects left to pollinate our crops.[38] There is also increasing competition between different societies and nations for diminishing supplies of fresh water. So far, we have postponed the Malthusian nightmare, but we have certainly not eliminated it permanently. There are also many other ways in which modern agriculture is reducing biodiversity by destroying the habitats of so many organisms for the sake of increasing human food production.[39]

That brings us to the final variable that affects the carrying capacity, K', of our contemporary human environments: life expectancy. That term refers to the number of years that the average human is expected to live in different parts of the world. It is relevant because the longer we live, the more energy and matter resources we consume and the more detritus we generate during our lives. Increased life expectancy adds to the impact or ecological footprint that each of us has on our environments. At the population level, if birth rates remain constant but life expectancy doubles in a human population, that greatly increases the size of the population. The size of the effective human population (N') will increase too. If it does, it will further increase the thermodynamic costs that a human population imposes on its external environment.

We can now compare the life expectancy of the average human being on Earth in 1700 BCE with the average life expectancy of human beings today. It has clearly increased, but by how much? I am not a demographer, so I cannot answer my own question. In Britain, however, where the Industrial Revolution began, as well as in most other countries, life expectancy has more than doubled since 1700.[40] This matters. Increased life expectancies are adding to the thermodynamic impact that we are all imposing on our environments.

Why have life expectancies increased so rapidly, almost worldwide? That cannot be explained exclusively by more secure food supplies or better living conditions. It is also due to the successes of the biomedical sciences in keeping us alive longer. The biomedical sciences have increased life expectancies in two main ways. They have reduced infant mortality in many countries, and they have increased longevity. Relative to the revised logistic growth equation that applies to human populations, it means that although the biomedical sciences have not directly contributed either to K' or r' in this equation, they have directly contributed to N'. They have done so by greatly increasing the effective size of most human populations on Earth.

The biomedical sciences can also claim to be reducing the size of N' and the rate of increase in our human populations (r') by developing contraceptive technologies. In the mid-twentieth century, it was expected that a phenomenon known as the "demographic transition" would occur. A demographic transition was first noticed in the industrialized countries of Europe toward the end of the nineteenth century.[41] It was expected to generalize to other countries as they industrialized. The availability of contraceptives, combined with increasing wealth and the increasing confidence of parents that any children they did have would survive into adulthood, was widely expected to halt the growth in human populations. This did happen in some prosperous countries, such as Japan, South Korea, and Germany. But it has not happened worldwide. Something more is still needed to stabilize human population numbers.

Another effective way of stabilizing human population growth is female emancipation. Female emancipation requires educating women up to at least the same level as men, as well as an increase in their economic independence, and allowing women much greater control over their own reproduction. At present, however, in some cultures, female emancipation is often opposed by long-standing cultural traditions and by prescientific beliefs and pronatalist religions.

Globally, the size of the effective human population on Earth (N') is still growing, partly thanks to the biomedical sciences. Therefore, the total amount of energy and matter, or R_p resources, that we are taking from our external environments and the total amount of detritus that we are dumping back into our environments are also increasing. In spite of the welcome achievements of the biomedical sciences in alleviating so much human suffering and the extra longevity that they have given humanity, the biomedical sciences are inadvertently making our problems worse. They are increasing rather than reducing our ecological footprint, simply by increasing N' in the logistic growth equation. They are unintentionally contributing to the looming threat of global warming despite their benign intentions.

Where are we going wrong? The main implication of the preceding analysis is that the construction of the two artificial trophic levels by our human niche-constructing activities enabled a considerable increase in the effective size of our human population. It also led to considerable increases in human wealth, although not for everyone. These two artificial trophic levels now seem to be indispensable to humanity. They have also introduced a positive feedback cycle that keeps ratcheting up the demands that we are making on our environments. Human societies quickly use up the benefits of the extra resources that we gain from our two artificial trophic levels. We do that by increasing our material wealth, by increasing the size of our populations, or both.

These outcomes then require a further increase in the productivity of our agricultural and industrial activities. This increases the thermodynamic demands that we make on our environments. One consequence of that is that the world's natural ecosystems may soon no longer be able to sustain our two artificial trophic levels due to the damage that they are inflicting on the natural ecosystems in which they occur. But we still depend on the natural ecosystems, as well as on the artificial trophic levels, for our lives. These two artificial trophic levels are neither stable nor sustainable. If we are to stabilize them, we are going to have to do more than just conserve biodiversity. We shall have to change the way that we manage them so they are no longer destroying the Earth's natural systems. Relative to the bioenergetic requirements of life, we may have to reduce the amount of energy and matter resources that we are taking from our environments, as well as the amount of detritus that we are returning to them. Relative to the bioinformatics requirements of life, we will certainly have to invest more in acquiring additional adaptive know-how to enable us to manage the detritus, or GDD, that so closely tracks our expanding global national GDPs.

It should not be impossible for us to achieve these adjustments, assuming that we have not already passed a point of no return. But it will be difficult. For example, in principle, we could reduce our resource take from our environments, or we could reduce the size of our populations. But most of us still seem to want greater, rather than less, material wealth (perhaps understandably). Also, collectively, we still seem to be unable to limit our reproductive activities enough to stabilize the size of the world's human population.

A SECOND ADAPTIVE LAG

In the same way as with the negative future, threatened by our invention of nuclear weapons, I suggest that we will not be able to avert the negative future threatened by global warming, exclusively through our cultural socioeconomic and geopolitical activities, at level 3. We will also have to take our past population-genetic evolution at level 1 into account, recognizing that we need to confront at least one more adaptive lag. This primarily refers to how we are dealing with our detritus, in particular how we are dealing with the greenhouse gases that we are generating by our machines.

This brings to mind a quote from Arthur Conan Doyle's hero, Sherlock Holmes: "Why didn't the dog bark in the night?" The corresponding question is: Why didn't our ancestors at the start of the Industrial Revolution establish the principle that whoever was responsible for the industries that were polluting the natural environment should have cleared up the mess, or pay for others to do so? There are some obvious reasons why this did not happen. It would have made their industries less profitable. Also, the relevant technology may not have been available at the time. But I suggest there was also another more profound reason. It involves another adaptive lag, generated by the incompatibility between another ancient level 1 adaptation and another much more recent level 3 adaptation in human evolution.

This adaptive lag works in the opposite way as the adaptive lag discussed previously, in connection with the threat of nuclear war. Instead of being due to the presence of an ancient level 1 social adaptation that has now become maladaptive, this second adaptive lag is due to the absence of an ancient level 1 adaptation, which we now need to prevent a level 3 adaptation from becoming fatally maladaptive. Similar to almost all other organisms, natural selection favored humans who dumped their raw detritus straight back into their environments. Consequently, dumping our raw detritus back into our environments was the only adaptive trait that we inherited from our ancestors for dealing with our detritus. This trait used to be adaptive. It was a cheap way of dealing with our detritus, in terms of its energy and matter (R_p) costs. Also, for as long as our environments could recycle our detritus, it was not a threat to ourselves and was far less of a threat to other organisms.

Humans have probably always assumed that our external environments are unlimited sinks for our detritus. That was true at least until the start of the Industrial Revolution. But it is no longer true today. Because of our heavy use of industries, we are now dumping more detritus, including greenhouse gases, back into our environments than our environments can absorb. If we are to consolidate the increased carrying capacities of our human environments that we have gained from our two artificial trophic levels, we are going to have to invest in an artificial reduction of the detritus, or GDD, that is generated by our expanding GDPs. We will have to become our own artificial detritivores relative to our own detritus.

HOW COULD WE AVOID A NEGATIVE FUTURE?

How might we avoid inducing a mass extinction of life on Earth, either due to nuclear war or global warming? It is now possible to see that both of the ways in which humans could cause mass extinction have the same origin. Both stem from the potency of our cultural niche constructive and destructive activities. Is there anything else that these two different routes into a negative future have in common? Possibly yes. Both involve our scientific and technological activities at level 3. Are we misusing science and technology in ways that are threatening our own futures, as well as the future of all life on Earth? Possibly yes.

The Asymmetric Application of Science and Technology

The philosopher Bertrand Russell once proposed that there are two kinds of science. He called them "power science" and "understanding science." He stated that some science is motivated by a quest for power, but wisdom is derived primarily from understanding science.[42] No scientific discipline is exclusively either a power science or an understanding science. They are all a mix of both, but to varying degrees. Evolutionary biology and ecology are primarily understanding sciences. Since the discovery of the molecular basis of genes, however, molecular biology has largely become a power science. For instance, it led to the ability to change the genetic composition of plants and animals, through selective breeding, and gene editing. Conversely, physics is usually thought of as a power science. It deals directly with the forces of nature, including gravity and its own more recent discoveries of electromagnetism and the strong and weak nuclear forces, and hence to their applications in technology. In my view, cosmology and astrophysics are understanding sciences rather than power sciences. Almost everyone finds power sciences attractive. Those who do range from would-be or actual dictators to enthusiastic users of the latest powerful artifacts, inventions, or gadgets. Conversely, most of us are disturbed or even frightened (sometimes profoundly) by the discoveries of understanding science. Maybe this is because scientific enlightenment so often seems to downgrade our own self-esteem and self-importance.

Over the last several centuries, science has taught us that the Earth is not the center of the universe. It is just a satellite of the Sun, orbiting it once a year. Both Copernicus and Galileo got into trouble with their contemporary authorities for pointing that out. As we now know from the work of later scientists, Earth is only one of billions of planets orbiting other stars in our galaxy. Moreover, our galaxy is only one of innumerable galaxies in the universe. None of that flatters humanity. Then along came Darwin, with his ideas about evolution, including human evolution. That made matters worse. Are we really so closely related to gorillas and chimpanzees? It's a thought that disturbs some people. Science often upsets the most powerful people in our societies too. Historically, it has challenged the divine rights claimed by popes, bishops, emperors, kings, ayatollahs, and caliphs by asking awkward questions about the presumed sources of their authority.

Also, as we've just noted, humanity has invested heavily in those niche-constructing activities that gratify us by giving us more of what we want, whether it be more wealth or more social, military, economic, or political power. But we have largely failed to apply understanding sciences to help us to understand the limitations of our cultural niche-constructing activities relative to what the rest of life on Earth can sustain. Nor have we applied understanding sciences sufficiently to help us to understand ourselves, or our relationships with other humans, better than we do at the moment. Nevertheless, we could still change the ways in which we niche-construct. We are purposeful agents that in principle could adjust our purposes, and therefore our niche-constructing activities. We could invest in different kinds of niche construction with respect to the adaptive management of the two artificial trophic levels that we have constructed. For example, until global warming becomes self-perpetuating by generating runaway positive feedback, it is in our hands to slow it down or halt it. But we will have to adjust our purposeful activities to do so. That will mean investing less in trying to exploit nature by the application of power sciences and more in trying to understand our relationship with the rest of nature through understanding sciences, and then applying this new knowledge.

A POSITIVE FUTURE

After spending so long on that treadmill of doom, what about the alternative? Earlier, I suggested that humanity is currently standing on the threshold of two very different futures. One refers to the negative futures that we have just been considering, the other to an alternative possible positive future. What form might that take? The positive path to the future is far more tenuous, for three main reasons. There is a lack of relevant data, there are difficulties with making predictions in biology, and we need to change the way we think.

We lack sufficient data, partly because we still don't know whether life exists anywhere in the universe other than on our own planet. Also, our knowledge of the past evolution of life, even on our own planet, becomes sketchier the further we go back in time. Further, as evolutionary biologists know to their detriment, biology is not like physics. It is almost impossible to predict what evolution is going to do next, at least in a specific population. It is possible for us to understand how evolution works and, in the light of that understanding, we can rule out some things that definitely *cannot* happen. This is what we have just been doing. For example, evolution cannot disregard the second law of thermodynamics. But we cannot predict the many ways in which evolution can obey the second law. Einstein was also right—we really do need to change the way we think, not just to avoid nuclear war or global warming, but to achieve a positive future. I want to concentrate on this third point. Changing the way that we think is a necessary precursor to changing our values, our purposes, our behaviors, and the ways in which we are currently niche-constructing. What will it take for us to change the way that we think and act?

What influences determine the way that we think at the moment? Obviously, we are affected by innumerable influences, but I'm going to ignore most of them. It would be reckless and out of place here to attempt to find a comprehensive answer to that question. In any case, it is beyond my expertise to do so. I shall not consider the ways in which our thinking is influenced by the arts, literature, the humanities, the entertainment industries, or even by demagogues. Instead, I want to concentrate on how the understanding sciences, and primarily the sciences of evolutionary biology and ecology, might eventually start to change the way that we think. Beyond that, I will only consider other sources of influence that are either directly connected to evolutionary biology and ecology or are affected by them. I also want to consider how substituting the proposed extended theory of evolution for SET might help us change the way that we think, to the point where a positive future, rather than a negative future, beckons.

Everything that influences the way that we think and behave starts as incoming information. This information is acquired by our natural or artificial senses (e.g., telescopes, microscopes) from both our internal and external environments. In the context of the extended evolutionary theory, this becomes meaningful information acquired by organisms. Recall that they are population-genetic evolution at level 1, individual developmental processes at level 2, and sociocultural processes at level 3. The understanding sciences are a level 3 information-gaining process.

Regardless of which evolutionary level we acquire our information from, we translate it into adaptive know-how, or R_i. From this know-how, whether it is adaptive or not, we then derive our emotions, motives, purposes, and goal-oriented behaviors (see chapter 2). It follows that if we are to change the way that we think and behave, it is always going to require changes in the meaningful information that we derive from at least one of those three information-gaining processes.

It also follows that new information derived from any of these levels could initiate new changes in our thinking and behavior. One possible source of new information is science, in particular the understanding sciences. Sciences provide us with information that was never available to our ancestors. Could the understanding sciences, particularly our understanding of evolution and ecology, eventually change the way that we think? Let's consider what we have now learned about the past history of our universe and our own relationship to the rest of nature on the largest possible scale.

The universe began with a bang that occurred approximately 13.8 billion years ago. That said, the Big Bang did not occur in time and space. Time and space were created by the Big Bang. The Big Bang was a physical singularity, which provided an unimaginable amount of free energy available to do physical work. According to astrophysicists, the heat left over after the Big Bang must initially have been so intense that the earliest universe must have glowed with light, before becoming opaque and dark. As the universe started to cool, matter formed. Light then returned, emitted by stars in the first galaxies. Since the launch of the Giant Web Space Telescope in 2022, astronomers have been receiving new data that allows them to peer so far back in time that they can now see what looks to be the very first galaxies in the universe that formed after the big bang. Most appear to be very small galaxies compared to our own galaxy, and they also appear to be composed of only the lightest elements, primarily hydrogen and helium. For example, apparently these first galaxies contained only about 2 percent of the oxygen (hardly a heavy element itself) that is present in our own galaxy. Astronomers are starting to call them "green pea" universes because of their smallness and their color.[43] Subsequently, the Big Bang and the laws of physics and chemistry generated our physical universe, comprising the myriad of galaxies, black holes, dark matter, and dark energy, as well as the visible stars and their planets.[44]

On one planet, the Earth, orbiting a star, the Sun, in our own galaxy, life originated from abiotic matter about 4 billion years ago. Apparently, it did so by self-assembly under the direction of the laws of physics, chemistry, and chance (see chapter 6). The subsequent evolution of life on Earth then eventually led from simple prokaryotic cells without nuclei to more complicated eukaryotic cells with organized nuclei and internalized organelles, including mitochondria. This occurred more than 1.5 billion years ago and was a major advance toward greater complexity. It led to the evolution of multicellular organisms or metazoans. That was followed by the appearance of the first animals in the Ediacaran era, about 600 million years ago. Subsequently, multiple different forms and kinds of animals appeared during the Cambrian explosion, about 540 million years ago. Brains and minds then evolved in animals. From animal brains, intelligence and conscious minds evolved, culminating in the evolution of our own species about 300,000 years ago and to today's levels of conscious human intelligence. Conscious human intelligence has now given us the first glimmer of awareness of both the past history of our universe and the history of the evolution of life on Earth. It has also given us a degree of self-awareness. Paul Nurse provides a similar list of discoveries in his book *What Is Life? Understand Biology in Five Steps.*[45]

In particular, we have learned that we are probably the only species capable of being aware of the evolutionary process that produced all of life on Earth. We can also assume that we are the only species alive today with enough self-awareness to start asking serious questions about our own existence. We are the first organisms on Earth to become consciously aware of the evolutionary processes that are responsible for our existence, as well as the existence of all other life on Earth. Through us, evolution has become aware of itself. This could eventually change the way that we think. It might also enable us to act as responsible custodians of the rest of life on Earth.

But we will have to pay a price for this insight. Collectively, we will have to learn much more from the understanding sciences about evolution and how it works. This is not a trivial task. Many years ago, I attended a conference on evolution at Cambridge University. There were a lot of enthusiastic evolutionary biologists present. Toward the end of the conference, our enthusiasms were dampened by a witty talk from the philosopher of biology David Hull. Remember, he said that the vast majority of people in the world have never heard of evolution. Among those that have heard of it, the vast majority don't understand it. Among those that have some understanding of it, the vast majority don't like it. That was sobering. It may still be true. If so, it means that the vast majority of humans on Earth today must still have a very poor understanding about how we really do relate to the rest of nature.

Following the construction of our two artificial trophic levels by our cultural niche constructing activities at level 3, we will still have to learn a lot more from ecology, evolution, and environmental sciences, including the relationship of our two artificial trophic levels with the rest of nature. However, to borrow from Hull again, most people alive on Earth today are probably more preoccupied with the daily struggles of life to think enough about the implications of these discoveries of our understanding sciences. Many people are interested in, or alarmed by, these discoveries, but not to the point that they change the way that they think.

After a period of unprecedented weather and climate events, however, ranging from uncontrollable forest fires in Australia, California, and southern Europe and peat fires in Siberia, as well as droughts in the Horn of Africa and dramatic increases in the loss of polar ice,[46] many people may have begun to think differently. We may also have started to think differently after experiencing the COVID-19 pandemic, which may itself be partly due to the ways in which we are now farming animals and destroying the habitats of multiple other organisms, thereby increasing zoonotic encounters between ourselves and disease-carrying organisms such as bats. But collectively, we are still falling far short of the radical changes in human thinking that will be necessary to rectify our previous mistakes relative to global warming. For instance, the need for humanity to give up fossil fuels is now widely recognized, but the implementation of policies to achieve this is still being opposed.

There are also some new dangers arising from our latest technologies, which we now need to understand better to avoid another misuse of science. For example, thanks to our new ability to edit not only our own genomes, but potentially the genomes of almost all organisms on Earth, we will have to understand, better than we do now, precisely what we are doing when we edit any organism's genome. This prospect takes us back to humans having the option to "design" their own babies,[47] and potentially also the designer offspring of multiple other organisms. It also may introduce the prospect of humanity having the option of directing the future evolution of some or much of life on Earth.

UNDERSTANDING SCIENCE VERSUS RELIGIOUS BELIEFS

I now want to consider another major influence over the ways that we think: religion. There is no direct connection between our relatively ancient religious beliefs and the modern understanding sciences of evolution and ecology. But in today's world, the sciences undoubtedly affect our religious beliefs. The relationship between evolutionary biology and religion is tense. Many religions see evolution as a threat to their beliefs. The prescientific world generated numerous creation myths, animisms, shamanisms, and later the doctrinal world religions.[48] They have a lot in common. Typically, they assume that the world, the universe, and life on Earth were created by a supernatural deity or deities. Many religions also claim that human creation was a special case. For example, Christianity claims that humans, unlike all other organisms, were created in the image of God.

Evolutionary biology provides a different kind of explanation for life. It accounts for the existence of life on Earth, including human life, in terms of the laborious, four-billion-year-old natural processes of evolution, rather than being the spontaneous products of a supernatural, intelligent creator. Does this difference between evolutionary biology and religious beliefs make them irreconcilable?

Let's go back to our prescientific selves again. Many people in the past were often able to find solace, their ethics, their meaning, and their purpose from the stories, legends, and religious beliefs that prevailed in their societies. I am not a religious person, but I can offer one suggestion about how it might be possible to reconcile science and religion. It stems from the seventeenth-century Dutch philosopher Baruch Spinoza. Apparently, Spinoza was indifferent to whether we call God nature or nature God. He came up with a memorable phrase, *Deus, sive natura*. In English, that translates to "God or Nature." Spinoza was criticized by his contemporaries. He was accused of pantheism. I'll ignore these criticisms. But might Spinoza's indifference to whether we call God nature or nature God still work for us, with one crucial proviso?

It could work, provided that the concept of God is repeatedly (and, if necessary radically) revised every time we learn more about nature from science. Failure to update our religions, in the light of new knowledge about nature from our sciences, may not rob religious beliefs of their emotional appeal, but it would threaten them with intellectual vacuity. It could leave us with a maladaptive cultural schizophrenia that could tear our world apart. It certainly would not help us avoid a human-induced mass extinction.

What would our religions have to give up to incorporate what the understanding sciences of evolution and ecology are now telling us? We will have to accept that humans are the products of the same natural processes of evolution that produced all other organisms on Earth. We owe our existence to the same laws of physics and chemistry, and to the same biological processes of evolution, as all other living organisms. We shall also have to accept that nature doesn't care about us, any more than it cares about any of its other creatures. What matters in nature is whether organisms can satisfy the demands made on them by the second law of thermodynamics in ways that allow them to survive and reproduce. If an organism cannot do that, too bad. Nature will evolve other organisms that are better adapted. Humans will also have to give up the idea that we are cared for by a benign, loving deity. Nature is not benign, and natural selection is not a benign way of designing organisms.

Perhaps the hardest idea that we may have to accept is that we are mortal, and as transitory as all other organisms. The evolutionary process could not work without the constant turnover of various mortal individual organisms, in successive generations of evolving populations. Death is part of life. Different organisms have different lifespans, and for an animal, ours is relatively generous. But we are still mortal.

What could we gain by updating our religious beliefs in the light of evolutionary biology and the other natural sciences? We would gain more truthful answers to those age-old questions than have ever been available to humans before. *Who are we?* We are social primates with ape and hominin ancestors. *Where did we come from?* We came from the same processes of evolution that are responsible for the existence of all organisms on Earth. *What are we doing here?* We are active, purposeful agents in the evolutionary process, and we contribute to the future evolution of our own species, as well as to many other species. *Where are we going?* We seem to be heading into one or the other of those two alternative possible futures that we have just been discussing—a negative future or a positive future. A positive future includes accepting our collective responsibility for the custodianship of life on Earth.

A FINAL QUESTION

This raises a final question. In the absence of a god who cares about us, and in the presence of an indifferent nature, do our lives matter? *Can* they matter? It is illuminating to consider two of the most extreme possible human reactions to the discoveries of modern science. One possible reaction is that, relative to the vastness of the universe, our individual human lives are meaningless. A second possible reaction is the exact opposite—that human lives are more meaningful than it has ever been possible for us to realize before.

I'll begin at the pessimistic end of this spectrum. It was anticipated by the words that Shakespeare put into the mouth of Macbeth after the death of his wife, when he was contemplating his imminent downfall and death: "Life is a tale told by an idiot, full of sound and fury, signifying nothing." (act V, scene V). Might that be true? Was Macbeth right? What does the scientific evidence tell us? It tells us that at least with respect to the evolution of life on Earth, Macbeth was wrong. Our lives are meaningful, if only because what we do and how we live affect not just our own evolution, but the evolution of countless other organisms on Earth. NCT demonstrates this point. But it raises a further question. Suppose that the evolutionary process itself is a tale told by an idiot, full of sound and fury, signifying nothing. What then? This is a far more difficult question to answer. Could Macbeth still be right?

If the evolutionary process itself showed no evidence of proceeding in any particular direction, but rather appeared to be just a random walk, then Macbeth might be right. It would not be possible for humans to impose purpose or direction on a fundamentally directionless process, whatever we did. Our individual human lives might then seem futile once again. But what does the evidence tell us this time? There is weak evidence that since the origin of life, the subsequent evolution of life on Earth has not been a random walk. It appears that the evolution of life since its origin has been traveling from simplicity to greater complexity and greater thermodynamic efficiency. Evolution happens because organisms are purposeful agents, so it cannot be a random walk. There have been many interruptions to this overall directionality. These interruptions have been caused by a series of mass extinctions, some very severe. So far, however, the evolutionary process has always recovered and resumed its travels toward greater overall complexity.

There is considerable evidence that the evolution of life on Earth has proceeded through a series of major transitions in evolution, and again niche construction lies at the heart of it (see chapter 8). Here, I'll focus on the major evolutionary transitions in bioinformatics.[49] Each successive major transition relates to an increase in efficiency of an information- or R_i-gaining process. According to Maynard-Smith and Szathmáry, the last of these major transitions refers to the appearance of humanity, human societies, and human language. However, I suggest that there has been another major transition in evolution, which was not mentioned by Maynard-Smith and Szathmáry. It was introduced by Darwin, and it corresponds with his discovery of the process of natural selection and with our subsequent, more comprehensive understanding of evolution. I suggest that this qualifies as another major transition in the bioinformatics of evolution. It corresponds to the moment when the evolutionary process becomes aware of itself, at least in the conscious, intelligent minds of one of its own creatures—humans.

This last hypothetical major transition in the bioinformatics of evolution takes us to the opposite end of the spectrum of possible human reactions to the discoveries of science. Potentially, we are the first organisms on Earth with the mental capacity to understand how the evolutionary process works. We also know that, thanks to the potency of our cultural niche-constructing activities, we are bound to contribute, either positively or negatively, to the future evolution of life on Earth, including the future of our own species and innumerable other species as well.[50]

Although evolutionary biology cannot eliminate mortality, it can supply us with a helpful analogy. All the individual cells in the bodies of multicellular organisms, including our own, are mortal. Most of them have shorter or far shorter lives than their host metazoans. But from our human point of view, the individual cells in our bodies are far from being purposeless, despite the brevity of their individual lives. Their purpose is to contribute to something greater than themselves—namely, us. Collectively, the cells in our body are responsible for who we are and how long we live. By analogy, I suspect that we may be contributing to something incomparably greater than ourselves by our mortal lives. But unlike the individual cells in our bodies, we may slowly come to know a great deal more about what we may be contributing to. If we can avoid self-destruction and human-induced mass extinction, and if we don't destroy science, there is every chance that our collective understanding of what we may be contributing to will increase.

What will we do with our responsibilities? Might we merely cause another mass extinction? That would be a dismal outcome. Or might we cause evolution to proceed further in the direction it was already traveling, toward still greater awareness of itself in the conscious, intelligent minds of the human species or a successor species? This idea touches on a point made by Max Tegmark about a future conscious intelligence waking up nature (see chapter 7).

There are several well-known, unresolved problems in both evolutionary biology and the physical sciences. The outstanding questions in evolutionary biology are: How did life originate on Earth? Does life exist anywhere else in the universe?

That brings me back to Schrödinger, whose "What is life?" question inspired this book. It is fitting to close this discussion by returning to him, as the second section in his 1944 book was called "Mind and Matter." Why "mind"? Schrödinger's early involvement in quantum mechanics may have something to do with it. In quantum mechanics, it is possible to measure either the velocity or the location of an elementary particle, but not both simultaneously. The observer has to choose which of these two things to measure, at any one time. Schrödinger was interested in the active, purposeful agency of living organisms in all respects, including with relation to quantum mechanics.

In his 1944 book, he even anticipated the logic of NCT, years before it emerged in biology. In a startling epilogue, only included in a later edition of his book, Schrödinger proposed that every intelligent, conscious person has some capacity to control the motions of atoms according to the laws of nature and to anticipate the consequences of his or her actions. On this basis, Schrödinger drew an extraordinary inference: "Hence, I am God Almighty."[51] He admitted that to Christians, this "sounds both blasphemous and lunatic." But he justified his statement in terms of the Upanishads of Hinduism, in which every mortal, individual self becomes a component in the omnipresent, eternal self. In Hindu philosophy, that is far from blasphemy. I have been suggesting that in spite of our mortality, our lives matter because, depending on how we live our lives, we are contributing to "waking up evolution" and "waking up nature." Did Schrödinger mean more than that? Possibly yes.

According to Schneider and Sagan (2005), there was a sequel. Schrödinger originally submitted his *What Is Life?* manuscript, without his epilogue, to an Irish publisher associated with Trinity College Dublin. He added the epilogue as an afterthought. But when his Irish publishers received it, they refused to publish the book unless he withdrew the epilogue, on the grounds that it was blasphemous. Schrödinger promptly took back his book and eventually sent it to a more secular English publisher, Cambridge University Press, which was happy to publish it. Schrödinger's experience with his Irish publisher suggests that enlightened blasphemy could be another requirement for reconciling science and religion, if ever they are to be reconciled.

Where have we ended up? At the moment, a negative future seems far more probable than a positive one. But one of the glories of evolution is its seemingly endless capacity to innovate and come up with surprises. Perhaps we may still contribute to a positive future, in which we accept the custodianship of the future evolution of life on Earth. We may still surprise ourselves. *Quo vadis?*

Notes

CHAPTER 1

1. Schrödinger 1944.
2. Vidral 2010.
3. Boltzman 1974.
4. Vidral 2010.
5. Vidral 2010.
6. Gibbs 1876.
7. Odling-Smee 1988.
8. Pross, personal communication, 2019.
9. Odling-Smee et al. 2003.
10. Odling-Smee et al. 2003; Laland et al. 2019.
11. Schrödinger 1944.
12. Schrödinger 1944, p. 73.
13. Gibbs 1876.
14. Darwin 1859.
15. Bennet 1987.
16. Huxley 1942.
17. Kauffman 2019.
18. Ravelli 2018.
19. Odling-Smee et al. 2003; Davis 2019.
20. Jaynes 1996.
21. For example, see Szilard 1929; Brillouin 1951; Bennet 1987; Jaynes 1996.

22. Odling-Smee et al. 2003.

23. Odling-Smee 1988; Odling-Smee et al. 2003.

24. Odling-Smee et al. 2003; Laland et al. 2012; Laland, Odling-Smee, and J. Endler 2017.

25. Odling-Smee et al. 2003; Laland et al. 2017.

26. Odling-Smee 1988.

27. Mayr 1982.

28. Odling-Smee 2007, 2010.

29. Liao and Zhang 2008.

30. Odling-Smee et al. 2003.

31. Cooper 1983.

32. See von Neumann 1956, 1966; Burke 1966; Cooper 1983; Vedral 2010.

33. Wolfram 2002.

34. See von Neumann 1956, 1966; von Neumann and Burks 1966; Vedral 2010.

35. Cooper 1983.

36. Turing 1937.

37. Vedral 2010, p. 44.

38. Cooper 1983.

CHAPTER 2

1. I owe this anecdote and the others in this chapter to Soni and Goodman 2017.

2. Soni and Goodman, 2017, p. 204.

3. Wilczek 1999.

4. In a subset of cases, animals can transmit just declarative knowledge, not procedural know-how, through social learning (Laland 2017).

5. Vedral 2010.

6. The biological utilities that organisms assign to incoming information from their environments are analogous to the subjective expected utilities, described by decision theory, that are usually but not invariably relative to humans.

7. Gibson 1950, 1966.

8. Contemporary ecological psychologists have developed the affordance concept further. The philosopher of biology Denis Walsh (2015) is a contemporary advocate of the affordances idea. Lynn Chiu (2019), another philosopher of biology, is also a supporter of the affordance concept. She makes an additional point, suggesting that the fluctuating biological utilities of

organisms correspond to a third kind of niche construction, in addition to the perturbational and relocational niche construction, as proposed by Odling-Smee et al. (2003). Chiu (2019, p. 301) calls it "mediational niche construction."

9. Possibly the most useful approach is still decision theory, which offers a way of combining arguably objective probabilities with the fluctuating subjective, expected utilities of organisms, relative to events in their local external environments. It also offers one way of calculating or at least of estimating the resulting biological utilities of organisms. We will return to these issues later in this chapter.

10. Watson and Szathmáry 2016.

11. Shannon emphasized that the person receiving a message may not be the person for whom it is intended, and the receiver must pass it to the target individual or destination. The destination may or may not subsequently use the knowledge that it has gotten from the receiver.

12. Laland 2017.

13. Tebbich et al. 2001, 2010; Grant 1999.

14. Grant 1999.

15. Laland, Odling-Smee, and Myles 2010.

16. Griffiths and Stotz 2013.

17. Vedral 2018; Carroll 2019.

18. Carroll 2019.

19. Carroll 2019.

20. For further information, I recommend Carroll 2019.

21. That could change for the simple reason that the world of quantum physics is not yet sufficiently well understood to rule out the relevance of quantum mechanics and qubits to evolution with any degree of authority. For instance, a large number of physicists are trying to apply quantum physics to the construction of a quantum computer. If and when they succeed, quantum computers should be many times more powerful than our most powerful conventional computers today. This may be relevant to biologists. For many years now, scientists have been comparing human brains to conventional computers. One of the first to do so was von Neumann again, but many others have followed in his footsteps. When quantum computers arrive on the scene, that may make comparisons between human brains and quantum computers irresistible. Do we already carry quantum computers inside our heads? At the moment, neuroscientists assume that we don't. But that might change. What about genomes? Do genomes have any quantum properties that are relevant to how we understand evolution? Evolutionary biologists assume that they don't. It seems too unlikely and too mind-befuddling to consider. But this too might change when we know more about quantum physics than we do today. An even more fundamental question concerns the origin of life. Did quantum mechanics contribute to the origin of life on Earth? We have no idea, but it can't be ruled out.

22. Odling-Smee et al. 2003; Griffiths and Stotz 2013.

23. Endler 1986.

24. Hamilton 1963.

25. In fact, this point is contestable. Some researchers, including Hamilton himself, have interpreted relatedness in a broader way than simply shared genes. For instance, relatedness might apply to individuals who are made more similar through acquiring the same shared cultural knowledge. To the extent that inclusive fitness is conceived in this broader manner, and that researchers recognize the similarity among individuals that can arise through shared environmental conditions that can come about through niche construction, then inclusive fitness can be useful here.

26. Odling-Smee et al. 2003.

27. Odling-Smee et al. 2003, 2013; Odling-Smee 2010.

28. These interactions between ecological inheritance systems and genetic inheritance systems then open up various hypotheses. For example, might niche-constructing organisms partly determine their degree of specialization versus generalization, relative to their niche interactions with their environments? Over successive generations, specialists may dig themselves deeper and deeper into their niches by specialized niche-constructing activities. Generalists may do the opposite. Like the Galapagos woodpecker finch, they may develop or maintain a high degree of flexibility in the ways that they niche-construct. As far as I know, this question has not been asked before.

29. Brown, Hall, and Sibly 2018.

30. Chaitin 2006; see also Schneider and Sagan 2005.

31. Chaitin 2006, p. 56.

32. Chaitin 2006.

33. Chaitin 2006, p. 56.

34. Chaitin 2006, p. 57.

35. Gerhart and Kirschner 1997, 2007; Popper 1966.

36. Gerhart and Kirschner 1997, 2007; Campbell 1960, 1974.

37. Odling-Smee 1988.

38. Hume 1739.

CHAPTER 3

1. Paley 1802.

2. Richard Dawkins's book *The Blind Watchmaker* (Dawkins 1986) depicts how contemporary evolutionary biology, in the form of neo-Darwinism, replies to Paley.

3. Lyell 1833.

4. Scott-Phillips et al. 2014; Futuyma 2017.

5. Odling-Smee et al. 2003.

6. Odling-Smee 1988; Odling-Smee et al. 2003, 2013; Sultan 2015.

7. Odling-Smee et al. 2003, 2013; Sultan 2015.

8. Plotkin and Odling-Smee 1979, 1981; Watson and Szathmáry 2016; Watson and Thies 2020. Also see chapter 7 of this book.

9. Williams 1966.

10. Reeve and Sherman 1993.

11. Hume 1739.

12. Slobodkin and Rapoport 1974. Note that interactions between purposeful systems can also be modeled by two-person games, which date back to much earlier work by von Neumann and Morgenstern (1944), with important later contributions from Nash and Maynard-Smith (Bhattachary 2021).

13. Here, I draw on a paper on adaptation by the morphologist Walter Bock (1980) because, to the best of my knowledge, he was one of the first biologists to clearly distinguish between the state of being adapted in the present and the historical origins of their adaptations. The former conception is most relevant to the adaptive niche management problems of organisms.

14. Odling-Smee 1988.

15. Slobodkin and Rapoport 1974.

16. Odling-Smee et al. 2003, chapter 4.

17. Ashby 1956, 1960.

18. Ashby 1956, 1960.

19. Odling-Smee 1988.

20. Odling-Smee 1988; Odling-Smee et al. 2003.

21. Odling-Smee 1988; Odling-Smee et al. 2003.

22. Dawkins 1982.

23. Odling-Smee 1988; Odling-Smee et al. 2003; Laland et al. 2019.

24. Hansell 1984.

25. Laland et al. 2017; Clark et al. 2019.

26. Mayr (1982, p. 828) describes the inheritance of acquired characteristics as "a chemical impossibility."

27. Bonduriansky and Day 2018. See also chapter 9 of this book.

28. Dickins and Rahman 2012.

29. Hamilton 1964; Gardner and West 2014.

30. Odling-Smee et al. 2003.

31. Puckett et al. 2018.

32. Couzin et al. 2005; Biro et al. 2006; Webster et al. 2017; Sasaki and Biro 2017.

33. For instance, see Seeley and Visscher 2004a, 2004b.

34. Turner 2000.

35. Hölldobler and Wilson 2009.

36. Levins 1968.

37. Environment influences on gene expression can also be mediated by gene methylation (Lyko et al. 2010).

38. Brakefield and Reitsma 1991.

39. Odling-Smee 1988.

40. Bock 1980.

41. Odling-Smee 2010.

42. Odling-Smee 2010, p. 182.

43. Odling-Smee 1988.

44. Odling-Smee 1988, 2010.

45. These examples come from Turner (2017), "Purpose and Desire."

46. Moore and Picker 1991.

47. Moore and Picker 1991.

48. Odling-Smee and Turner 2012.

49. Darwin 1881.

50. Turner 2000.

51. Anderson et al. 2017.

52. Turner 2000, pp. 106–107.

53. Turner 2000.

54. Turner 2000, p. 117.

55. Odling-Smee et al. 2003; Laland et al. 2017; Clark et al. 2020.

56. Turner 2017.

57. Turner 2017.

58. Hutchinson 1957.

CHAPTER 4

1. Williams 1966.

2. Whitehead 1978.

3. Feynman 1965; Odling-Smee 1988.

4. Gribbin 2005.

5. Bohm 1957; Vedra 2010; Rovelli 2021.

6. Plotkin and Odling-Smee 1981.

7. Popper 1979; Plotkin and Odling-Smee 1981; Hull 1981; Plotkin 1994; Dennett 1995.

8. Watson and Szathmáry 2016.

9. Odling-Smee 1988.

10. Lewontin 1983; Chiu 2019.

11. Antao et al. 2020.

12. Laland et al. 2017; Clarke et al. 2020.

13. Laland et al. 2017; Clarke et al. 2020.

14. Hansell 2000; Reid et al. 2002; Odling-Smee et al. 2003.

15. Laland et al. 2017; Clarke et al. 2020.

16. Laland et al. 2017; Clarke et al. 2020.

17. Odling-Smee et al. 2003.

18. Jones, Lawton, and Shachak 1997.

19. Clarke et al. 2020.

20. Davies 2015.

21. Odling-Smee et al. 2003.

22. Laland and Chiu 2020.

23. Odling-Smee et al. 2003.

24. Kondrashov 1993.

25. Uller et al. 2018.

26. Uller et al. 2018.

27. Uller et al. 2018.

28. Watson and Szathmáry 2016.

29. Allen et al. 2008.

30. Brakefield 2010.

31. Uller et al. 2018.

32. Gerhart and Kirschner 1997.

33. Russon 2003.

34. Hansell 1984; Odling-Smee et al. 2003.

35. Odling-Smee et al. 2003.

36. Laland et al. 2017; Clark et al. 2020.

37. Laland et al. 2017; Clark et al. 2020.

38. Odling-Smee et al. 2003; Laland 2014; Laland et al. 2017; Clark et al. 2020.

39. Turner 2000.

40. Pross 2012, p. 68.

CHAPTER 5

1. Gamow 1952; Guth 1997.

2. Schrödinger 1944.

3. Vedral 2010; Ravelli 2015, 2021; Schneider and Sagan 2005; Pross 2012; Lane 2015; Deamer 2019.

4. It is possible for endergonic energy–generating chemical reactions to occur in purposeless systems. It is also possible for autocatalytic reactions to occur in purposeless systems, which can greatly speed up chemical reactions. But none of that is enough to allow purposeless abiotic systems to resist the second law.

5. Damer and Deamer 2020; Deamer 2019.

6. Kauffmann 2019.

7. Lewontin 1983; Odling-Smee et al. 2003, 2013.

8. Odling-Smee et al. 2003.

9. Odling-Smee et al. 2003.

10. Kauffman 2019, p. 4.

11. Bhattacharya 2021.

12. This possibility of a universal theory of evolution was anticipated by Schrödinger himself in his "What is life?" question and by his supplementary question of what else is needed beyond the known laws of physics and chemistry to understand life. His questions not only referred to life on Earth, but potentially to life anywhere in the universe.

13. Rothschild and Mancinelli 2001.

14. For instance, see Phillip Pullman's *His Dark Materials* (1995).

15. I write "quasi-autonomous" natural selection pressures because although SET recognizes that organisms commonly modify natural selection (e.g., in cases such as habitat selection,

sexual selection, frequency dependent selection, and coevolution), it attributes no causal significance to this. Niche construction is treated as a product of evolution, not an evolutionary process—that is, as an effect rather than as a cause." SET recognizes that the sources of selection may be other organisms, but it treats such cases identically to cases where the source of selection is independent of the evolving population. Niche construction is viewed as determined by genes and fully explained by earlier natural selection.

16. Lewontin 1983.

17. Jones and Lawton 1995; Holt 1995; Jones, Lawton, and Shachak 1994, 1997.

18. Maynard Smith and Szathmáry 1995.

19. Maynard Smith and Szathmáry 1995.

20. Lane and Martin 2010.

21. Margulis 1967.

22. Post and Palkovacs 2009; Odling-Smee et al. 2013; Matthews et al. 2014; Sultan 2015; Laland et al. 2017; Clarke et al. 2019.

23. Schrödinger 1944.

24. Odling-Smee 1988, p. 82.

25. Rovelli 2021.

26. Odling-Smee 1988.

27. Godfrey-Smith 1996, p. 30.

28. Mayr 1961; West et al. 2011; Scott-Phillips et al. 2011; see also Laland et al. 2011.

29. Lewontin 1983; Odling-Smee 1988; Odling-Smee et al. 2003.

30. Dawkins 2006.

31. Lewontin 1983.

32. Godfrey-Smith 1996; Boltzmann 1974.

33. John Wheeler, quoted in Blundell 2015, 13; see also Rovelli 2015, 2021.

34. Godfrey-Smith 1996.

35. Godfrey-Smith 1996.

36. Odling-Smee 1988.

37. Krakauer made this remark at a conference about niche construction held at the Santa Fe Institute in 2014.

38. Odling-Smee 1988.

39. Odling-Smee 1988.

40. Odling-Smee 1988; Odling-Smee et al. 2003, see also chapter 4 of this book.

41. Laland et al. 2011, 2014, 2015.

42. Turner 2000, 2006; Laland, Odling-Smee, and Turner 2014.

43. West 1970; Abushama 1974; Sieber and Kokwaro 1982; Lys and Leuthold 1994.

44. Turner 2000, 2005; Laland et al. 2014.

45. Brodie 2005, p. 251.

46. Odling-Smee et al. 2003.

47. Lane and Martin 2010; Lane 2015.

48. Judson 2017.

49. Lenton, Pichler, and Weisz 2016.

50. San Roman and Wagner 2018.

51. Oxygen is a rich source of energy: the use of oxygen as an electron acceptor releases more energy per electron transfer than that of any other element except for chlorine and fluorine. See Catling et al. 2005.

52. Judson 2017.

53. Towe 1970.

54. Erwin 2005, p. 1752; Erwin and Tweedt 2012; Ward and Kirschvink 2015.

55. Ward and Kirschvink 2015; Hannah 2021.

56. Doug Erwin, quoted in Brannen 2017.

57. Hannah 2021.

58. Erwin, quoted in Brannen 2017.

59. Hannah 2021.

60. Ward and Kirschvink 2015.

61. There is one preliminary point. We need to distinguish between a species and an evolving population. A species may include more than one population. However, it is possible that an endangered species may be reduced to only a single population. That point is relevant to the current debate about the loss of biodiversity. It is possible for an endangered species to look less in danger of extinction than it actually is because there may be a number of different populations in the species that still exist in different regions of the world. Superficially, that may make the species appear to be in good health. But some of or all the populations in the species may be losing a very large percentage of their individual organisms very rapidly, thereby putting the species at a much higher risk of extinction than appears to be the case (Ceballos et al. 2017).

62. For instance, two other species might pool their R_i in a mutualist relationship, and then in combination outcompete a focal species for the same R_p resources.

63. See Kauffman 2019, 100.

64. Hannah 2021.

65. Hannah 2021, p. 173.

66. Ward and Kirschvink 2015.

CHAPTER 6

1. Cairns-Smith 1985; Fry 2002; Pross 2012; Lane 2015; Deamer 2012, 2019.

2. Hoyle 1983.

3. Mitchell 1961; Martin and Russell 2003; Lane 2015.

4. Deamer 2012, 2019; Damer and Deamer 2020.

5. Lane 2015.

6. Deamer 2019.

7. Mitchell 1961.

8. Martin and Russell 2003.

9. Mitchell 1979.

10. Eigen and Schuster 1977.

11. Pross 2012.

12. Deamer 2019.

13. Deamer 2019.

14. On February 1, 1871, Charles Darwin wrote a letter to Joseph Hooker that included some of his speculations on the spontaneous generation of life in "some warm little pond." To view this, see http://www.age-of-the-sage.org/darwin-quotes/warm-little-pond.html.

15. Damer and Deamer 2015, 2020; Deamer 2019.

16. Deamer 2012, 2019; Damer and Deamer 2020.

17. Maguire, Smokers, and Huck 2021; Cornell et al. 2019.

18. Kaufman 1993, p. 120.

19. Woese and Fox 1977, p. 5088; Woese 1998.

20. Bartel and Szostak 1993.

21. Lewontin 1970.

22. Laland, Odling-Smee, and Gilbert 2008; Gilbert, Bosch, and Ledon-Rettig 2015; Chiu and Gilbert 2015; Gilbert 2019.

23. Boogert, Paterson, and Laland 2006.

24. Lane 2015.

25. Gilbert 2019.

26. Damer and Deamer 2020.

CHAPTER 7

1. Laland et al. 2014, 2015.

2. SET does recognize that individual organisms can learn, of course, but it does not regard this as bootstrapping, as that learning is generally regarded as being under genetic control. This point is discussed later in this chapter.

3. Campbell 1960; Plotkin and Odling-Smee 1981; Kirschner and Gerhart 2005; Watson and Szathmáry 2016.

4. Kirschner and Gerhart 2005.

5. See Watson and Thies, 2019, for a discussion about the relationships that can occur among particles (in this case, individual organisms and different kinds of collectives, here evolving populations or sociocultural groups).

6. In saying this, I am riding roughshod over two points. First, individual metazoan organisms actually comprise collections of cells. Second, biologists have great difficulty in defining what is meant by the concept of individuality. It is often blurred, but for the sake of my present argument, I'm going to ignore both of these points here.

7. I will ignore the effects of somatic mutation and horizontal gene transfer here.

8. I use the term "memory" broadly, to encompass not just memories of events experienced by an individual (level 2), but also genetic registers of gene products that produced fit phenotypes in ancestral environments (level 1), memories of antibodies that proved effective in fighting off pathogens earlier in life (level 2), and cultural wisdom accrued in a population (level 3).

9. Kawai 1965.

10. Whiten et al. 1999.

11. Fisher and Hinde 1949.

12. Whitehead and Rendell 2015.

13. Hauser 1996.

14. Tomlinson 2015, 2018.

15. Olson and Torrance 2009.

16. Cavalli-Sforza and Feldman 1981.

17. Mesoudi 2011.

18. Hamilton 1964.

19. Hoppitt and Laland 2013.

20. Durham 1990; Richerson and Boyd 2005.

21. Durham 1990.

22. Mineka and Cook 1988.

23. Bateson and Martin 2000.

24. West-Eberhard 2003; Laland et al. 2011.

25. Bonduriansky and Day 2018.

26. Laland et al. 2015.

27. Amundson 2005; Muller 2021.

28. Segerstrale 2000.

29. Lewontin 1970.

30. Goldstein 2022.

31. Monroe et al. 2022; Stoltzfus 2019.

32. Campbell 1960; Hull, Langman, and Glenn 2001; Dennett 1995; Watson and Szathmáry 2016; Gerhart and Kirschner 1997; Plotkin and Odling-Smee 1979, 1981.

33. Plotkin and Odling-Smee 1979, 1981.

34. Waddington 1969, p. 122.

35. Burnet 1957, 1959.

36. Gerhart and Kirschner 1997.

37. The terminology currently in use seems to have dissuaded many other scientists working in other disciplines, as well as other psychologists, from wanting to know more about the processes of animal learning. It may also have discouraged investigations into the relationships between individual animal learning and the underlying processes of individual development and population-genetic evolution. For instance, psychologists who have investigated animal learning have seldom shown much interest in the evolutionary functions of learning or in the contributions that learning makes to the adaptations of individual organisms or to their fitness. They primarily have been interested in how learning works in the brains of animals rather than why or when learning occurs in the natural world, and in connecting the learning process to the neurosciences rather than to evolutionary biology.

38. Mayr 1961; Laland et al. 2011.

39. Heyes 1993, 2013.

40. Cavalli Sforza and Feldman 1981; Boyd and Richerson 1985; Henrich 2016.

41. Cavalli Sforza and Feldman 1981; Boyd and Richerson 1985; Henrich 2016.

42. Evershed et al. 2022.

43. Gerbault et al. 2011; Itan et al. 2009.

44. Laland, Odling-Smee, and Myles 2010.

45. For instance, see Popper 1966.

46. Goldstein 2022.

47. Walsh 2015; Uller and Helanterä 2019; Laland et al 2019.

CHAPTER 8

1. Reiners 1986.

2. Erwin and Valentine 2013.

3. Haller and Hendry 2014.

4. Erwin and Valentine 2013; Ward and Kirschvink 2015.

5. Grant and Grant 2014.

6. Alberti et al. 2017.

7. Alberti 2015; Derryberry et al. 2020; Otto 2018; Zeder 2017.

8. O'Neill et al. 1986.

9. May 1973; Thompson 1994.

10. Odling-Smee et al. 2003, chapter 5.

11. O'Neill et al. 1986; Roughgarden 1995.

12. Roughgarden 1995; Ricklefs and Miller 1999; Levin 2012.

13. O'Neill et al. 1986.

14. O'Neill et al. 1986; Reiners 1986; Jones and Lawton 1995; Holt 1995.

15. Odling-Smee et al. 2003, 2013.

16. Jones, Lawton, and Shachak 1994, 1997.

17. Naiman 1988; Wright and Jones 2006.

18. Lewontin 1982.

19. Odling-Smee 1988.

20. Odling-Smee et al. 2003, 2013.

21. Odling-Smee et al. 2003, 2013. It is also captured by the concept of environmentally mediated phenotypic associations, where the phenotypes are not necessarily fully determined by naturally selected genes (Odling-Smee et al. 2013).

22. Post and Palkovacs 2009; Hendry 2020.

23. Matthews et al. 2014.

24. Post and Palkovacs 2009; Hendry 2020.

25. Post and Palkovacs 2009; Loreau 2010; Kyfalis and Loreau 2008, 2011; Odling-Smee et al. 2013; Matthews et al. 2014; Lion 2018.

26. Auer et al. 2017.

27. Schrödinger 1944, p. 70.

28. Laland, Odling-Smee, and Endler 2017.

29. Clark et al. 2020.

30. Interestingly, there might be information in the organism-constructed remains such as the artifacts of a planet on which all life has gone extinct.

31. Ward and Kirschvink 2015.

32. Ward and Kirschvink 2015.

33. Kauffman 1993.

34. Weinberg 1993.

35. Schrödinger 1944.

36. Davies 2015.

37. Bejan and Zane 2012.

38. Harold 2014.

39. Davidson and Erwin 2006; Erwin and Valentine 2013.

40. Gould and Vrba 1982.

41. Odling-Smee et al. 2003.

42. Boogert, Paterson, and Laland 2006.

43. Bascompte and Jordano 2014.

44. Bascompte and Jordano 2014.

45. Watson et al. 2016.

46. Watson et al. 2016; Watson and Szathmáry 2016.

47. Judson 2017.

48. Margulis 1970.

49. Maynard-Smith and Szathmáry 1995.

50. Margulis 1970.

51. Lane 2015.

52. Lane 2015.

CHAPTER 9

1. Dawkins 1976.

2. Dawkins 1976.

3. Goldstein 2022.

4. Arguably, it is not really a major problem for the statistical approach of quantitative genetics that there are different mechanisms of inheritance, as it is the relationship between the fitness

of parents and the phenotypes of offspring that matters. One might reason that so long as it is possible to estimate parent-offspring similarity, all is well. However, there is now extensive theoretical evidence that extragenetic inheritance makes a substantial difference in evolutionary dynamics and can often take populations to alternative equilibria (Fogarty and Wade 2022; Gonzalez-Forero 2023). In other words, the mechanistic details of inheritance matter.

5. Bonduriansky and Day 2018.

6. Sultan, Moczek, and Walsh 2021; Bonduriansky and Day 2018; Jablonka and Lamb 2005.

7. Sultan et al. 2021; Bonduriansky and Day 2018; Jablonka and Lamb 1995, 2014; Jablonka and Raz 2009; Danchin et al. 2019; Uller 2019; Anastasiadi et al. 2021.

8. Mayr 1974.

9. Sultan 2015.

10. Gilbert and Epel 2009; Schmidt-Huntzel, et al. 2005; Carneiro et al. 2011; Aigner et al. 2000.

11. Bonduriansky and Day 2018.

12. Bonduriansky and Day 2018; Jablonka and Lamb 1995, 2014; Anastasiadi et al. 2021.

13. Cubas, Vincent, and Coen 1999.

14. Schmid et al. 2018; Bonduriansky and Day 2018; Jablonka and Lamb 1995, 2014.

15. For instance, see Bonduriansky and Day 2018; Jablonka and Lamb 1995, 2014.

16. Damer and Deamer 2019.

17. Bonduriansky and Day 2018.

18. Lane 2015.

19. Chiu and Gilbert 2015; Gilbert 2020.

20. Hundreds of published studies now attest to the importance of epigenetic inheritance in a wide variety of organisms, including humans. See Jablonka and Raz 2009; Heard and Martienssen 2014; Bonduriansky and Day 2018 and Anastasiadi et al. 2021 for reviews.

21. Heijmans et al. 2008; Tobi et al. 2018.

22. Durham 1991.

23. Durham 1991; O'Brien and Laland 2012.

24. Jablonka and Lamb 2005; Henrich 2016; Laland 2017.

25. Jablonka and Lamb 2005; Henrich 2016; Laland 2017.

26. Jablonka and Lamb 2005; Henrich 2016; Laland 2017.

27. Woese 1998.

28. Hehemann et al. 2010.

29. Monod 1972

30. Gould and Vrba 1982.

31. Shubin, Daeschler, and Coates 2004.

32. Judson 2017.

33. Maynard Smith and Szathmáry 1995.

34. Lane 2015.

35. Lewontin 1970.

36. Lane 2015.

37. Odling-Smee, Laland, and Feldman 2003.

38. Watson and Thies 2019.

39. Watson and Thies 2019.

40. Mukherjee et al. 2018.

41. Tomasello 2008; Bickerton 2009; Laland 2017.

42. Maynard Smith and Szathmáry 1995.

43. Judson 2017.

44. Wrangham 2009.

45. Wrangham 2009.

46. Aiello and Wheeler 1995.

47. Judson 2017.

48. Sober and Wilson 1998; Godfrey-Smith 2009; Okasha 2018.

49. Odling-Smee 1988; Odling-Smee et al. 2003; Laland et al. 2015.

50. Lewontin 1983; Odling-Smee et al., 2003, 2013; Matthews et al. 2014.

51. See Uller 2023 for a discussion of this point.

52. Riskin 2016.

53. Waddington 1969, p. 122.

54. Walsh 2019.

55. Uller 2023.

56. Pittendrigh 1958.

57. Lewontin 1983; Odling-Smee et al. 2003.

58. Gilbert and Epel 2009.

CHAPTER 10

1. See Witze 2020, and references therein.

2. Albert Einstein Quotes, Goodreads (https://www.goodreads.com/quotes/17014-the-release-of-atomic-power-has-changed-everything-except-our).

3. Boyd and Richerson 1985; Tomasello 1999; Dunbar 2022.

4. Holldobler and Wilson 1990; Wilson 2020.

5. Dunbar 2022.

6. Lindenfors, Wartel, and Lind 2021.

7. Richerson and Boyd 2005; Turchin 2016.

8. Richerson and Boyd 2005; Turchin 2016.

9. Boyd and Richerson 1985; Richerson and Boyd 2005; Boyd 2018.

10. Turchin 2016.

11. Jeffery 2008; Rohner et al. 2013; McGaugh et al. 2019; Kowalko 2020.

12. Pérez et al. 2009.

13. Wilson 1975.

14. Dawkins 1976.

15. Schneider and Sagan 2005.

16. Odum 1971; Reiners 1986; Odling-Smee et al. 2003, 2013; Duffy 2021.

17. Begon, Townsend, and Harper 2005.

18. Hölldobler and Wilson 1990; Wilson 2020.

19. Hölldobler and Wilson 1990; Wilson 2020; Odling-Smee et al. 2003.

20. Smith 2007a, 2007b; Smith and Zeder 2013; Zeder 2017.

21. Jarrige and Le Roux 2021.

22. Smith 2007a, 2007b; Smith and Zeder 2013; Zeder 2017; Hünemeier et al. 2012.

23. But see Ideen 2023.

24. Erwin 2005, 2008, 2012; Erwin et al. 2011.

25. Jones, Lawton, and Shachak 1994, 1997; Odling-Smee et al. 2003, 2013; Matthews et al 2014.

26. Boserup 1981.

27. Romer 1990, 1994.

28. Erwin 2008.

29. Solow 1956.

30. Romer 1990, 1994.

31. Duffy 2021.

32. Erwin 2008, see also Ideen 2023.

33. Duffy 2021.

34. Smil 2000; Syvitski et al. 2020.

35. Roughgarden 1979, 1998.

36. Borlaug 1970; Antonelli 2023.

37. Jarrige and Le Roux 2021; Duffy 2021.

38. Wagner et al. 2021; Klein et al. 2007; Aizen et al. 2009, 2019.

39. Antonelli 2023.

40. Roser, Ortiz-Ospina, and Ritchie 2013.

41. Kreager et al. 2015.

42. Russell 1948, 1961.

43. Rhoads et al. 2023.

44. Weinberg 1974.

45. Nurse 2020.

46. Kreibich et al. 2022.

47. Goldstein 2022.

48. Dunbar 2022.

49. Maynard-Smith and Szathmáry 1995.

50. Odling-Smee and Laland 2011.

51. Schrodinger 1944, p. 87.

References

Abushama, F. T. 1974. Water-relations of the termites *Macrotermes bellicosus* (Smeathman) and *Trinervitermes geminatus* (Wasmann). *Zeitschrift für Angewandte Entomologie* 75: 124–134.

Aiello, L. C., and P. Wheeler. 1995. The expensive-tissue hypothesis: The brain and the digestive system in human and primate evolution. *Current Anthropology* 36: 199–221.

Aigner, B., U. Besenfelder, M. Müller, and G. Brem. 2000. Tyrosinase gene variants in different rabbit strains. *Mammalian Genome* 11: 700–702.

Aizen, M. A., S. Aguiar, J. C. Biesmeijer, et al. 2019. Global agricultural productivity is threatened by increasing pollinator dependence without a parallel increase in crop diversification. *Global Change Biology* 25: 3516–3527.

Aizen, M. A., L. A. Garibaldi, S. A. Cunningham, and A. M. Klein. 2009. How much does agriculture depend on pollinators? Lessons from long-term trends in crop production. *Annals of Botany* 103: 1579–1588.

Alberti, M. 2015. Eco-evolutionary dynamics in an urbanizing planet. *Trends in Ecology & Evolution* 30: 114–126.

Alberti, M., C. Correa, J. M. Marzluff, et al. 2017. Global urban signatures of phenotypic change in animal and plant populations. *Proceedings of the National Academy of Sciences* 114: 8951–8956.

Allen, C. E., P. Beldade, B. J. Zwaan, and P. M. Brakefield. 2008. Differences in the selection response of serially repeated color pattern characters: Standing variation, development, and evolution. *BMC Evolutionary Biology* 8: 94. https://doi.org/10.1186/1471-2148-8-94.

Amundson, R. 2005. *The Changing Role of the Embryo in Evolutionary Thought: Roots of Evo-Devo.* Cambridge: Cambridge University Press.

Anastasiadi, D., C. J. Venney, L. Bernatchez, and M. Wellenreuther. 2021. Epigenetic inheritance and reproductive mode in plants and animals. *Trends in Ecology & Evolution* 36: 1124–1140.

Anderson, F. E., B. W. Williams, K. M. Horn, et al. 2017. Phylogenomic analyses of Crassiclitellata support major Northern and Southern Hemisphere clades and a Pangaean origin for earthworms. *BMC Evolutionary Biology* 17: 123.

Antonelli, A. 2023. Indigenous knowledge is key to sustainable food systems. *Nature* 613: 239–242.

Ashby, W. R. 1956. *An Introduction to Cybernetics*. London: Chapman & Hall.

Ashby, W. R. 1960. *Design for a Brain: The Origin of Adaptive Behavior*. London: Chapman & Hall.

Bartel, D. P., and J. W. Szostak. 1993. Isolation of new ribozymes from a large pool of random sequences. *Science* 261: 1411–1418.

Bateson, P., and P. Martin. 2000. *Design for a Life*. London: Vintage.

Begon, M., C. R. Townsend, and J. L. Harper. *Ecology: From Individuals to Ecosystems*. 4th ed. Oxford, UK: Wiley, Blackwell.

Bejan, A. and J. Peder Zane. 2012. *Design in Nature: How the Constructal Law Governs Evolution in Biology, Physics, Technology, and Social Organization*. New York: Doubleday Books.

Bennet, S. S. S. R. 1987. *Name Changes in Flowering Plants of India and Adjacent Regions*. Dehra Dun, India: Triseas Publishers.

Bhattacharya, V. 2021. An empirical model of R&D procurement contests: An analysis of the DOD SBIR program. *Econometrica* 89: 2189–2224.

Bickerton, A. 2009. *Adam's Tongue*. New York: Hill and Wang.

Biro, D., D. J. T. Sumpter, J. Meade, and T. Guilford. 2006. From compromise to leadership in pigeon homing. *Current Biology* 16: 2123–2128.

Blundell, K. 2015. *Black Holes: A Very Short Introduction*. Oxford: Oxford University Press.

Bock, W. J. 1980. The definition and recognition of biological adaptation. *American Zoologist* 20: 217–227.

Bohm, D. 1957. *Causality and Chance in Modern Physics*. London: Routledge.

Boltzmann, L. 1974. The second law of thermodynamics. In *Theoretical Physics and Philosophical Problems: Selected Writings*, ed. B. McGuinness. Dordrecht, Netherlands: Springer, pp. 13–32.

Bonduriansky, R., and T. Day. 2018. *Extended Heredity. A New Understanding of Inheritance and Evolution*. Princeton, NJ: Princeton University Press.

Boogert, N. J., D. M. Paterson, and K. N. Laland. 2006. The implications of niche construction and ecosystem engineering for conservation biology. *BioScience* 56: 570–578.

Boyd, R., and P. J. Richerson. 1985. *Culture and the Evolutionary Process*. Chicago: University of Chicago Press.

Brakefield, P. M. 2010. Radiations of Mycalesine butterflies and opening up their exploration of morphospace. *The American Naturalist* 176: S77–S87.

Brakefield, P. M., and N. Reitsma. 1991. Phenotypic plasticity, seasonal climate and the population biology of Bicyclus butterflies (Satyridae) in Malawi. *Ecological Entomology* 16: 291–303.

Brillouin, L. 1951. Maxwell's demon cannot operate: Information and entropy. *Journal of Applied Physics* 22: 334–337.

Brodie, E. D. III. 2005. Caution. Niche construction ahead. *Evolution* 59: 249–251.

Brown, J. H., C. A. S. Hall, and R. M. Sibly. 2018. Equal fitness paradigm explained by a trade-off between generation time and energy production rate. *Nature Ecology & Evolution* 2: 262–268.

Burnet, F. M. 1957. A modification of Jerne's theory of antibody production using the concept of clonal selection. *Australian Journal of Science* 20: 67–69.

Burnet, F. M. 1959. *The Clonal Selection Theory of Acquired Immunity.* Nashville: Vanderbilt University Press.

Cairns-Smith, A. G. 1985. *Seven Clues to the Origin of Life: A Scientific Detective Story.* Cambridge: Cambridge University Press.

Campbell, D. T. 1960. Blind variation and selective retention in creative thought as in other knowledge processes. *Psychological Review* 67: 380–400.

Campbell, D. T. 1974. Downward causation in hierarchically organised biological systems. In *Studies in the Philosophy of Biology: Reduction and Related Problems*, ed. F. J. Ayala and T. Dobzhansky. London: Macmillan Education UK, pp. 179–186.

Carneiro, M., S. Afonso, A. Geraldes, et al. 2011. The genetic structure of domestic rabbits. *Molecular Biology and Evolution* 28: 1801–1816.

Carroll, S. M. 2019. *Something Deeply Hidden: Quantum Worlds and the Emergence of Spacetime*: New York: Dutton.

Catling, D. C., C. R. Glein, K. J. Zahnle, and C. P. McKay. 2005. Why O_2 is required by complex life on habitable planets and the concept of planetary "oxygenation time." *Astrobiology* 5: 415–438.

Cavalli-Sforza, L. L., and M. W. Feldman. 1981. *Cultural Transmission and Evolution: A Quantitative Approach.* Princeton, NJ: Princeton University Press.

Ceballos, G., P. R. Ehrlich, and R. Dirzo. 2017. Biological annihilation via the ongoing sixth mass extinction signaled by vertebrate population losses and declines. *Proceedings of the National Academy of Sciences* 114: E6089–E6096.

Chaitin, G. 2006. *Meta Math: The Quest for Omega.* Bloomsbury, UK: Atlantic Books.

Chiu, L. 2019. Decoupling, commingling, and the evolutionary significance of experiential niche construction. In *Evolutionary Causation, Biological and Philosophical Reflection*, ed. T. Uller, and K. N. Laland. Cambridge, MA: MIT Press, pp. 299–321.

Chiu, L., and S. F. Gilbert. 2015. The birth of the holobiont: Multi-species birthing through mutual scaffolding and niche construction. *Biosemiotics* 8: 191–210.

Clark, A. D., D. Deffner, K. N. Laland, F. J. Odling-Smee, and J. Endler. 2020. Niche construction affects the variability and strength of natural selection. *The American Naturalist* 195: 16–30.

Cooper, N. G. 1983. From Turing and von Neumann to the present. *Los Alamos Science* Fall. 22–27.

Cornell, C. E., R. A. Black, M. Xue, et al. 2019. Prebiotic amino acids bind to and stabilize prebiotic fatty acid membranes. *Proceedings of the National Academy of Sciences* 116: 17239–17244.

Couzin, I. D., J. Krause, N. R. Franks, and S. A. Levin. 2005. Effective leadership and decision-making in animal groups on the move. *Nature* 433: 513–516.

Cubas, P., C. Vincent, and E. Coen. 1999. An epigenetic mutation responsible for natural variation in floral symmetry. *Nature* 401: 157–161.

Damer, B., and D. W. Deamer. 2015. Coupled phases and combinatorial selection in fluctuating hydrothermal pools: A scenario to guide experimental approaches to the origin of cellular life. *Life* 5: 872–887.

Damer, B., and D. W. Deamer. 2020. The hot spring hypothesis for an origin of life. *Astrobiology* 20: 429–452.

Danchin, E., A. Pocheville, O. Rey, B. Pujol, and S. Blanchet. 2019. Epigenetically facilitated mutational assimilation: Epigenetics as a hub within the inclusive evolutionary synthesis. *Biological Review* 94: 259–282.

Darwin, C. R. 1881. *The Formation of Vegetable Mold through the Action of Worms, with Observations on Their Habits*. London: John Murray.

Darwin, C. R. 1968 [orig. 1859]. *On the Origin of Species by Means of Natural Selection, or the Preservation of Favoured Races in the Struggle for Life*. Reprint of the first edition. London: Penguin Books.

Davidson, E. H., and D. H. Erwin. 2006. Gene regulatory networks and the evolution of animal body plans. *Science* 311: 796–800.

Davies, N. 2015. *Cuckoo: Cheating by Nature*. London: Bloomsbury Publishing.

Dawkins, R. 1976. *The Selfish Gene*. Oxford: Oxford University Press.

Dawkins, R. 1982. *The Extended Phenotype*. Oxford: Oxford University Press.

Dawkins, R. 1986. *The Blind Watchmaker*. New York: W. W. Norton & Company.

Dawkins, R. 2006. *The God Delusion*. London: Black Swan.

Deamer, D. W. 2012. *First Life: Discovering the Connections between Stars, Cells, and How Life Began*. Berkeley: University of California Press.

Deamer, D. W. 2019. *Assembling Life: How Can Life Begin on Earth and Other Habitable Planets?* Oxford: Oxford University Press.

Dennett, D. 1995. *Darwin's Dangerous Idea: Evolution and the Meanings of Life*. London: Penguin.

Derryberry, E. P., J. N. Phillips, G. E. Derryberry, M. J. Blum, and D. Luther. 2020. Singing in a silent spring: Birds respond to a half-century soundscape reversion during the COVID-19 shutdown. *Science* 370: 575–579.

Dickins, T. E., and Q. Rahman. 2012. The extended evolutionary synthesis and the role of soft inheritance in evolution. *Proceedings of the Royal Society B* 279: 2913–2921.

Duffy, J. Emmett. 2021. *Ocean Ecology: Marine Life in the Age of Humans*. Princeton, NJ: Princeton University Press.

Dunbar, R. 2022. *How Religion Evolved: And Why It Endures*. London: Pelican.

Durham. W. H. 1991. *Coevolution: Genes, Culture, and Human Diversity*. Stanford, CA: Stanford University Press.

Eigen, M., and P. Schuster. 1977. A principle of natural self-organization. *Naturwissenschaften* 64: 541–565.

Endler, J. 1986. *Natural Selection in the Wild: Monographs in Population Biology 21*. Princeton, NJ: Princeton University Press.

Erwin, D. H. 2005. Seeds of diversity. *Science* 308: 1752–1753.

Erwin, D. H. 2008. Macroevolution of ecosystem engineering, niche construction and diversity. *Trends in Ecology & Evolution* 23: 304–310.

Erwin, D. H. 2012. Novelties that change carrying capacity. *Journal of Experimental Zoology Part B: Molecular and Developmental Evolution* 318: 460–465.

Erwin, D. H., M. Laflamme, S. M. Tweedt, E. A. Sperling, D. Pisani, and K. J. Peterson. 2011. The Cambrian conundrum: Early divergence and later ecological success in the early history of animals. *Science* 334: 1091–1097.

Erwin, D. H., and S. Tweedt. 2012. Ecological drivers of the Ediacaran-Cambrian diversification of Metazoa. *Evolutionary Ecology* 26: 417–433.

Erwin, D. H., and J. W. Valentine. 2013. *The Cambrian Explosion: The Construction of Animal Biodiversity*. Greenwood Village, CO: Robert and Company Publishers.

Evershed, R. P., G. D. Smith, M. Roffet-Salque, et al. 2022. Dairying, diseases and the evolution of lactase persistence in Europe. *Nature* 608: 336–345.

Feynman, R. 1965. *The Character of Physical Law*. Cambridge, MA: MIT Press.

Fisher, J. B., and R. A. Hinde. 1949. Opening of milk bottles by birds. *British Birds* XLII: 347–357.

Fogarty, L., and M. J. Wade. 2022. Niche construction in quantitative traits: Heritability and response to selection. *Proceedings of the Royal Society B* 289: 20220401.

Fry, I. 2002. *Emergence of Life on Earth: A Historical and Scientific Overview*. New Brunswick, NJ: Rutgers University Press.

Futuyma, D. J. 2017. Evolutionary biology today and the call for an extended synthesis. *Interface Focus* 7: 20160145.

Gamow, G. 1952. *The Creation of the Universe*. New York: Viking.

Gardner, A., and S. A. West. 2014. Inclusive fitness: 50 years on. *Philosophical Transactions of the Royal Society B: Biological Sciences* 369: 20130356.

Gerbault, P., A. Liebert, Y. Itan, et al. 2011. Evolution of lactase persistence: An example of human niche construction. *Philosophical Transactions of the Royal Society of London B* 366: 863–877.

Gerhart, J. C., and M. W. Kirschner. 1997. *Cells, Embryos & Evolution*. Hoboken, NJ: Wiley.

Gerhart, J. C., and M. W. Kirschner. 2007. The theory of facilitated variation. *Proceedings of the National Academy of Sciences* 104: 8582–8589.

Gibbs, J. W. (May 1876—July 1878). On the equilibrium of heterogeneous substances. *Transactions of the Connecticut Academy of Arts and Sciences* 3: 441–458.

Gibson, J. J. 1950. *The Perception of the Visual World*. Boston: Houghton-Mifflin.

Gibson, J. J. 1966. *The Senses Considered as Perceptual Systems*. Boston: Houghton-Mifflin.

Gilbert, S. F. 2019. Evolutionary transitions revisited: Holobiont evo-devo. *Journal of Experimental Zoology Molecular and Developmental Biology* 332: 307–314.

Gilbert, S. F. 2020. *Holobionts Can Evolve by Changing Their Symbionts and Hosts*. Stanford, CA: Stanford University Press.

Gilbert, S. F., and D. Epel. 2009. *Ecological Developmental Biology: Integrating Epigenetics, Medicine, and Evolution*. Sunderland, MA: Sinauer Associates.

Gilbert, S. F., T. C. Bosch, and C. Ledon-Rettig. 2015. Eco-Evo-Devo: Developmental symbiosis and developmental plasticity as evolutionary agents. *Nature Review Genetics* 16: 611–622.

Godfrey-Smith, P. 1996. *Complexity and the Function of Mind in Nature*. Cambridge, MA: Cambridge University Press.

Godfrey-Smith, P. 2009. *Darwinian Populations and Natural Selection*. New York: Oxford University Press.

Goldstein, D. B. 2022. *The End of Genetics Designing Humanity's DNA*. New Haven, CT: Yale University Press.

Gonzalez-Forero, M. 2023. How development affects evolution. *Evolution* 77: 562–579.

Gould, S. J., and E. Vrba. 1982. Exaptation: A missing term in the science of form. *Paleobiology* 8: 4–15.

Grant, P. R. 1999. *Ecology and Evolution of Darwin's Finches*. Princeton, NJ: Princeton University Press.

Grant, P. R., and B. R. Grant. 2014. *40 Years of Evolution: Darwin's Finches on Daphne Major Island*. Princeton, NJ: Princeton University Press.

Griffiths, P. E., and K. Stotz. 2013. *Genetics and Philosophy: An Introduction*. Cambridge: Cambridge University Press.

Guth, A. 1997. *The Inflationary Universe: The Quest for a New Theory of Cosmic Origins*. Reading, MA: Addison Wesley.

Haller, B. C., and A. P. Hendry. 2014. Solving the paradox of stasis: Squashed stabilizing selection and the limits of detection. *Evolution* 68: 483–500.

Hamilton, W. D. 1963. The evolution of altruistic behavior. *The American Naturalist* 97: 354–356.

Hamilton, W. D. 1964. The genetical evolution of social behaviour. I. *Journal of Theoretical Biology* 7: 1–16.

Hannah, M. 2021. *Extinctions: Living and Dying in the Margin of Error*. Cambridge: Cambridge University Press.

Hansell, M. H. 1984. *Animal Architecture and Building Behaviour*. London: Longman.

Hansell, M. H. 2000. *Bird Nests and Construction Behaviour*. Cambridge: Cambridge University Press.

Harold, F. M. 2014. *In Search of Cell History: The Evolution of Life's Building Blocks*. Chicago: University of Chicago Press.

Hauser, M. D. 1996. *The Evolution of Communication*. Cambridge, MA: MIT Press.

Heard, E., and R. A Martienssen. 2014. Transgenerational epigenetic inheritance: Myths and mechanisms. *Cell* 157: 95–109.

Hehemann, J. H., G. Correc, T. Barbeyron, W. Helbert, M. Czjzek, and G. Michel. 2010. Transfer of carbohydrate-active enzymes from marine bacteria to Japanese gut microbiota. *Nature* 464: 908–912.

Heijmans, B. T., W. T. Elmar, D. S. Aryeh, et al. 2008. Persistent epigenetic differences associated with prenatal exposure to famine in humans. *Proceedings of the National Academy of Sciences* 105: 17046–17049.

Hendry, A. P. 2020. *Eco-Evolutionary Dynamics*. Princeton, NJ: Princeton University Press.

Henrich, J. 2016. *The Secret of Our Success: How Culture Is Driving Human Evolution, Domesticating Our Species, and Making Us Smarter*. Princeton, NJ: Princeton University Press.

Heyes, C. M. 1993. Imitation, culture and cognition. *Animal Behaviour* 46: 999–1010.

Heyes, C. M. 2013. What can imitation do for cooperation? In *Cooperation and Its Evolution*, ed. K. Sterelny, R. Joyce, B. Calcott, and B. Fraser. Cambridge, MA: MIT Press.

Hölldobler, B., and E. O Wilson. 1990. *The Ants*. Cambridge, MA: Harvard University Press.

Hölldobler, B., and E. O. Wilson. 2009. *The Superorganism: The Beauty Elegance and Strangeness of Insect Societies*. New York: W. W. Norton.

Holt, R. D. 1995. Linking species and ecosystems: Where's Darwin? In *Linking Species and Ecosystems*, ed. C. G. Jones and J. Lawton. London: Chapman & Hall, pp. 273–279.

Hoppitt, W. J. E., and K. N. Laland. 2013. *Social Learning: An Introduction to Mechanisms, Methods, and Models*. Princeton, NJ: Princeton University Press.

Hoyle, F. 1983. *The Intelligent Universe*. London: Michael Joseph.

Hull, D. L. 1981. Units of evolution: A metaphysical essay. In *The Philosophy of Evolution*, ed. U. J. Jensen, and R. Brighton, UK: Harvester Press, pp. 23–44.

Hull, D. L., R. E. Langman, and S. S. Glenn. 2001. A general account of selection: Biology, immunology, and behavior. *Behavioral and Brain Sciences* 24: 511–528.

Hume, D. 1739. *A Treatise of Human Nature*. London: Clarendon Press.

Hünemeier, T., C. E. G. Amorim, S. Azevedo, et al. 2012. Evolutionary responses to a constructed niche: Ancient Mesoamericans as a model of gene-culture coevolution. *PLOS ONE* 7: e38862.

Hutchinson, G. E. 1957. Concluding remarks. *Cold Spring Harbor Symposia on Quantitative Biology* 22: 415–427.

Huxley, J. 1942. *Evolution. The Modern Synthesis.* London: Allen and Unwin.

Itan, Y., A. Powell, M. A. Beaumont, J. Burger, and M. G. Thomas. 2009. The origins of lactase persistence in Europe. *PLOS Computational Biology* 5: e1000491.

Jablonka, E., and M. J. Lamb. 1995. *Epigenetic Inheritance and Evolution: The Lamarckian Dimension.* Oxford: Oxford University Press.

Jablonka, E., and M. J. Lamb. 2005. *Evolution in Four Dimensions.* Cambridge, MA: MIT Press.

Jablonka, E., and M. J. Lamb. 2014. *Evolution in Four Dimensions.* Revised ed. Cambridge, MA: MIT Press.

Jablonka, E., and G. Raz. 2009. Transgenerational epigenetic inheritance: Prevalence, mechanisms, and implications for the study of heredity and evolution. *Quarterly Review of Biology* 84: 131–176.

Jarrige, F., and T. Le Roux. 2021. *The Contamination of the Earth.* Cambridge, MA: MIT Press.

Jaynes, D. B. 1996. Improved soil mapping using electromagnetic induction surveys. *Proceedings of the Third International Conference on Precision Agriculture* 169–179.

Jeffery, W. R. 2008. Emerging model systems in evo-devo: Cavefish and microevolution of development. *Evolution & Development* 10: 265–272.

Jones, C. G., and J. H. Lawton. 1995. *Linking Species and Ecosystems.* New York: Chapman & Hall.

Jones, C. G., J. H. Lawton, and M. Shachak. 1994. Organisms as ecosystem engineers. *Oikos* 69: 373–386.

Jones, C. G., J. H. Lawton, and M. Shachak. 1997. Positive and negative effects of organisms as physical ecosystem engineers. *Ecology* 78: 1946–1957.

Judson, O. P. 2017. The energy expansions of evolution. *Nature Ecology & Evolution* 1: 0138.

Kauffman, S. A. 1993. *The Origins of Order: Self-Organization and Selection in Evolution.* New York: Oxford University Press.

Kauffman, S. A. 2019. *A World beyond Physics: The Emergence and Evolution of Life.* New York: Oxford University Press.

Kawai, M. 1965. Newly-acquired pre-cultural behavior of the natural troop of Japanese monkeys on Koshima islet. *Primates* 6: 1–30.

Kirschner, M. W., and J. C. Gerhart. 2005. *The Plausibility of Life: Resolving Darwin's Dilemma.* New Haven, CT: Yale University Press.

Klein, A.-M., B. E. Vaissière, J. H. Cane, et al. 2007. Importance of pollinators in changing landscapes for world crops. *Proceedings of the Royal Society B: Biological Sciences* 274: 303–313.

Kondrashov, A. S. 1993. Classification of hypotheses on the advantage of amphimixis. *Journal of Heredity* 84: 372–387.

Kowalko, J. 2020. Utilizing the blind cavefish *Astyanax mexicanus* to understand the genetic basis of behavioral evolution. *Journal of Experimental Biology* Feb 7; 223(Pt Suppl 1): jeb208835.

Kreager, P., B. Winney, S. Ulijaszek, and C. Capelli. 2015. *Population in the Human Sciences: Concepts, Models, Evidence*. Oxford: Oxford University Press.

Kreibich, H., A. F. Van Loon, K. Schröter, et al. 2022. The challenge of unprecedented floods and droughts in risk management. *Nature* 608: 80–86.

Laland, K. N. 2017a. *Darwin's Unfinished Symphony: How Culture Made the Human Mind*. Princeton, NJ: Princeton University Press.

Laland, K. N. 2017b. The origins of language in teaching. *Psychonomic Bulletin & Review* 24: 225–231.

Laland, K. N., and L. Chiu. 2020. Evolution's engineers. *Aeon*. https://aeon.co/essays/organisms -are-not-passive-recipients-of-evolutionary-forces.

Laland, K. N., F. J. Odling-Smee, and J. Endler. 2017. Niche construction, sources of selection and trait coevolution. *Interface Focus* 7: 20160147.

Laland, K. N., F. J. Odling-Smee, and M. W. Feldman. 2019. Understanding niche construction as an evolutionary process. In *Evolutionary Causation, Biological and Philosophical Reflections*, ed. T. Uller and K. N. Laland. Cambridge, MA: MIT Press, pp. 127–152.

Laland, K. N., F. J. Odling-Smee, and S. F. Gilbert. 2008. Evodevo and niche construction: Building bridges. *Journal of Experimental Zoology Part B* 310: 549–566.

Laland K. N., F. J. Odling-Smee, W. Hoppitt, and T. Uller. 2012. More on how and why: Cause and effect in biology revisited. *Biology & Philosophy* 28: 719–745.

Laland, K. N., F. J. Odling-Smee, and S. Myles. 2010. How culture shaped the human genome: Bringing genetics and the human sciences together. *Nature Reviews Genetics* 11: 137–148.

Laland, K. N., F. J. Odling-Smee, and S. Turner. 2014. The role of internal and external constructive processes in evolution. *The Journal of Physiology* 592: 2413–2422.

Laland, K. N., K. Sterelny, F. J. Odling-Smee, W. Hoppitt, and T. Uller. 2011. Cause and effect in biology revisited: Is Mayr's proximate-ultimate dichotomy still useful? *Science* 334: 1512–1516.

Laland, K. N., T. Uller, M. W. Feldman, et al. 2014. Does evolutionary theory need a rethink? Yes. *Nature* 514: 161–164.

Laland, K. N., T. Uller, M. W. Feldman, et al. 2015. The extended evolutionary synthesis: Its structure, assumptions and predictions. *Proceedings of the Royal Society B* 282: 20151019.

Lane, N. 2015. *The Vital Question*. London: Profile books.

Lane, N., and W. Martin. 2010. The energetics of genome complexity. *Nature* 467: 929–934.

Lenton, T. M., P. P. Pichler, and H. Weisz. 2016. Revolutions in energy input and material cycling in Earth history and human history. *Earth System Dynamics* 7: 353–370.

Levin, S. A. 2012. *The Princeton Guide to Ecology*. Princeton, NJ: Princeton University Press.

Levins, R. 1968. *Evolution in Changing Environments*. Princeton, NJ: Princeton University Press.

Lewontin, R. C. 1970. The units of selection. *Annual Review of Ecology and Systematics* 1: 1–18.

Lewontin, R. C. 1982. Organism and environment. In *Learning, Development and Culture*, ed. H. C. Plotkin. Chichester, UK: Wiley, pp. 151–170.

Lewontin, R. C. 1983. Gene, organism, and environment. In *Evolution from Molecules to Men*, ed. D. S. Bendall, Cambridge: Cambridge University Press, pp. 273–285.

Liao, B. Y., and J. Zhang. 2008. Null mutations in human and mouse orthologs frequently result in different phenotypes. *Proceedings of the National Academy of Sciences* 105: 6987–6992.

Lindenfors, P., A. Wartel, and J. Lind. 2021. 'Dunbar's numbers' deconstructed. *Biology Letters* 17: 20210158.

Lion, S. 2018. Theoretical approaches in evolutionary ecology: Environmental feedback as a unifying perspective. *The American Naturalist* 191: 21–44.

Loreau, M. 2010. *From Populations to Ecosystems*. Princeton, NJ: Princeton University Press.

Lyell, C. 1833. *Principles of Geology: Being an Attempt to Explain the Former Changes of the Earth's Surface, by Reference to Causes Now in Operation*, Vol. 3. London: John Murray.

Lyko, F., S. Foret, R. Kucharski, S. Wolf, C. Falckenhayn, and R. Maleszka. 2010. The honey bee epigenomes: Differential methylation of brain DNA in queens and workers. *PLOS Biology* 8: e1000506.

Lys, J. A., and R. Leuthold. 1994. Forces affecting water imbibition in Macrotermes workers (Termitidae, Isoptera). *Insectes Sociaux* 41: 79–84.

Maguire, O. R., I. B. A. Smokers, and W. T. S. Huck. 2021. A physicochemical orthophosphate cycle via a kinetically stable thermodynamically activated intermediate enables mild prebiotic phosphorylations. *Nature Communications* 12: 5517.

Margulis, L. 1967. On the origin of mitosing cells. *Journal of Theoretical Biology* 14: 225–274.

Margulis, L. 1970. *Origin of Eukaryotic Cells: Evidence and Research Implications for a Theory of the Origin and Evolution of Microbial, Plant, and Animal Cells on the Precambrian Earth*. New Haven, CT: Yale University Press.

Martin, W., and M. J. Russell. 2003. On the origins of cells: A hypothesis for the evolutionary transitions from abiotic geochemistry to chemoautotrophic prokaryotes, and from prokaryotes to nucleated cells. *Philosophical Transactions of the Royal Society B* 358: 59–83.

Matthews, B., L. De Meester, C. G. Jones, et al. 2014. Under niche construction: An operational bridge between ecology, evolution, and ecosystem science. *Ecological Monographs* 84: 245–263.

May, R. M. 1973. *Stability and Complexity in Model Ecosystems*, Vol. 1. Princeton, NJ: Princeton University Press.

Maynard Smith, J., and E. Szathmáry. 1995. *The Major Transitions in Evolution*. Oxford, UK: W. H. Freeman.

Mayr, E. 1961. Cause and effect in biology. *Science* 134: 1501–1506.

Mayr, E. 1974. Behavior programs and evolutionary strategies. *American Scientist* 62: 650–659.

Mayr, E. 1982. *The Growth of Biological Thought. Diversity, Evolution, and Inheritance*. Cambridge, MA: Harvard University Press.

McGaugh, S. E., S. Weaver, E. N. Gilbertson, et al. 2019. Evidence for rapid phenotypic and behavioural shifts in a recently established cavefish population. *Biological Journal of the Linnean Society* 129: 143–161.

Mesoudi, A. 2011. *Cultural Evolution: How Darwinian Evolutionary Theory Can Explain Human Culture and Synthesize the Social Sciences*. Chicago: University of Chicago Press.

Mineka, S., and M. Cook. 1988. Social learning and the acquisition of snake fear in monkeys. In *Social Learning: Psychological and Biological Perspectives*, ed. B. G. Galef Jr. and T. R. Zentall. Hillsdale, NJ: Erlbaum, pp. 51–73.

Mitchell, P. 1961. Coupling of phosphorylation to electron and hydrogen transfer by a chemi-osmotic type of mechanism. *Nature* 191: 144–148.

Monod, J. Y. 1972. *Chance and Necessity: An Essay on the Natural Philosophy of Modern Biology*. New York: Random House USA.

Monroe, J. G., T. Srikant, P. Carbonell-Bejerano, et al. 2022. Mutation bias reflects natural selection in Arabidopsis thaliana. *Nature* 602: 101–105.

Moore, J. M., and M. D. Picker. 1991. Heuweltjies (earth mounds) in the Clanwilliam district, Cape Province, South Africa: 4000-year-old termite nests. *Oecologia* 86: 424–432.

Mukherjee, I., R. R. Large, R. Corkrey, and L. V. Danyushevsky. 2018. The Boring Billion, a slingshot for complex life on earth. *Scientific Reports* 8: 443.

Muller, G. B. 2021. Evo-devo's contributions to the extended evolutionary synthesis. In *Evolutionary Developmental Biology, A Reference Guide*, ed. L. Nuño de la Rosa and G. B. Müller. Cham, Switzerland: Springer International Publishing, pp. 1127–1138.

Naiman, R. J., C. A. Johnston, and J. C. Kelley. 1988. Alteration of North American streams by beaver: The structure and dynamics of streams are changing as beaver recolonize their historic habitat. *BioScience* 38: 753–762.

Nurse, P. 2020. *What Is Life? Understand Biology in Five Steps*. Oxford, UK: David Fickling Books.

O'Brien, M. J., and K. N. Laland. 2012. Genes, culture and agriculture: An example of human niche construction. *Current Anthropology* 53: 434–470.

Odling-Smee, F. J. 1988. Niche-constructing phenotypes. In *The Role of Behavior in Evolution*, ed. H. C. Plotkin. Cambridge, MA: MIT Press, pp. 73–132.

Odling-Smee, F. J. 2007. Niche inheritance: A possible basis for classifying multiple inheritance systems in evolution. *Biological Theory* 2: 276–289.

Odling-Smee, F. J. 2010. Niche construction. In *Evolution: The Extended Synthesis*, ed. G. B. Müller and M. Pigliucci. Cambridge, MA: MIT Press, pp. 175–208.

Odling-Smee, F. J., and K. N. Laland. 2011. Ecological inheritance and cultural inheritance: What are they and how do they differ? *Biological Theory* 6: 220–230.

Odling-Smee, F. J., D. H. Erwin, E. P. Palkovacs, M. W. Feldman, and K. N. Laland. 2013. Niche construction theory: A practical guide for ecologists. *Quarterly Review of Biology* 88: 3–28.

Odling-Smee, F. J., K. N. Laland, and M. W. Feldman. 2003. *Niche Construction. The Neglected Process in Evolution*. Princeton, NJ: Princeton University Press.

Odum, E. P. 1971. *Fundamentals of Ecology Philadelphia*. Philadelphia: Saunders.

Okasha, S. 2018. *Agents and Goals in Evolution*. Oxford: Oxford University Press.

Olson, D., and N. Torrance, N. 2009. *Cambridge Handbook of Literacy (Cambridge Handbooks in Psychology)*. Cambridge: Cambridge University Press.

O'Neill, R. V., D. L. DeAngelis, J. B. Waide, and T. F. H. Allen. 1986. *A Hierarchical Concept of Ecosystems*. Princeton, NJ: Princeton University Press.

Otto, S. P. 2018. Adaptation, speciation and extinction in the Anthropocene. *Proceedings of the Royal Society B* 285: 20182047.

Paley, W. 1802. *Natural Theology: Or, Evidences of the Existence and Attributes of the Deity, Collected from the Appearances of Nature*. London: R. Faulder.

Pérez, V. I., R. Buffenstein, V. Masamsetti, et al. 2009. Protein stability and resistance to oxidative stress are determinants of longevity in the longest-living rodent, the naked mole-rat. *Proceedings of the National Academy of Sciences* 106: 3059–3064.

Pittendrigh, C. S. 1958. Adaptation, natural selection, and behavior. In *Behavior and Evolution*, ed. A. Roe and George Gaylord Simpson. New Haven, CT: Yale University Press, 390–416.

Plotkin, H. C. 1994. *Darwin Machines and the Nature of Knowledge*. London: Penguin.

Plotkin, H. C., and F. J. Odling-Smee. 1979. Learning, change, and evolution: An enquiry into the teleonomy of learning. In *Advances in the Study of Behavior*, ed. J. S. Rosenblatt, R. A. Hinde, C. Beer, and M. C. Busnel. Oxford, UK: Academic Press, pp.1–41.

Plotkin, H. C., and F. J. Odling-Smee. 1981. A multiple-level model of evolution and its implications for sociobiology. *Behavioral and Brain Sciences* 4: 225–235.

Popper, K. R. 1966. *Of Clouds and Clocks. An Approach to the Problem of Rationality and the Freedom of Man*. Washington, DC: Washington University.

Popper, K. R. 1979. Review of E. J. Steele's *Somatic Selection and Adaptive Evolution. Times Literary Supplement* 1979: 5.

Post, D. M., and E. P. Palkovacs. 2009. Eco-evolutionary feedbacks in community and ecosystem ecology: Interactions between the ecological theatre and the evolutionary play. *Philosophical Transactions of the Royal Society B* 364: 1629–1640.

Pross, A. 2012. *What Is Life? How Chemistry Becomes Biology*. Oxford: Oxford University Press.

Puckett, J. G., A. R. Pokhrel, and J. A. Giannini. 2018. Collective gradient sensing in fish shoals. *Scientific Reports* 8: 7587.

Pullman, P. 1995. *His Dark Materials*. London: Everyman.

Reeve, H. K., and P. W. Sherman. 1993. Adaptation and the goals of evolutionary research. *The Quarterly Review of Biology* 68: 1–32.

Reid, J. M., W. Cresswell, S. Holt, R. J. Mellanby, D. P. Whitfield, and G. D. Ruxton. 2002. Nest scrape design and clutch heat loss in pectoral sandpipers (*Calidris melanotos*). *Functional Ecology* 16: 305–312.

Reiners, W. A. 1986. Complementary models for ecosystems. *American Naturalist* 127: 59–73.

Rhoads, J. E., I. G. B. Wold, S. Harish, et al. 2023. Finding peas in the early universe with JWST. *The Astrophysical Journal Letters* 942: L14.

Riahi, I. A. 2023. Macroevolutionary origins of comparative development. *The Economic Journal* (upcoming).

Richerson, P. J., and R. Boyd. 2005. *Not by Genes Alone: How Culture Transformed Human Evolution*. Chicago: University of Chicago Press.

Ricklefs, R. E., and G. L. Miller. 1999. *Ecology*. New York: W. H. Freeman & Co Ltd.

Riskin, J. 2016. *The Restless Clock*. Chicago: University of Chicago Press.

Rohner, N., D. F. Jarosz, J. E. Kowalko, et al. 2013. Cryptic variation in morphological evolution: HSP90 as a capacitor for loss of eyes in cavefish. *Science* 342: 1372–1375.

Romer, P. M. 1990. Endogenous technological change. *Journal of Political Economy* 98: S71–S102.

Romer, P. M. 1994. The origins of endogenous growth. *Journal of Economic Perspectives* 8: 3–22.

Roser, Max, Esteban Ortiz-Ospina, and Hannah Ritchie, 2013. "Life Expectancy." *Our World in Data*, https://ourworldindata.org/life-expectancy.

Rothschild, L. J., and R. L. Mancinelli. 2001. Life in extreme environments. *Nature* 409: 1092–1101.

Roughgarden, J. 1979. *Theory of Population Genetics and Evolutionary Ecology: An Introduction*. Stanford, CA: Stanford University Press.

Roughgarden, J. 1995. *Theory of Population Genetics and Evolutionary Ecology: An Introduction*. Hoboken, NJ: Prentice-Hall.

Roughgarden, J. 1998. *Primer of Ecological Theory*. Upper Saddle River, NJ: Prentice-Hall.

Rovelli, C. 2015. *Seven Brief Lessons on Physics*. London: Penguin Books.

Rovelli, C. 2018. *The Order of Time*. London: Allen Lane.

Rovelli, C. 2021. *Helgoland: The Strange and Beautiful Story of Quantum Physics*. London: Penguin Books.

Russell, B. 1948. B.B.C. broadcast transcript, published as "Science as a Product of Western Europe," The Listener 39 (May 27 1948), 865–6 Repr. as "Nature and Origin of Scientific Method," The Western Tradition, a Series of Talks Given in the B.B.C. European Programme, 1949.

Russell, B. 1961. *History of Western Philosophy*, 2nd ed. London: George Allen & Unwin Ltd.

Russon, A. E. 2003. Innovation and creativity in forest-living rehabilitant orangutans. In *Animal Innovation*, ed. S. M. Reader and K. N. Laland. Oxford: Oxford University Press, pp. 279–306.

Saini, A. 2019. *Superior: The Return of the Race Science*. London: Fourth Estate.

San Roman, M., and A. Wagner. 2018. An enormous potential for niche construction through bacterial cross-feeding in a homogeneous environment. *PLOS Computational Biology* 14: e1006340.

Sasaki, T., and D. Biro. 2017. Cumulative culture can emerge from collective intelligence in animal groups. *Nature Communications* 8: 15049.

Schmid, M. W., C. Heichinger, D. C. Schmid, et al. 2018. Contribution of epigenetic variation to adaptation in Arabidopsis. *Nature Communications* 9: 4446.

Schneider, E. D., and D. Sagan. 2005. *Into the Cool Energy Flow, Thermodynamics, and Life*. Chicago: University of Chicago Press.

Schrödinger, E. 1944. *What Is Life?* Cambridge: Cambridge University Press.

Scott-Phillips, T. C., T. E. Dickins, and S. A. West. 2011. Evolutionary theory and the ultimate–proximate distinction in the human behavioral sciences. *Perspectives on Psychological Science* 6: 38–47.

Scott-Phillips, T. C., K. N. Laland, D. M. Shuker, T. E. Dickins, and S. A. West. 2014. The niche construction perspective: A critical appraisal. *Evolution* 68: 1231–1243.

Seeley, T. D., and P. K. Visscher. 2004a. Group decision making in nest-site selection by honey bees. *Apidologie* 35: 101–116.

Seeley, T. D., and P. K. Visscher. 2004b. Quorum sensing during nest-site selection by honeybee swarms. *Behavioral Ecology and Sociobiology* 56: 594–601.

Shubin, N. H., E. B. Daeschler, and M. I. Coates. 2004. The early evolution of the tetrapod humerus. *Science* 304: 90–93.

Sieber, R., and E. D. Kokwaro. 1982. Water intake by the termite macrotermes michaelseni. *Entomologia Experimentalis et Applicata* 31: 147–153.

Slobodkin, L. B., and A. Rapoport. 1974. An optimal strategy of evolution. *Quarterly Review of Biology* 49: 181–200.

Smil, V. 2000. Energy in the twentieth century: resources, conversions, costs, uses, and consequences. *Annual Review of Energy and the Environment* 25: 21–51.

Smith, B. D. 2007a. Niche construction and the behavioral context of plant and animal domestication. *Evolutionary Anthropology* 16: 188–199.

Smith, B. D. 2007b. The ultimate ecosystem engineers. *Science* 315: 1797–1798.

Smith, B. D., and M. A. Zeder. 2013. The onset of the Anthropocene. *Anthropocene* 4: 8–13.

Sober, E., and D. S. Wilson. 1998. *Unto Others: The Evolution and Psychology of Unselfish Behavior*. Cambridge, MA: Harvard University Press.

Solow, R. M. 1956. A contribution to the theory of economic growth. *The Quarterly Journal of Economics* 70: 65–94.

Soni, J., and R. Goodman. 2017. *A Mind at Play: How Claude Shannon Invented the Information Age*. New York: Simon & Schuster.

Stoltzfus, A. 2019. Chapter 3: Understanding bias in the introduction of variation as an evolutionary cause. In *Evolutionary Causation, Biologigal and Philosophical Reflections*, ed. T. Uller and K. N. Laland. Cambridge, MA: MIT Press, pp. 29–62.

Sultan, S. E. 2015. *Organism and Environment: Ecological Development, Niche Construction and Adaptation*. Oxford: Oxford University Press.

Sultan, S. E., A. P. Moczek, and D. Walsh. 2021. Bridging the explanatory gaps: What can we learn from a biological agency perspective? *BioEssays* 44: 2100185.

Syvitski, J., C. N. Waters, J. Day, et al. 2020. Extraordinary human energy consumption and resultant geological impacts beginning around 1950 CE initiated the proposed Anthropocene Epoch. *Communications Earth & Environment* 1: 32.

Szilard, L. 1929. Über die Entropieverminderung in einem thermodynamischen System bei Eingriffen intelligenter Wesen. *Zeitschrift fur Physik* 53: 840–856.

Tebbich, S., K. Sterelny, and I. Teschke. 2010. The tale of the finch: Adaptive radiation and behavioural flexibility. *Philosophical Transactions: Biological Sciences* 365: 1099–1109.

Tebbich, S., M. Taborsky, B. Fessl, and D. Blomqvist. 2001. Do woodpecker finches acquire tool-use by social learning? *Proceedings of the Royal Society of London. Series B: Biological Sciences* 268: 2189–2193.

Tegmark, M. 2017. *Life 3.0*. London: Allen Lane.

Thompson, J. N. 1994. *The Coevolutionary Process*. Chicago: University of Chicago Press.

Tobi, E. W., R. C. Slieker, R. Luijk, et al. 2018. DNA methylation as a mediator of the association between prenatal adversity and risk factors for metabolic disease in adulthood. *Science Advances* 4: eaao4364.

Tomasello, M. 1999. *The Cultural Origins of Human Cognition*. Cambridge, MA: Harvard University Press.

Tomasello, M. 2008. *Origins of Human Communication*. Cambridge, MA: MIT Press.

Tomlinson, G. 2015. *A Million Years of Music*. New York: Zone Books.

Tomlinson, G. 2018. *Culture and the Course of Human Evolution*. Chicago: Chicago University Press.

Towe, K. M. 1970. Oxygen-collagen priority and the early metazoan fossil record. *Proceedings of the National Academy of Sciences* 65: 781–788.

Turchin, P. 2016. *Ultrasociety: How 10,000 Years of War Made Humans the Greatest Cooperators on Earth*. Chaplin, CT: Beresta Books.

Turing, A. M. 1937. On computable numbers, with an application to the entscheidungs problem. *Proceedings of the London Mathematical Society* 2.42: 230–265.

Turner, J. S. 2000. *The Extended Organism: The Physiology of Animal-Built Structures*. Cambridge, MA: Harvard University Press.

Turner, J. S. 2005. Extended physiology of an insect-built structure. *American Entomologist* 51: 36–38.

Turner, J. S. 2017. *Purpose and Desire: What Makes Something "Alive" and Why Modern Darwinism Has Failed to Explain It.* San Francisco: HarperOne.

Uller, T. 2019. Evolutionary perspectives on transgenerational epigenetics. In *Transgenerational Epigenetics, Second Edition*, ed. T. O. Tollefsbol. London: Academic Press, pp. 333–350.

Uller, T. 2023. Agency, goal orientation, and evolutionary explanations. In *Evolution "On Purpose": Teleonomy in Living Systems, Vienna Series in Theoretical Biology*, ed. P. Corning, S. A. Kauffman, D. Noble, J. A. Shapiro, R. I. Vane-Wright, and A. Pross. Cambridge, MA: MIT Press, preprint 10.31219/osf.io/49qrs.

Uller, T., and H. Helanterä. 2019. Niche construction and conceptual change in evolutionary biology. *British Journal for the Philosophy of Science* 70: 351–371.

Uller, T., A. P. Moczek, R. A. Watson, P. M. Brakefield, and K. N. Laland. 2018. Developmental bias and evolution: A regulatory network perspective. *Genetics* 209: 949.

Vedral, V. 2010. *Decoding Reality.* Oxford: Oxford University Press.

Vedral, V. 2018. *From Micro to Macro: Adventures of a Wandering Physicist.* Singapore: World Scientific.

von Neumann, J. 1956. Probabilistic logics and synthesis of reliable organisms from unreliable components. In *Automata Studies in Annals of Mathematical Studies* 34, ed. C. E. Shannon and J. McCarthy. Princeton, NJ: Princeton University, pp. 43–98.

von Neumann, J., and A. W. Burks. 1966. *Theory of Self-Reproducing Automata.* Urbana: University of Illinois Press.

von Neumann, J., and O. Morgenstern. 1944. *Theory of Games and Economic Behavior.* Princeton, NJ: Princeton University Press.

Waddington, C. H. 1969. Paradigm for an evolutionary process. In *Towards a Theoretical Biology 2. Sketches*, ed. C. H. Waddington. Edinburgh: Edinburgh University Press.

Wagner, D. L., E. M. Grames, M. L. Forister, M. R. Berenbaum, and D. Stopak. 2021. Insect decline in the Anthropocene: Death by a thousand cuts. *Proceedings of the National Academy of Sciences* 118: e2023989118.

Walsh, D. M. 2015. *Organism, Agency, and Evolution.* New York: Cambridge University Press.

Walsh, D. M. 2019. The paradox of population thinking: First order causes and higher order effects. In *Evolutionary Causation, Biological and Philosophical Reflections*, ed. T. Uller and K. N. Laland. Cambridge, MA: MIT Press, pp. 227–246.

Ward, P., and J. Kirschvink. 2015. *A New History of Life: The Radical New Discoveries about the Origins and Evolution of Life on Earth.* New York: Bloomsbury Publishing.

Watson, J. D., and F. H. C. Crick. 1953. Molecular structure of nucleic acids: A structure for deoxyribose nucleic acid. *Nature* 171: 737–738.

Watson, R. A., and E. Szathmáry. 2016. How can evolution learn? *Trends in Ecology & Evolution* 31: 147–157.

Watson, R. A., and C. Thies. 2019. Are developmental plasticity, niche construction and extended inheritance necessary for evolution by natural selection? The role of active phenotypes in the minimal criteria for Darwinian individuality. In *Evolutionary Causation, Biological and Philosophical Reflections*, ed. T. Uller and K. N. Laland. Cambridge, MA: MIT Press, pp. 197–226.

Watson, R. A., R. Mills, C. L. Buckley, et al. 2016. Evolutionary connectionism: Algorithmic principles underlying the evolution of biological organisation in evo-devo, evo-eco and evolutionary transitions. *Evolutionary Biology* 43: 553–581.

Webster, M. M., A. Whalen, and K. N. Laland. 2017. Fish pool their experience to solve problems collectively. *Nature Ecology & Evolution* 1:0135.

Weinberg, S. 1993. *Dreams of a Final Theory: The Search for the Fundamental Laws of Nature*. New York: Vintage Books.

West, S. A., C. El Mouden, and A. Gardner. 2011. Sixteen common misconceptions about the evolution of cooperation in humans. *Evolution and Human Behavior* 32: 231–262.

West, W. F. 1970. The Bulawayo Symposium papers: No 2. termite prospecting. *Chamber of Mines Journal* 30: 32–35.

West-Eberhard, M. J. 2003. *Developmental Plasticity and Evolution*. New York: Oxford University Press.

Whitehead, A. N. 1978. *Process and Reality: An Essay in Cosmology*, Corrected edition, ed. D. R. Griffin and D. W. Sherburne. New York: The Free Press.

Whitehead, H., and Rendell, L. 2015. *The Cultural Lives of Whales and Dolphins*. Chicago: University of Chicago Press.

Whiten, A., J. Goodall, W. C. McGrew, et al. 1999. Cultures in chimpanzees. *Nature* 399: 682–685.

Wilczek, F. 1999. Getting its from bits. *Nature* 397, 303–306.

Williams, G. C. 1966. *Adaptation and Natural Selection*. Princeton, NJ: Princeton University Press.

Wilson, E. O. 1975. *Sociobiology: The New Synthesis*. Cambridge, MA: Harvard University Press.

Wilson, E. O. 2020. *Tales from the Ant World*. New York: Liveright.

Witze A. 2020. How a small nuclear war would transform the entire planet. *Nature* 579: 485–487.

Woese, C. R. 1998. The universal ancestor. *Proceedings of the National Academy of Sciences* 95: 6854–6859.

Woese, C. R., and G. E. Fox. 1977. The concept of cellular evolution. *Journal of Molecular Evolution* 10: 1–6.

Wolfram, S. 2002. *A New Kind of Science*. Champaign, IL: Wolfram Media.

Wrangham. 2009. *Catching Fire: How Cooking Made Us Human*. London: Profile Books.

Wright, J. P., and C. G. Jones. 2006. The concept of organisms as ecosystem engineers ten years on: Progress, limitations, and challenges. *BioScience* 56: 203–209.

Zeder, M. A. 2017. Domestication as a model system for the extended evolutionary synthesis. *Interface Focus* 7: 20160133.

Index

Page numbers followed by *f* indicate figures.

Fungal mycorrhizae and roots of plants, 253–254
Future of humanity
  optimistic view, 294–298, 331, 334
  pessimistic view, 292–294, 334
Future of Life Institute (FLI), 195, 196, 213, 230

Galapagos woodpecker finch, 50, 235, 338n28
Galileo, Galilei, 323
Gambler's Ruin, 77. *See also* Adaptive niche management game
Game theory, 75. *See also* Adaptive niche management game
GDD. *See* Gross domestic detritus (GDD)
GDP. *See* Gross domestic product (GDP)
Gene-based memories, 12, 146, 202
Gene-cultural coevolution, 224
Gene editing, 323, 328
Gene methylation, 340n37
Generalist species, 311
Generalization vs. specialization, 338n28
Generation of variety, 25
Genetic evolution, 201
Genetic inheritance
  adaptive know-how (R_i), 57, 144
  based on inheritance of specific sequences of DNA, 266
  evolution, 46–51
  genetic mutations, 129
  niche construction, 58
  standard evolutionary theory (SET), 261
Genetic memories, 13, 114, 118
Genetic mutations, 129
Genomes, 337n21
Geochemical epoch, 162, 163, 259
Geophysical activities, 243, 244
Gerhart, John, 216
Giant Web Space Telescope, 326
Gibson, James, 45
Giraffe, 67, 211

Global positioning system (GPS), 309–310
Global warming, 145, 171, 301–308, 313, 319, 324
God, 67, 154–155, 282, 329
Goldilocks zone of the Sun, 18
Goldstein, David, 227, 228
GPS. *See* Global positioning system (GPS)
Grant, Peter and Rosemary, 235
Grazing path, 259, 302
Greater thermodynamic efficiency, 135, 136, 257, 279
Great oxidization event, 162
Greenhouse gases, 313, 321, 322
Green pea universe, 326
Green Revolution, 317
Gross domestic detritus (GDD), 312, 322
Gross domestic product (GDP), 312, 322

Hadean era, 243, 244
Hamilton, Bill, 56, 130, 207, 338n25
Hamlet (Shakespeare), 259
Harvester termite, 100, 101
Heat death of the universe, 287
Herbicide, 305
Here-and-now environments
  adapting or failing to survive, 311
  discrete abiota, 122
  dumping detritus into environment as consequence of here-and-now economic activities, 312
  elsewhere-and-later environments, 83, 310, 311
  immediate here-and-now gratification, 312
  "world" in four-dimensional space, 110
Heterotrophs, 302
Heuristic niche construction, 132–137
Heuweltjies, 100, 101
HGT. *See* Horizontal genetic transmission (HGT)
High-centering, 283
Himalayan rabbit, 263–264
Hinduism, 334

Niche construction theory (NCT) (cont.)
  evolutionary theory being about organism-
    environment coevolution, 70
  first biotic-by-abiotic interactions, 245
  forerunner for future universal theory of
    evolution, 172
  knowledge-gaining capabilities that
    allow organisms to improve their fit to
    environment, 74
  niche construction being central aspect of
    process of adaptation, 84, 85
  orderliness, 241
  organism-environment coevolution, 281
  relativistic reference device, 158
  self-control of inductive gambles, 86
  supplementary inheritances, 90
  supplementary knowledge-gaining pro-
    cesses, 89
  two-way-street interactive niche
    relationships, 241
Niche inheritance, 25, 288
Niche management game. See Adaptive niche
  management game
Nitrification, 249
Nitrogen cycle, 249
Noise, 199
Noncoding RNA, 265
Nonergodic evolutionary process, 145
Nonexcludable goods, 310
Nongenetic inheritance systems, 263,
  266–269
Nonrivalrous goods, 309
Novel change, 144–145
Novel environmental event, 85
Novel genetic mutations and recombinations,
  215, 216
Nuclear DNA, 266, 267
Nuclear war, 292–301
Nurse, Paul, 327

Obligatory epigenetic process, 264
Occam's razor, 257

Odling-Smee, John, 178. See also *Niche
  Construction* (Odling-Smee)
Oil, 312
One cause/two effects dilemma, 284–285
O'Neill, Robert, 235
On the Origin of Species (Darwin), 155, 233,
  271
Open environment, 17, 18, 143
Opportunistic niche construction, 50,
  144–145, 168
Opportunity costs, 106
Orangutans, 132
Order, 12
Order for free, 187, 244
Orderliness, 241
Order of Time, The (Ravelli), 13
Organism-environment coevolution
  evolution, 143, 172, 242, 260
  Lewontin, Richard, 242, 260
  more comprehensive theory of evolution,
    289
  niche construction theory (NCT), 281
  relativism, 156, 158
  standard evolutionary theory (SET),
    281
  universal theory of evolution, 172
Organisms
  active, purposeful, goal-seeking agents in
    their own evolution, 5, 31
  adaptations allowing them to perceive or
    sense their environments, 81
  adapting to multiple selection pressures
    simultaneously at each place and
    moment during their lives, 82
  "design" to their environments, 67
  determining their own futures, 137
  dynamic kinetic stability (DKS), 9
  external environments, 21–22
  fundamental purposes of living organisms,
    81
  interactions between abiota and organisms,
    122–124